THE GREAT ENERGY TRANSITION

A qualitative assessment of the external impacts of fossil, nuclear and renewable energy resources within the context of the rising total cost of fossil fuels and nuclear energy and the declining costs of renewable energy and how existing and emerging technologies can solve humanity's resource challenges today.

Mark Townsend Cox

ISBN 978-1-964462-16-5 (Paperback)
ISBN 978-1-964462-18-9 (Hardback)
ISBN 978-1-964462-17-2 (Ebook)

Inquiries and Book Orders should be addressed to:

Leavitt Peak Press
17901 Pioneer Blvd Ste L #298, Artesia, California 90701
Phone #: 2092191548

I dedicate this book to the men in the 3rd Battalion, The Parachute Regiment who were killed or wounded in the Battle for Mount Longdon, in the closing chapter of the Falklands Conflict in June of 1982. The 3rd Battalion, The Parachute Regiment was the first of five British infantry battalions to engage the Argentinian Army on high ground to the West of Port Stanley on the East Island of the Falkland Islands. Being first, the plan was to attack a mountain called Mt. Longdon with a surprise night attack. I was the Lieutenant in command of 5 platoon, B Company. B Company was the point company for the attack on Mt. Longdon. The battle started after midnight and persisted until the Argentinian unconditional surrender, 2 days later, on the 12th of June 1982.

5 Platoon emerged from the battle with relatively low casualties, compared to 4 and 6 Platoons, commanded by my friends, Lieutenants Andrew Bickerdyke (whose leg I put a tourniquet on after receiving a bullet wound) and John Shaw, who recently retired as a Major General after commanding in Iraq and being the assistant Chief of the UK Defense Staff.

I specifically mention those who died or were wounded in my platoon who participated in that action.

Corporal McLaughlin died. He was my 3rd section commander, who was my elder and exemplified the professionalism and competence you would expect of an ex-SAS special forces soldier. He was a born leader. In the days prior to the battle and with what strange moods occasion soldiers on active duty, we stood shoulder to shoulder, facing Mt. Longdon looking from the Estancia mountainside, singing "Always look on the Bright Side of Life" from Monty Python where we both fumbled the whistling part. Private John Crow, who was attached to my platoon from the battalion anti-tank platoon was killed by enemy fire.

The wounded include Corporals Ian Bailey (who earned the Military Medal), Graham Heaton and Phil Skidmore (attached from anti-tanks), Lance Corporal Lenny Carver, privates Grant Grinham, Dominic Grey, Peter Hindmarsh, Frank Regan, Mark Meredith and Andy Steadman for a total of 2 dead and 10 wounded.

Foreword

THE GREAT ENERGY TRANSITION by MARK TOWNSEND COX is an unusual book by a one of a kind author. Mark's amazingly diverse background is in itself unusual: military, investment, energy and entrepreneurial with a refreshing attitude of "it can be done". He brings out the problems, the facts and figures but he does it in a delightfully fresh way; it's upbeat (largely) and tries not to shy away from real problems.

So much is said and written today about the changing world and we all certainly need to take notice. Issues that we all face, wherever we live and whatever we do cannot be hidden. I confess that I am sometimes amongst those who often complain that we are under such a constant barrage of negatives we have almost had enough! The "GREAT ENERGY TRANSITION" has cheered me greatly: it puts a positive spin on a great many of my concerns and this book is a pleasure to read and a valuable volume to quote from. Many facts clearly presented.

Mark offers a forward-looking narrative with examples that show quite convincingly how traditional sources of energy are about to be eclipsed (partial if not total) by renewables and that this actually augurs well for global prosperity and improvements in our degraded biosphere, Dirty air, polluted rivers and oceans, loss of species are concerns and Mark looks at these with examples and references. He has seriously researched his presentation with countless examples. The feeling I had at the end of the book was one of being recharged with hope.

Well done, Mark, a seriously useful and positive contribution.

Richard E. Leakey, FRS.
Chair, Board of Trustees, Turkana Basin Institute Chair,
Board of Trustees, Kenya Wildlife Service
Professor of Anthropology, Stony Brook University

Dedication

I started writing this book in 2013 to put words to what has become a very strong vision of a world that can be much more resilient and stronger if it chooses to adopt sustainable practices. We can have our cake and eat it. We know that we are facing challenges to provide food, water, shelter and energy to the existing 7.4 billion people who are here now. This number is due to grow and peak at about 9 billion in 2100. With this vision, all of us together can have a drastically reduced footprint and yet enjoy a booming economy that allows individuals to flourish and respond to challenges in the healthiest way. It's not a zero-sum game. The sustainable energy revolution that is under way will bring us all more efficient ways to use more plentiful, sustainable energy.

Years of weekends at my desk accumulating material for the thesis have made me appear hostile to friends, but I am hoping that it will be worth it for its message. Thank you, Olivia Huntington, in the Upper West Side of Manhattan for your constant and active efforts to leave me alone to work on this book despite no evident sign of its completion, and to the Huntington family in general who have made this possible.

Also, I want to make a special mention of my "internal editor in chief" and business partner, Olushola "Shola" Ashiru, who carefully read each chapter and whose comments and corrections I unflinchingly adopted. Professor Geoffrey Heal of Columbia University's business school also read an early draft and gave me some good orientation advice which I followed. Later in the production I welcomed the help of Shellka Arora, later to become my wife, for her excellent ability to find details that all previous eyes missed and of Andrew Mongar, an air-conditioning entrepreneur and former scientific advisor to the first "scientist prime minister" in the UK, Margaret Thatcher, for making pertinent observations which I acted on. Finally, Professor Satyajit Bose, of Columbia University who teaches courses in the Sustainability Management program, for his effective filter on my egregious wandering pen. I also want to thank Christina Palaia of Emerald Editorial Services for her kind and patient advice.

I dedicate the book to my wonderful parents, Joan Lillian Cox and Lt. Colonel John Joseph Geoffrey Cox, MBE, OBE, of whom I'm so proud. My father passed away in November 2015, as I was writing, and I was so glad I had put up a draft of the book online so that he could see it two weeks before he left us.

Contents

Preface

*"I think that the world is in the middle of a huge
transition that we have to make to renewable
energy. We have to transition away from fossil fuels
very, very quickly."* —Josh Fox, director of the
Oscar-nominated documentary *Gasland*, 2010.

A gunshot reverberated in the hot air about 300 yards away. I looked up
and saw thousands of Africans with sticks charging toward me, yelling
with passion. Billowing dust clouds increased the scale of the spectacle.
I was transfixed. It fascinated me. My eyes were locked on the oncoming
human wave. They were giving it everything and only seconds remained
before they threw themselves, their nets and calabashes headlong into
the Matan Fada River, which mercifully separated us, the audience, from
this mass of farmers and workers from all over Western Africa eager to
catch the largest Nile perch they could in return for a purse of 1 million
Naira (then about $7,600), a free bus they could operate and a trip to
Mecca to fulfill the Hajj, the once-in-a-lifetime pilgrimage to Mecca.

The four-day Argungu Fishing Festival is in Kebbi State, Northern
Nigeria, and was started in 1934 to celebrate peace between the Nigerian
states of Kebbi and the Caliphate of Sokoto. The festival has now
become a tourist attraction, sporting an ambiance unlike the calm of any
other fishing festival known. I witnessed this in March of 1979 when I
was lucky enough to spend a year in Nigeria.

Being in Nigeria was an incredible privilege. I was what's still
known as an "army brat," a child of a military family. My father, a
career officer in the British Army's Parachute Regiment, was sequen-
tially posted to different military postings in different Commonwealth
countries, as were many others. My three siblings and I shared an amazing childhood and never lived in the same house or part of the world
for more than two years at a time. When old enough, we all were sent

to boarding schools in the UK, since it was too difficult to go to day schools during overseas postings. Our family lived in a total of 17 different locations during our upbringing.

We all remember the day our boxes would arrive at a new house somewhere exotic like Cyprus or Pakistan containing the furniture and fixtures that make a family function. I sometimes hear it said that a child growing up requires stability and that too many disruptions can cause psychological imbalance, but I felt nothing but excitement about flying to far-off lands to meet our parents for school holidays. It was adventure and excitement. Home was wherever Mom and Dad were to be found. The Nigerian experience was later in this cycle of country experiences but key in my story because it's where I first had my eyes opened to renewable energy.

In 1979, when I was 22 years old, I boarded a British Caledonian aircraft (as was) to fly to Kaduna, a city in northern Nigeria, to spend the summer holidays with my family. At the time, the United Kingdom trained Nigerian Army officers, and my father was a lieutenant-colonel training officer there. I languished away the summer in the swimming pool at the Durbar Hotel and frequented parties interspersed with trips into the countryside to do amazing and never repeated activities like the relaxing in the geological wonder of the hot water Wikki Springs in Yankari National Park and witnessing the spectacular Argungu Fishing Festival mentioned above.

The summer played itself out. One of the other colonels was hurt in a parachute jump. I and a friend were miraculously unhurt in a car accident when I allowed him to drive one evening at sunset after a party. He couldn't hold the Peugeot to the curve as we raced around the tree-lined, tight- bending Coronation Crescent. We rolled, miraculously right through a gap in the neatly interspersed trees that lined the road and ended upside down. We climbed out of the open windows, disoriented but luckily unscathed. The tape deck was still playing Mike Oldfield's "Tubular Bells." We looked helplessly at the surreal sight and panicked about its implications. The wheels were still spinning and the music was playing loudly, as though the musicians were so consumed by their music that they hadn't noticed.

People who were there helped us push the car back onto its wheels. There was a significant dent in the roof. I sat in the driver's seat and turned the starter motor over and over. Before long, our desperate efforts to restart the engine bore fruit. The handicapped vehicle limped back to my parents' garage and a difficult confession. The upshot was that my friend and I had to pay for repairs to the car. A hefty sum for unemployed young students.

When the autumn arrived, I was preparing to return to London and the question mark of a career. Instead, an offer of a job turned up right there in Kaduna: Would I like to work for a year as a stand-in teacher at the local Sacred Heart School? I would have a class of about 30 multinational eight- year-olds. I agreed. It would keep me "at home" for a while longer and allow me time to ponder my next steps. It was a useful "gap year" and I could also pay for the damage to the car.

As the winter term ended, my headmistress, Dorothy Ajijola, an Englishwoman who had married a Nigerian, asked me if I had any ideas for a summer science project. I decided in this equatorial location that it might be a good idea to make a couple of 3-foot mirror solar parabolic dishes. I designed a parabolic curve and had my class make copies, which we then joined in the center to make frames for two dishes. Then we used papier- mâché to create as smooth a bowl in the center as we could and glued kitchen foil, shiny side up, onto the bowl surface. On the big day, the bowls stretched the children's patience, but finally the water in a bottle began to boil. The eight-year-olds had produced a disruptive, essential commodity in a location where there was a tangible need. This was an extraordinary thing to accomplish and I have never forgotten that the sun provides 5,000 times more energy annually than human beings consume.

After my year in Nigeria, I joined the British Army and attended the Royal Military Academy, Sandhurst, in 1980 and was posted as a lieutenant to the Third Battalion, the Parachute Regiment. I had an eye-opening initial exercise in the Oman, where I set up camp for and met the soldiers in my platoon for the first time. Twenty-one years earlier, in 1961, my father was a captain in the same battalion of the British Army, "3 Para." The battalion dug in a defensive position on the Iraqi border to help dissuade an Iraqi invasion of Kuwait long before the two U.S.-led Iraqi engagements of 1990 and 2003. So much has happened in the region since that time. It is amazing that the gulf has indeed remained open to tanker traffic all these years.

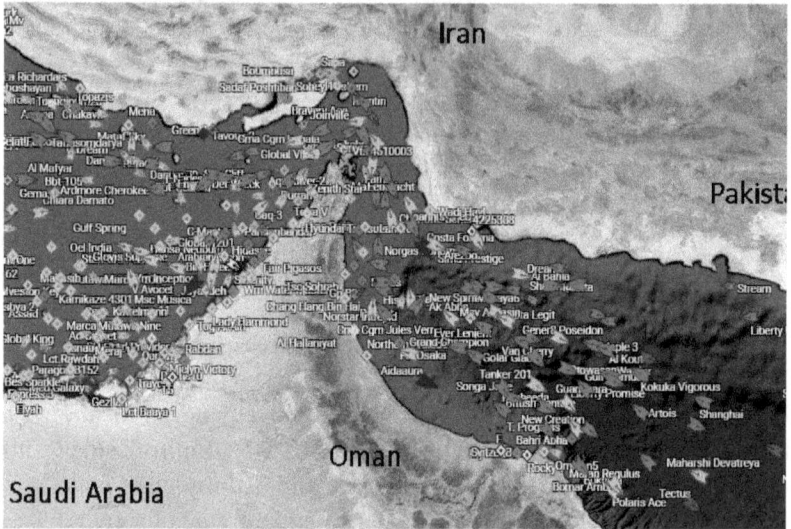

Figure 1: The Gulf of Oman and Gulf of Hormuz on 11 March 2016, showing a line of oil tankers winding in and out of the Persian Gulf to collect oil from Iraq and Saudi Arabia and the Gulf States and take it to the world. Source: Google Maps, Marinevesseltraffic. com.

Immediately after my unit returned from the Oman, General Galtieri, the leader of the military junta that governed Argentina from 1981 to 1982, ordered the invasion of the Falkland Islands, a British territory in the South Atlantic.

The Falkland Islands were known to have geology that supported the presence of oil, and in the full knowledge of the finite nature of existing, proven global oil resources, many exploration companies were on the lookout for new deposits. Oil most likely was not at the top of Margaret Thatcher's priorities when she confronted General Galtieri, nor was it spoken about in the military mission, but it remains one of the background issues in the conflict.

My unit was on what was called standby, so we were among the first to react to the threat. An idea that was briefly considered was to parachute our battalion of 650 men into Port Stanley airport on East Falkland as quickly as possible to let Galtieri know that his decision was a bad one. A tragic decision by General Montgomery toward the end of World War II, to establish a bridgehead, ahead of the advancing Allied front at Arnhem, as shown in the film *A Bridge Too Far*, possibly locked that plan out of consideration. Instead, we traveled the 8,000 miles south in the S.S. *Canberra*, a beautiful, white passenger liner, also known as the "Great White Whale," along with the rest of the task force ships accompanied by the "still- communist" USSR shadow ships trailing behind us.

The *Canberra* has since scrapped, but I still have my first-class cabin key, number 41. It was very collegiate and organized in the officers' mess. We stopped in the mid-Atlantic at volcanic Ascension Island, where, for several days, ships, helicopters and small boats busily "cross-decked" or redistributed weapons, men and war materiel, before we all headed south again. We trained, with the soldiers performing exhaustive deck running for fitness and perfecting weapons drills. Since the oil in the deep water under the Falklands is part of the fossil energy issue, I mention this conflict in greater detail in chapter 5.

After my brief, exciting military service, I worked in the city of London learning the stockbroker business and eventually came to the United States in 1987. In 2003, I created my own socially responsible investment (SRI) hedge fund to invest in publicly quoted renewable energy technology companies. Over these years, it became obvious that there is a major transition from destructive energy to sustainable energy going on. This book provides a perspective that helps put that evolution into context.

Introduction: Context and Journey

Introduction: Context and Journey

*"If you want to build a ship, don't drum up peo-
ple to collect wood and don't assign them tasks and
work, but rather teach them to long for the immen-
sity of the sea."* —Antoine de Saint Exupéry

Cox father and son were both involved with the military protection of valuable overseas energy resources. Oil trade protection is a major part of the U.S. and Western nations' military role and a task that keeps U.S. naval carrier groups on duty around the world. It also provides energy supply insurance, and although it is extremely expensive for governments to maintain, it's also good for ongoing military preparedness and training. This protection cost is a geopolitical, external cost of oil and, as I mention in chapter 5, accounts for trillions of dollars. It just may be, as this book strongly suggests, a cost that need not exist. This will in part be due to the different nature of renewable energy logistics. There is no cost, feedstock agreement or logistics of any kind to a ton of wind or a barrel of solar power. The external costs of fossil fuels always made renewable energy cheaper, but we pay only the internal, lower cost at the pump, so fossil fuels have been more economic to the consumer until this point. Most of the benefit from renewable energy has come from the fall in the fossil fuel external costs as they are replaced by renewables. Today, however, falling equipment costs for renewable energy formats have made renewable energy competitive with fossil fuels on an absolute basis as well. This is known as grid parity, where obtaining a watt of renewable energy, whether solar, wind or from other renewable sources, becomes as cheap as the power already distributed on the electricity grid from coal, gas and oil. It looks clear that renewable energy costs are set to continue falling as economies of scale, new technology and efficiency records continue to tumble, even beyond the need for the small quantity of subsidies they receive.

Sustainability suggests an absence of damage to any part of the environment or other aspect of human affairs. It's also a term which means "affordable" in every sense. "Return to the mean" is a common Wall Street idea that unusual events are actually bubbles that eventually end and that normalcy will return. I maintain that this time, the transition to sustainable energy is not a bubble and instead is evidence of a major infrastructural transition that will take 30 to 50 years to get to 80–100%

global sustainable practice and, in the process, boost economic growth and result in an improved quality of life.

In establishing the investment process of the New Energy Fund, I felt it was important to highlight the external costs, the externalities of fossil fuels and nuclear power. I share a vision about what it would take to make an orderly transition to a state of sustainability, a condition of humankind where all our needs, food, water, energy and shelter, can be obtained with minimal or even a positive impact on the environment, even with an expanding population. This would effectively amount to zero human footprint.

The measure of the fossil fuel externalities examined in this book tells us loud and clear that a world without fossil fuels would be far preferable to one in which we keep them if we had renewable energy instead. It's not to say that fossil fuels have not done us good service, merely that they have outgrown their usefulness now. In the face of growing global population and energy consumption, this challenge and the search for its solution have become an immediate global preoccupation, and yet, the private sector often still behaves in a singularly detached and complacent manner about this subject. People just flick the light switch. At its worst, it has taken the form of a political polemic about what's more expensive: putting in billions of dollars of new technologies and learning as we go or realizing too late that we are in an unthinkable economic destabilization that becomes an existential risk and too expensive to fix. This book's vision is optimistic due to the mass of evidence supporting the idea that we are already on an unstoppable path toward a major economic rebirth that has the potential to benefit all. Is this too optimistic? Not at all; read on.

Introduction to Key Terms and Concepts

I juggled with a suitable acronym to collect the "bad" energy resources in a single word: coal, oil, nuclear and gas. These combine neatly to make CONG. The nuclear represented by the N refers to the most common nuclear technology, pressurized water reactors or boiling water reactors (PWRs/BWRs). Choosing this nuclear design was a bad decision made worse by the string of serious accidents and proliferation risk which have resulted in the peaceful use of the atom being burdened with regulatory safety measures which are very expensive to implement. As you will see in chapter 5, my opinion about nuclear power changed as I studied the subject for this book, leading me to see the N in CONG could also stand in for the "natural" in natural gas.

I identified eight fossil fuel externalities that cover the lion's share of the subject. *Externalities* are costs accrued by the use of an energy source that are not paid directly by the buyer of the energy. Filling your car with gasoline is affordable, but I am making the point that the accrual of all eight external costs, if internalized and added to the cost of a gallon of gasoline, would push the price to a shocking level.

Taken in a larger context, a country, faced with the higher externality-laden price, should have this same shocked reaction. In the case of some externalities such as depletion or geopolitics, it can be said that these are not real externalities, but for the purposes of this book I treat them as such. Depleting a valuable resource such as a fish species or an oil resource can be seen as a cost to a community that depends on those commodities and that will have to make significant adaptations when the resource is gone. There is also no question that geopolitics is the subject of international relations literally poisoned by international energy stresses, resulting in wars, refugees and huge armament expenses and trade protection efforts.

I started presenting these externalities while raising capital for my first fund, the New Energy Fund, LP, in 2003 to make the case for investing in renewable energy companies that had solutions. These are the eight externalities:

THE EIGHT EXTERNALITIES OF CONG

CLIMATE	PRICE VOLATILITY
CONG DEPLETION	INNAPROPRIATE SUBSIDIES
GEOPOLITICS	POLLUTION
HEALTH	WASTE

In this book, I explore each of the above externalities. Most of any public, media-led discussion about our current resources of fossil fuels involves just one consequence at a time, say, pollution, or geopolitics, and increasingly perhaps climate. It's rare to see these negatives discussed together, and all of them, never. The moment you see the impact of each, it's easier to recognize the magnitude of their cumulative impact.

Instead, the discussion is still all about the high perceived cost of renewable energy. All the other bad things about the use of CONG don't come out at every telling. When solar power is spoken of, there is little mention of the fact that its energy resource, the sun, is perhaps the only real free lunch and provides almost 5,000 times more energy than humankind currently uses annually.

Also, renewable equipment costs are falling, while CONG fuel prices remain volatile. There is rarely a comprehensive statement of the full cost we pay because of our reliance on CONG. This book can't get quantitative about the whole impact but begins to articulate and demonstrate the alternatives. There is, justifiably, a vision of a relatively low-cost renewable energy paradigm and its beneficial impacts on the wider economy. Over time, the externalities of fossil fuels have accumulated. The real picture is much larger, more alarming and compelling. Despite falling use of coal and oil, the world's high volume of consumption is greater than ever and set to continue to grow. Taken all together, this view is the most powerful indictment of the continued use of fossil fuels and the most powerful support for the transition to renewable, sustainable alternatives. Those sticking to the CONG playbook may be ignorant of these facts or at worst paid to perpetrate falsehoods about them, which I demonstrate in chapter 3, in a discussion of climate denial and its antecedents.

The Great Energy Transition Chapter Summaries

Introduction. This part of the book introduces my background and how I became aware of the larger issues of sustainable energy. It also lists each chapter's contribution to the thesis of humanity's big adventures in applying its hard-fought-for new understanding of science to take us all off the "human footprint," a term that describes all the negatives of any economic inputs that we put up with, specifically the fossil fuel externalities. It describes coal, oil, nuclear and gas (CONG), the fossil fuels, as a collection of finite resources of expensive, polluting energy and shows how this is now giving way to almost infinite quantities of cheap, clean, sustainable energy.

Chapter 1. Humans and Technology. This chapter sets the scene of the great energy transition by showing the historical global energy context. There is a significant transition from CONG to good energy already well under way. It has already reached so significant a point that the use of the word *alternative* to describe its energy product is no longer accurate since renewable energy has become mainstream.

The year 2015 was the first that global renewable energy capacity installations were larger than CONG capacity installations. This chapter also makes the point, backed up by later chapters and references, that there is so little external cost in renewable energy, properly applied, that the transition is in everybody's favor. If we eliminate the fossil fuel externalities, there will be a huge positive impact on the economy and everyone's lives. You **can** have your cake and eat it.

All this means that you can drive a powerful electric sport utility vehicle (SUV) powered by solar panels installed in your garden and on your roof, or from a green grid, or even from a CONG-supplied grid without guilt. Soon, you will be able to fly as many times as you want without the almost habitual refrain from critics that you imploded your environmental credentials by taking a flight with all its implied, hypocritical, damaging emissions. Renewable technology makes all this possible with lower absolute energy costs. Many, including environmentalists, have not yet realized that this phase of the adventure of human progress can literally lead to a new age, the Resiliocene (yes, there is a competitive effort to name the new era, with *Anthropocene* close behind.... Okay, ahead!), if we can somehow articulate it and actually, physically manifest it over the short period we have ahead before the many tipping points make it a one-way trip. There is an appropriate urgency and need to articulate this vision of a better world so that as much information as possible is made available. Like Al Gore's most recent 2016 TED talk,[1] there is plenty of reason for optimism and hope. The remainder of the book focuses on why this is true.

Chapter 2. Renewable Energy Externalities. This is an examination in as honest a way as possible of the downsides of renewable energy technologies. Bird and bat deaths, noise and many other negatives associated with wind, solar and hydro power are examined. It turns up some surprisingly favorable facts from the field. Early excitement about hydroelectric energy locked in the worst of its damaging ecological impacts with extraordinary stories of failed multi-billion-dollar dams in places like China, flooding, silting and drowning large areas and large numbers of people and animals in the name of progress and only sometimes taming the water challenge. In the end, green externalities turn out to be virtually immaterial in the face of CONG.

Chapter 3. Climate Denial. Climate denial is a big externality, presenting humanity with an existential threat. This is such a dynamic human issue. Science observes and measures the facts. If you look at the context within which the scientists work, you can understand and have confidence in their results. Science is true whether you believe it or not because of the discipline of throwing every hypothesis open to testing, reproduction or rejection by skeptically minded academic professionals who only work on good evidence. Science is slowly built from nothing, has no shame in not knowing and explores phenomena with an inexorable logic that explains what's going on clearly and often with results we don't want to hear. If we are to be intellectually honest, it's the only road to go down. That's why the truth about the climate is "inconvenient" since it demonstrates that humans are responsible and can fix it and that there are tipping points that will occur soon that will make it all but impossible to return to the way the Earth was before the industrial revolution. We found out that installation of a renewable energy paradigm is not enough; we also need geoengineering to take out the CO2 to obviate the worst climate impacts. All the solutions exist, however, and can be implemented, but we need the political will, which, happily and increasingly, almost all countries in the world possess today. Someone tell the USA!

Chapter 4. CONG Externalities: Depletion. Whatever one's theory about where fossil fuels come from and, yes, there are people out there who believe the resource is constantly being topped up, there is no doubt that at the rate at which we are using the resource, it is getting increasingly expensive to find and extract. I look at the reasonable amounts of each energy source which have been proven to remain and how our consumption of it impacts how long we can enjoy—or be cursed by—it.

Chapter 5. CONG Externalities: Geopolitics. Unequally distributed resources turn countries into economic "haves" and "have-nots." In this relatively long chapter, I explore various impacts of having and lacking fossil fuel resources as an economy. There are a lot of stories and, since I was there, I also include the 1982 Falkland's Conflict, where one of the underlying themes was actually about potential oil resources that lie under the islands. We also take the nuclear story to account to figure out why it's got lots of promise to combat climate change but why it probably is too expensive, given irrational fear born of continuing accidents, to perform. However, it became clear that being limited to using the main nuclear fuel, uranium 235, and not being able to embrace breeder reactors due to the risk of proliferation and doggedly sticking to early technology without modernization have combined to cause stagnation in developed nuclear markets, exemplified by the U.S. situation, where the Nuclear Regulatory Commission (NRC) holds the industry in a chokehold. It prevents new research and is unable to step out of its own way to find solutions to higher safety costs. This is all the more extraordinary when you consider that the United States was the source of much of the technological innovation that created the industry in the first place.

Chapter 6. CONG Externalities: Health. It might seem surprising to include health as a factor that is negatively impacted by fossil fuels, but a closer look will surprise you with how much money comes out of an economy to cope with it.

Chapter 7. CONG Externalities: Price Volatility. Again, the unequally distributed resources mean that those without them are prepared to pay for them, and in fact, the prices of CONG have bounced around in a way that makes energy, something a good, growing economy needs as a stable feedstock, very volatile, interfering with strategic planning.

Chapter 8. CONG Externalities: Pollution. The original problem with using oil and gas, especially in the early decades of its use, was that spilling it didn't appear to matter, although of course it did. All the chemicals in fossil fuels have turned out not only to have an impact on human health but also to be responsible for this first and major phase of the sixth major species extinction, which is currently under way.

Chapter 9. CONG Externalities: Subsidies. The hue and cry about renewable energy receiving far too many subsidies is in fact one of those totally incorrect myths. It turns out that fossil fuels have been used during the period of history when many countries organized their tax codes and so tax deductions, grants and other subsidies are baked into many tax systems. In the United States, this is very much the case and we have look at different studies that describe the billions in subsidies enjoyed by fossil fuels today, just at a time when we ought to tilt support in favor of sustainability.

Chapter 10. CONG Externalities: Waste. As time has progressed, the efficiency with which a vehicle can use a gallon of gasoline, for example, has improved dramatically. Since CONG fuel sources are finite, every inefficient use, that is, all the historical uses, have been marked by significant waste of these precious resources. Humanity has thrown away so much energy. Renewable energy, by definition, is almost infinite and consequently not so impacted by the efficiency issue. There will always be more, and it turns out we have learned the efficiency lesson and every generation of new technology is showing greater and greater efficiency at the same time as becoming cheaper. Its interesting that the internal combustion engine today is about the same efficiency 15% - 25% as the solar panel currently is as well.

     Chapter 11. Conclusion. This short chapter summarizes the huge amount of damage, the externalities, that the current volumes of fossil fuels still in use are actually and irrevocably having on our economies. Empowered by the facts of the book, readers will want to make changes in their own lives and do their best to encourage and accelerate the transition to a sustainable environment.

What's at Stake

> *"We know from the study of evolution that, again and again, various branches of animal stock have become over-specialized, and that over-specialization has led to their extinction. Present- day Homo sapiens is in many physical respects still very unspecialized– ... But in one thing man, as we know him today, is over-specialized. His brain power is very over-specialized compared to the rest of his physical make-up, and it may well be that this over-specialization will lead, just as surely, to his extinction. ... if we are to control our future, we must first understand the past better."* —Louis S. B. Leakey, *Adam's Ancestors*

As our current high fossil and nuclear energy subsidies and externalities become better known and impact us ever harder, it becomes clear that renewable energy technologies are a much-needed bargain with few discernible risks. A sustainable energy paradigm represents a major improvement in the human condition. This book defends the solar and wind and many other clean energy technologies that have appeared in the last three decades and demonstrates that they can effectively solve all the externality problems. I aim to articulate, in lay terms, how much actual energy is available to us in renewable, sustainable energy forms and how new technologies have made it possible to exploit them. The vision is clear. A family or community can easily have an electricity-based lifestyle that is efficient, with distributed sources of renewable energy to run houses, hot water, transportation, communications and so forth.

The book highlights the CONG externalities, the frequently unseen negative sides of current fossil fuel use and compares them to those of renewable energy. A sustainable human community with all its economic advantages is the prize. It's a vision, supported by rigorous, skeptical, scientific evidence, of a world where we can aspire to an even better quality of life than the hugely improved one that we experience today and certainly leaps and bounds ahead of the preindustrial lifestyle. It demonstrates how fossil fuels hurt as well as help us, whereas renewable energy simply helps us.

A common view of a world without fossil fuels is a poor one with a poverty of energy, where meager current for a dim lamp is generated with a squeaky wind turbine. After looking at the evidence, I think the hair shirt and cold shower environmentalist department can stand down. Instead of austerity, we can improve our lot considerably after fossil fuels have gone.

We have a world that is, in fact, totally capable of welcoming the expected rise in population from 7.4 billion to perhaps 9 to 11 billion this century, with all the implied demand for food, energy, shelter and water. It turns out that we are currently using the wrong, finite, expensive resources of coal, gas and oil. Instead, we need to exploit the resources we know to be almost limitless, clean, sustainable and increasingly affordable: solar, wind, new nuclear, hydro, thermal, efficiency, waste to energy and other new entries to the sustainable fold.

Today, no other forms of energy capacity are being installed faster than renewable energy technologies. Financing methods have evolved to match the industry led by individuals such as Jigar Shah, with his company SunEdison, who realized that the revenue stream from the energy produced by solar panels would pay for the equipment, making it possible to offer free installations and cheaper power to residential and business customers. Companies like Exelon and NRG in the United States and E.On in Germany have also learned how renewable energy can be put to work profitably.

Today, in 2018, financial investors favor tried and tested, and sufficiently economic technologies as opposed to the deeply economic but untried disruptive new technologies that are nonetheless slowly appearing and that are also very appealing. For example, an up-and-coming concentrated-solar- with-storage configuration transfers solar thermal energy to an insulated chamber full of ceramics that resembles a kiln, but a kiln that takes in heat from the sun and the ceramic media inside keep the heat until it is needed. This allows us to generate solar electricity for up to a week without sunshine by storing the sun's energy in solid media so that it can be drawn upon at night or during cloudy weather. Because this energy configuration is not placed in service yet, it's also not bankable, even though it works. In the world of energy, where machinery is expected to operate faultlessly for 30 years at minimal cost, the bar for new technologies is raised high. We are already over 50 years into the solar industry and so we know now that these technologies will work for the long term. This underlines how important it is to make a start.

Richard Leakey discovered a 196,000-year-old *Homo sapiens* fossil in the Omo valley of Ethiopia, the earliest modern human remains at the time to be found and now eclipsed by the finds in Jebel Irhoud, Morocco, that date us back to 315,000 years ago. Brain size is often linked to the ability of the human being, born with no other real assets in strength, flight, camouflage or cunning that other species became specialized in. The Leakey family were witness to the increase in cranial capacity, or brain size of hominids, likely due to many causes, among which was diet but also adaptation to changing environment, notably climate. Finding food, planning for a group of humans to survive over long cold periods and communicating were learned skills abetted by the wetware (brains) to make it happen. In the last 10,000 years, brain size actually diminished but today is on the rise again, with the resurgence of food and good conditions. The question remains whether the increase in our brain size will be sufficient, species wide, to overcome the challenges man has set himself in this era, notably with the trend toward sustainability.

The tone of this book is optimistic despite the discussion of externalities, because of the huge economic promise of a sustainable economy, and it's very clear that the progress of renewable energy and sustainability technology to date has far outpaced expectations of even just one decade ago. This is a David and Goliath story, where expectations are slowly but inexorably switching to what appeared to be the less likely horse, away from fossil fuels, to the benefit of all.

Chapter 1: Humans
and Technology

Humans and Technology

Figure 2: Two men in the act of breaking a Jacquard loom. Source: An 1844 engraving from the Penny Magazine.

They said Ned Ludd was an idiot boy
That all he could do was wreck and destroy, and
He turned to his workmates and said: Death to
Machines
They tread on our future and they stamp on our
dreams.
—Robert Calvert, from the album *Freq*, 1985

We are in the early stages of a significant transition in the way we obtain energy. History tells us that there are consequences of changes in the use of technology that have significant influence on the quality of life (QOL). A quick look at some relevant historical events helps to put these changes into context. A previous major turning point was the industrial revolution that brought us huge increases in productivity and quality of life and led, not in a straight line, to the global civilization we enjoy today. When man makes a change in the world, there appear to be good and bad consequences of these changes. The development of innovative textile machines brought growth to the textile industry and the economy that led to a follow-on revolution in the energy, shipping, manufacturing and distribution activities of an emerging industrial society.

It was not all positive, however. The impact of new technologies often had a bad effect on local populations. Wartime encourages innovation of new technologies that can provide an impetus toward a competitive edge over an enemy, with more powerful, fuel-efficient engines and more accurate, longer-range weapons and so forth. Innovation came as much out of necessity in geopolitical struggles as from the parallel increase in knowledge in the fields of science and engineering. The demand for weapons required standard methods of manufacturing them; demand for sheets and cloth for military and civilian clothing inspired the introduction of cheap-to-operate, wide-frame automated looms.

This all led to a reduction in the need for artisanal skilled labor. You would think, in a fair world, that the introduction of labor-saving machinery that puts people out of work would create wealth that would be used to compensate those losing their jobs in some manner. In the big picture, and much later in many countries, social security and health care safety nets evolved. In practice and depending on the government, it meant the owners of the productive assets got to make that decision and generally kept the rewards to themselves if they could. Today there is a significant new wave of automation, with robots manufacturing many things, from vehicles and planes to electronic components and even food. This plays a major role in continuing to cut the cost of manufacturing and creates value, which translates into increased profits, increased pay and higher corporate value.

On 20 December 1811, at the start of the industrial revolution, in Leicester, England, the *Nottingham Review,* a local newspaper, published an article about textile industry labor conditions. Weavers and craftspeople were being put out of business by the new large mills, which were operated for long hours by workers whose lives were almost reduced to slavery. Mill owners were only interested in producing as much as possible for the least possible cost. Human welfare was at the bottom of the list of priorities, resembling the worst excesses of medieval feudal society. Growing mill businesses deprived ordinary craftspeople of incomes and lifestyle. It could only end one way.

Ned Ludd was a weaver from Anstey, near Leicester. Sometime in 1799 he disappointed his employer and was given a whipping for laziness. In a fit of rage, he smashed two of the textile machines, an act that cemented his name into history and gave rise to a rebellious British working-class movement called the Luddites. The event inspired folklore, songs and tales and was also reminiscent of another, earlier legendary English hero, Robin Hood. The movement picked up steam, so to say. Between 1811 and 1817, the British Army was often more engaged battling the Luddites all over Britain than it was fighting the Napoleonic Wars on the European Continent.

In 1813, just a year after General Ross beat the Americans in the Battle of Bladensburg and sacked Washington, DC, there was a show trial in York, United Kingdom. Sixty Luddite defendants were put in the dock, many not even associated with the Luddite cause. Everyone in that trial was either executed or sent into penal exile. This draconian reaction appeared at least temporarily to calm the rebellious storm.

The unrest was also seen in another area of industry, agriculture, which was experiencing the introduction of labor-saving machines. As new harvesting and planting machines replaced the bucolic harvest image of seasonal workers cutting sheaves of wheat and barley, those same workers lost their farm incomes. The negative social effects of this gave rise to the start of friendly societies, or unions.

The Tolpuddle Martyrs were the initial example of a group of farm workers from Tolpuddle, Dorset, in the United Kingdom, whose leaders assembled legally in 1832 and founded the Friendly Society for Agricultural Labourers. Local landowners reacted fast. Six ringleaders were put on trial, found guilty and deported to Australia's penal colonies for seven years. Public outcry against this unfair treatment resulted in the first ever recorded protest march and the collection of 800,000 signatures, a staggering achievement when you consider it was done manually. As a result, all but one of the Tolpuddle Martyrs were released and many moved to Canada. Luddite events echoed these early examples and spread throughout Europe and later to the United States and the rest of the world.

In today's world, this movement still exists and is represented by a not-so- passive resistance to change, visible as anti-globalization, anti-consumerism and reactions to a new wave of ever more complex, "inhuman" computerization—the reduction of complex personalities to numbers. A human laborer is something to be "dealt" with. These thoughts recently resurfaced in the Occupy movement that swept the United States in 2011 and now are evident in the science denial, fake news, antivaccination and alternative medicine movements and are enshrined as an anti-modernization philosophy termed neo-Luddism.

Martin Heidegger,[2] the German philosopher, articulated in 1953 that human alienation is a reaction to the exploitation of natural resources. He said it altered the human "way of being" in a grotesque manner, reducing us to "not-beings" and an associated abandonment of awe and wonder. He wrote of a yearning to return to an earlier, romanticized, populist, agrarian simplicity. The reduction of human beings in modern capitalism to exploitable machines, human capital capable of producing yield and production, emulated slavery. It was too demeaning not to result in a strong reaction from the exploited person.

Figure 3: The bucolic version of human productivity, rudely smashed by the industrial revolution and mass production. Source: The Harvesters by Pieter Bruegel the Elder ca. 1525- 1569. Brussels. The Rogers Fund, 1919. Public Domain.

The concept of returning to a simpler, more romantic time is wide-spread and is balanced in the nerd, scientific mind by a positive alternative vision and a more elegant future existence that offers a higher quality of life as a result of embracing technology.

During my long summer holidays from university in 1977 and 1978, I worked in the vineyards of southeastern France, in the Languedoc, in a small town called Saint Chinian. I experienced the hard work, but bucolic nature of the seasonal farm worker. I was taken on as part of a grape-picking team that had about eight people in it, made up of some Spanish seasonal pickers, a couple of local people, a German and me. We lived in a barn on the vineyards and got up with the sun. We had a bucket and a pair of clippers to cut the bunches of grapes from the vine. At first, we got a few cut fingers as we fumbled around under the leafy vines, but eventually we got the hang of cutting grapes quickly.

Once full, the buckets were emptied into a rapidly filling larger container in a truck. At lunchtime, we had a break to eat tomatoes, baguette bread, *saucisson sec*, soft cheese, apples and bars of nougat. We all sat together on our buckets and drank bottles of early wine that, although only partially fermented, were nonetheless delicious. If we reached the end of a row of vines at the exact same time as a neighboring team, those bottles were brought out again and shared. Everyone would hug and there were pleasant introductions in the sunshine. In the evening, the team would take their day's activity and fatigue, and after dinner with more of the wine, get to bed early. There was a clear sense of purpose, happiness and participation in something worthwhile in the sunshine and, for a university student, it was well paid.

The idea that technology is the enemy of the people echoes a similar antipathy to capitalism. The ability to effectively mobilize a workforce toward an end was nonetheless commonplace. Take the example of armies fighting with discipline toward a goal or masons building a cathedral. This form of large-scale human mobilization had not been as common in the industrial workplace except in the growing coal mining industry.

The wider issue calls into question what we are here for and how we live and with what aspirations. I am, however, going to leave all that to better trained observers and I will concentrate on the story of technology and its sustainability. It does seem self-evident, though, that as we find ways of making our own energy and our own water using sunlight and the humid air, and growing our own food, the attraction of self-sufficiency serves as a connection to this "simpler world" and, with a bit of luck, can absorb much of the need for a return to simplicity in coming years. The desire to "return to nature" is deep. Indeed, even "going on holiday" is at bottom merely a reversion to this elemental human nomadic freedom and a temporary escape to paradise from an otherwise oppressive modern age, where we all need a job and income to support a family and pay for food and housing. The trend toward more control by householders of their own distributed energy supplies is attractive because it also plays into this helpful "self-realization."

A more recent and famous modern American Luddite was Theodore "Ted" John Kaczynski He was accepted to Harvard at the age of 16 and became the youngest professor hired at the University of California, Berkeley. After only two years at Berkeley, he sacrificed an academic career that could have taken him all the way to the top of his profession. Instead, he fostered a deep indignation about human so-called progress and its damage to the planet. He embraced a life of self-sufficiency in the woods. Upset by a development close to his dwelling, he became radicalized and initially indulged in small- scale sabotage. The indignation he felt grew, and he determined to redress society not by reform but by revolution. He decided to use letter bombs to make his point. His bombs were aimed at universities and airlines, and so the FBI called him the "Un"iversity and "A"irline bomber, or Unabomber, which stuck in the popular imagination.

From 1978 to 1995, he mailed letter bombs nationwide, killing 3 and injuring 23. He offered to stop his campaign if the *New York Times* and the *Washington Post* both printed his 35,000-word manifesto, "Industrial Society and Its Future."[3] The opening words were, *"The industrial revolution and its consequences have been a disaster for the human race."* Eventually, on September 19, 1995, both newspapers did publish the manifesto. Kaczynski explained that human beings have become divorced from their connection with nature and from natural human behavior by technology and the power in the "industrial-technological system." He called for a revolution against technology and included steps to accomplish this. He sent no further bombs, but his writing style and idiosyncrasies were finally recognized by his brother, David. The FBI obtained a warrant and raided Kaczynski's hut and found the damning evidence, his original manifesto.

Like pro-lifer zealots who kill abortion doctors, Kaczynski's violent reaction went way over the top, but he nonetheless tapped into a real theme of the modern industrial alienation of human beings from nature. He was given eight concurrent life sentences without the possibility of parole. All his possessions were auctioned, and the proceeds sent to his victims. Later, he managed to update his personal details on the Harvard alumni association website, listing his eight life sentences as awards and his current occupation as prisoner. Somehow, in his mind, in a possible world of the future where humans indeed became simpler and more agrarian, he would be released and celebrated as a hero and political prisoner of the inhuman capitalist system. Unfortunately for him, going violent was the last straw for the rest of us. The question is, How did he place such a low value on the positive side of technological progress in fields such as medicine, transportation and communications?

Today, some Neo-Luddites continue to promote violence as a means of achieving the changes they seek, although the reality of their impact is generally, and thankfully, more of political and intellectual engagement. The 2011 Japanese Tōhoku earthquake, the fifth largest earthquake ever recorded, caused the tsunami that brought about the Fukushima nuclear disaster. This event also inspired a horrifying and violent example of Luddism. On the seventh of May 2012, Roberto Adinolfi, CEO of Finmeccanica in Italy, was shot and injured in Genoa by an anarchist group called the Olga Cell of the Informal Anarchist Federation International Revolutionary Front.[4] They were worried that a European Fukushima was only a matter of time. Three of their members were arrested for plotting to bomb IBM's Swiss headquarters in Zurich, and they later sent a parcel bomb which exploded in the offices of the Swiss nuclear lobby group, Swissnuclear. They wrote:

> Science in centuries past promised us a golden age, but it is pushing us toward self-destruction and slavery. [...] With our action we give back to you a small part of the suffering that you scientists are bringing to the world.

To be sure, there have been huge technological disasters for the human race. Technology is a double-edged sword, and even though its worst effects are often a result of good intentions and not malice, there are, of course, examples of technology used with malicious intent.

Acid Rain	Loss of Community
Anthropocentrism	Loss of Language and Culture
Anxiety Disorders	Mania
Attention Deficit Disorder (ADD)	Medical Mistakes (440,000 fatalities in the US in 2012)
Asbestos	Mercury Poisoning
Climate Change and Global Warming	Natural Resource Depletion
Consumerism	Nuclear Waste
Domestic Abuse	Overpopulation and Overcrowding
Economic and Political Inequality	Ozone Depletion
Environmental Degradation	PCBs
Eugenics	Pesticides
Depression	Pollution
Genetic Deterioration due to relaxation of Natural Selection	Powerlessness
Genetically Modified Organisms (GMO)	Propaganda and Psychological Manipulation
Globalization	Psychological Disorders
Guns and Transportation Accidents	Social Alienation
Heart Disease	Species Extinction
Immigration	Stress
Impersonal Computerization	Surveillance Technology
Lifestyle Diseases from sugar, processed foods, meat	Thalidomide and other bad drugs
Incompetence and complexity	The Military Industrial Complex
Industrial capitalism	The Rat Race
Corporate Social Control	Unemployment

Figure 4: Some of the problems associated with technology. Source: The author.

Figure 4 is a very incomplete, but nonetheless long list of bad things that can happen because of technology that doesn't even mention warfare or weapons.

We have a paradox. Industrialization gave us sanitation, electricity and running water, which led to better health, disposable time and wealth for some. Much of the world's population, the billions without energy or food or clean living conditions, understandably aspires to achieve these basic necessities that are a critical foundation stone of a higher quality of life.

The introduction of some technologies has been dramatically positive. Developments as apparently simple as the washing machine freed up women (Swedish statistician Hans Rosling's TED talk[5]) and allowed them better access to education, the workforce, and careers and induced major social change. On the other hand, billions of women still do laundry manually.

An interesting development in today's world of automation is the growing use of technology to weld, walk, assemble, paint, drive, mix drinks or chemicals and myriads of other things. I know of a capacitor manufacturing company that will save significant amounts of money normally spent making the capacitors with a robot arm, able to pick up components and assemble them exactly, and to do it 24 hours a day, without a break, food, salary, illness, accommodation, clothing, counseling or union membership. It simply consumes a measure of capital and electricity, allowing productivity to soar. This explains why the significant increase in manufacturing revenues in the United States is matched by a decline in manufacturing jobs. Jobs in major industrial countries such as the United States, Germany and Japan have been exported to lower-cost countries like India and China and increasingly to domestic robots in the developed nations. Half of the United States' current jobs are at risk of automation by 2050.

The increase in income to the manufacturer and the decrease in incomes to a huge part of the community mean a jobless economy and increasing wealth inequality. Of course, in the passage of time, European countries among others have inaugurated single payer health care, free education and a significant safety blanket for jobless workers, all measures that redress the initial lawless exploitation of labor. Today a more thorough form of safety blanket is being debated worldwide: a living, minimum, guaranteed income to every citizen whether they are working or not. This concept is literally and figuratively "alien" to the capitalist mind-set.

In Switzerland, they were debating a system to pay a minimum wage. If you had an income already, you could expect another $2,600 (U.S. dollars equivalent) deduction in taxes, but if you and your wife were jobless, at least there was $5,200 coming in every month, enough for shelter and food, with a bit more for children. In June 2016, Switzerland's parliament defeated a measure that would have provided $2,600 to every legal resident of the country: 76.9% voted against the measure. In the United States, a 2010 report[6] by the White House for Congress said that the probability of a worker making less than $20 per hour losing their job to automation was 83% and if you earned $40 per hour, it was lower at 31%. Another term for this is "universal basic income," which is being seriously considered by some countries, whose progress will be closely watched.

Today, we are much more aware that technological innovations can have secondary consequences. The negative consequences are called *costs* or *externalities*. Some were listed earlier in this book. I focus on the externalities of coal, oil, nuclear and gas (CONG) and, though they have fueled our current civilization, how they also deteriorate the wider quality of human life. I believe that persisting with CONG is possibly one of the worst things humanity can do over the medium term, even though CONG is, nonetheless, a vital bridge to get us to sustainability. The effects of CONG's externalities today are compounding. What a shame it would be if the global community was so newly sensitive, conservative and fearful of unforeseen consequences that they imposed checks on all new and promising innovations that actually fix many of the problems. This, of course, would be tantamount to making the current problems permanent and therefore... terminal.

Good Versus Bad Capitalism

"*If something exists, it must be possible*" —Amory Lovins, about the possibility of implementing resilient, existing solutions.

Henry David Thoreau wrote a tale of man and nature in balance in his *Walden Pond*. The book's popularity was mainly due to the sense that this balance was at risk of disappearing. The French concept of the Noble Savage, epitomized by the artist Gaugin's long residence in Tahiti in the South Pacific, was also an exploration of this theme. In some ways, this labor lifestyle has not gone away. Business had a reputation of despoiling the biosphere and undervaluing human capital, taking what it needed to make and sell its products. This became a very deeply rooted mindset. The despoliation, also known as the tragedy of the commons, was highlighted by authors like Rachel Carson, who wrote *Silent Spring*[7] in 1962 as a reaction to the relatively naïve and careless way that industries simply used poisons in the environment or dumped chemical wastes, causing pollution of all sorts in the absence of any regulatory control.

One of the subjects discussed in her book was an odorless, colorless liquid that was widely used as a pesticide without any regard to side effects such as disrupting animal life or its role as a carcinogen. It was called dichlorodiphenyltrichloroethane (DDT). DDT quickly became the main culprit.

At the time, "good" capitalism was deemed, by definition, not to be "for" workers or nature. The purpose of business was defined as existing for the maximum return to owners or shareholders and thus exploitative of workers and the environment. It was "supposed" to be bad. Carson's work reinforced the stereotype. The character Michael Corleone in *The Godfather* famously said, "*It's nothing personal. It's strictly business.*" This was a core sentiment of business everywhere. Success was something you worked for no matter the cost. If you want a friend, went the idea, get a dog.

In the latter half of the 20th century, there was much evidence that business could also be an agent for good, even though this was deemed a radical idea. It took time to gain a foothold. In 1974, a Bangladeshi with a PhD in economics from Vanderbilt University in the United States, Mohamed Yunus,[8] decided to lend out small sums of money to 42 poor families so that they could avoid predatory lending rates and make things that could be sold in a market. Grameen Bank (Bank of the Villages) emerged in 1976 and became a very successful business, initiating a revolution in business models that showed you could do well by doing good. Yunus was awarded the 2005 Nobel Peace Prize for the impact he was making, and, I'll namedrop, I met him at a meeting at the United Nations Foundation in New York in 2007, when I was invited to attend by Ted Turner and his son-in-law.

In 1993, Paul Hawkin published *The Ecology of Commerce*[9] in which he suggested the reverse of the established capitalist paradigm. Business had indeed plundered the Earth, but only business could reverse the process. He said that business could be good for people and the environment and be profitable. Interestingly, despite already being a successful author with two previously well-received books, every major business publisher in the United States boycotted the book. They were loath to risk highlighting a work that openly betrayed "well understood" "fiduciary" responsibilities, namely, the idea that shareholders can only do well when the companies they are invested in are exploiting the natural world. Despite this, the book became a best seller and its message was gradually heard by businesses all over the world.

There have been many positive results. One impactful result of the book concerned Ray Anderson, the founder and CEO of a company called Interface that makes office carpet tiles. His business depended on oil and fossil energy for its energy and raw materials. After reading Hawkins's book, he determined to change this state of affairs. He changed his supply chain and successfully reduced his production costs by $400 million, which paid for all the process improvements. By 2015, Interface had cut CO_2 emissions by 82%, with the goal of 100% by 2020. The old business paradigm would say that he was probably losing money and foolish to think that he could both do good and be profitable. In fact, he became the world's dominant producer of recycled and emissions-free carpet tiles, competitively priced, very attractive, convenient and a profitable, billion-dollar business to boot. The process inspired his staff, changed his factories, transformed his supply chain and even spread goodwill in the market. There is a very well- presented TED talk[10] by this humble, humorous and effective philosopher- CEO.

The Growing Impact of Efficiency as an Indicator of Energy Maturity

> *"If one day, my words are against science, choose science."* —Mustafa Kemal Atatürk, secular leader of Turkey after the fall of the Ottoman Empire.

Industry in general has often resisted government calls for improvements in efficiency. The most obvious example of this is the auto industry's claims that the consumer wanted to buy cars with power and safety like sport utility vehicles (SUVs) that consumed gasoline at a rate of 1 gallon for every 15 miles (mpg) at the same time the average mpg in Europe and Japan was over 40, and each gallon more expensive too. In 1975, after the Arab oil embargo, the U.S. Congress embraced the Corporate Average Fuel Economy (CAFE) standards to make every drop of available fuel go further. Some stern discussion about the future of CAFE efficiency standards in the United States and the rescue package for GM during the 2008–9 market collapse imposed a better level of commitment to car efficiency standards. The Obama administration's efforts to clean up coal-fired power stations has resulted in much state resistance to the Environmental Protection Agency's (EPA) designation of CO2 as a pollutant and, of course, all the implied issues about closing coal-fired power stations in pursuit of cleaner environments, despite the fact that there are now far more solar and wind jobs in America than in the coal industry.[11]

On 8 August 2016, in a unanimous decision, the Chicago-based 7th Circuit U.S. Court of Appeals rejected an industry-backed request to overturn a 2014 ruling that set energy efficiency standards, specifically aimed at standards for commercial refrigerators. For the first time, the court used the concept of the "social cost of carbon." The Obama administration used this standard to estimate the cost per metric ton of carbon dioxide emitted into the atmosphere and put it at $36. When the DOE did its cost-benefit analysis, it was in pursuit of an energy conservation measure and the environmental costs of emitted carbon needed to be taken into account. This was the first instance in the United States of a federal court using an externality as a justification for a ruling and as a tool for internalizing the impact of the otherwise emitted carbon on climate change. The Air-Conditioning, Heating and Refrigeration Institute, which led the industry litigation, naturally said it was "disappointed" with the ruling. Industry and congressional Republicans, business interests and energy companies attacked the ruling as bad mathematics and irredeemably flawed.

Robert Socolow and Stephen Pascala of Princeton University produced a relevant paper[12] in 2004 that demonstrated that if we wanted to reduce our greenhouse gas emissions over the coming 50 years, there were about 13 approaches, which they called "wedges," including various efficiency approaches that could be used and for which the resources exist, each of which would have its own impact on reducing the greenhouse gas emissions. They are:

Improve fuel economy   Nuclear fission
Reduce reliance on cars   Carbon capture and storage
More efficient buildings   Wind electricity
Improve power plant efficiency   Photovoltaic electricity
Decarbonization of electricity and fuels   Biofuels
Substitution of natural gas for coal   Forest management
Agricultural soils management

Socolow and Pascala quantified this by saying each wedge would save 25 gigatons of carbon (Gtc) over 50 years, meaning that only 7 of them would be required to stabilize CO2 emissions at 500 parts per million. Most of these involve the squeezing of existing energy sources to obtain more energy from less fuel, an efficiency drive that effectively answers the challenge.

In 2009, *Time* magazine named Amory Lovins[13] one of its 100 most influential people in the world. The reason was for his four decades of work as an energy consultant. He is a Harvard-trained physicist who is passionate about the use of energy efficiency as one of the key paths forward. He argues that customers don't want kilowatt-hours of energy; rather, they want energy "services" such as lighting, heating, cooling and spinning shafts. If you can get a task done with less energy, that's an improvement. His analysis also suggests a different energy business plan than the current utility subscriber model. As a free enterprise thinker, rather than an environmentalist, his impact has been immense and has resulted in his being a consultant to states, countries and corporations around the world as they feel their way forward to an emissions-free energy paradigm.

The Arab oil embargo in 1973 created an audience for his ideas about energy resource management, and he started publishing books about the subject in the early 1970s. He established the Rocky Mountain Institute (RMI) in 1982, built to demonstrate energy-saving methodologies and to show that a high standard of living was possible using very little energy. Today RMI has a staff of 85 and an annual budget of $13 million. He points out that nuclear fission as it has been implemented in the USA is really just another intermittent source and that even coal power can be considered intermittent too. Of Fukushima, he said, *"An earthquake-and-tsunami zone, crowded with 127 million people is an unwise place for 54 reactors."*

Lovins introduced the concept of "negawatts," or watts saved by not being consumed, a delightful and useful metric for efficiency. LED bulbs and lighter vehicles allow global warming to be addressed not at a cost but at a profit. There is no cheaper source of energy than the energy not consumed.

Another of his projects was to design a car that was lighter mainly due to being made of carbon-fiber body panels. The hypercar project, in conjunction with BMW, resulted in the BMW i3, which Lovins currently drives, along with Volkswagen's 313-mpg XL1. Both cars have dramatically reduced weight using carbon-fiber body panels and dramatically improved overall efficiency using electric motors. Lovins points out that these measures, efficiency gains and reduced energy use, could combine to produce a faster growing economy and $5 trillion in savings without any changes in taxes, subsidies or laws. Private enterprise is eager to reap the savings offered by existing technology and improve living standards as it does so. Any technological steps forward just add grist to this energy mill.

Impact Investing and Markets

"Present thinking holds that man has a time window of five to ten years before the need for hard decisions regarding changes in energy strategies

might become critical." —James Black, Exxon climate scientist in 1978.

In the aftermath of the 2008 mortgage-inspired stock market collapse in the United States, which also impacted global equity markets, investors' appetite for risk dried up to an extreme. The renewable energy public equity markets have experienced a long boom/bust cycle. In the early 2000s, the market grew steadily from just about 50 stocks in the year 2000 to over 2,000 companies by 2008. Some big energy engineering companies, such as GE and Siemens, found themselves transformed by the renewable energy revolution. By 2008, fully 40% of Siemens's earnings were accounted for by renewable energy activities such as wind and solar. GE Wind emerged successfully from the bankruptcy of energy company Enron. The start of renewable energy exchange-traded funds (ETF) saw the first solar and wind ETFs appearing. The 2008 market collapse witnessed even large, successful, liquid stocks like First Solar (FSLR) soundly punished.

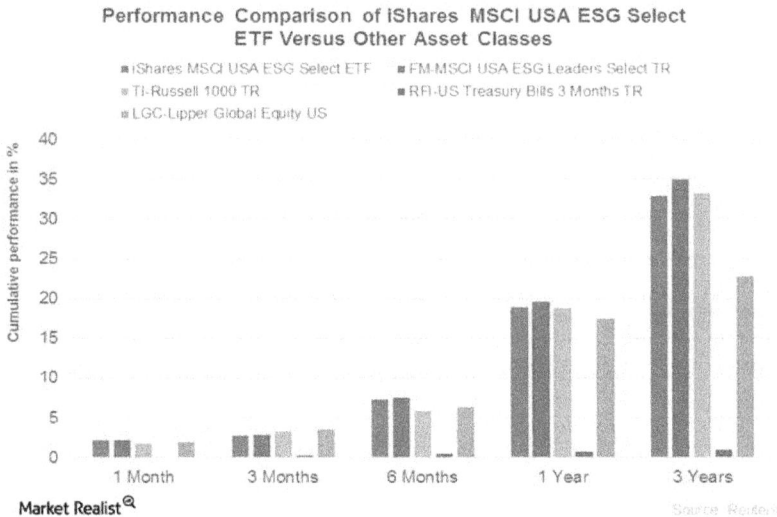

Performance Comparison of iShares MSCI USA ESG Select ETF Versus Other Asset Classes

■ iShares MSCI USA ESG Select ETF ■ FM-MSCI USA ESG Leaders Select TR
■ TI-Russell 1000 TR ■ RFI-US Treasury Bills 3 Months TR
■ LGC-Lipper Global Equity US

Market Realist℗

Figure 5: It's often maintained that SRI funds do better than regular market baskets. This data would appear to prove that point. Published in September 2017, it shows that this basket of SRI or ESG stocks has indeed outperformed over the last 3 years. Source: Market Realist, Amanda Lawrence. September 20, 2017.

First Solar is a manufacturer of cadmium telluride thin-film solar panels that had an advantageously low price per watt compared to the cost of silicon solar panels. The stock dropped in 2008 from $310 per share all the way to $11.43 by June 2012, even though it would have been difficult to find another company with as strong a balance sheet, as impressive a product and as solid and visible a growth of revenues and earnings.

In 2010 and 2011, to strengthen strategic Chinese export industries, the Chinese government added to the global solar industry's struggles by awarding their own solar companies $40.7 billion,[14] effectively bailing them out for weakness caused by the global market crash and giving them more competitivity in the European, Asian and U.S. markets. This was done precisely at the time when new funding was all but impossible after the market crash in the United States had sent its tsunami of wealth destruction around the world. Sales of Chinese panels in Germany, at the time the leading solar market, caused indignation there about paying feed-in-tariff (FIT) subsidies to local developers that favored the cheaper but also newly high-quality Chinese panels. This accelerated the decline in the German FITs and caused a German company, Solar World, to challenge the Chinese in the World Trade court system, a case that continues today. The news in late 2016 was that Solar World was hacked by Chinese cyberwarfare military units looking for information about the case. The Chinese market has gradually moved from solar exports to domestic installations as well, and their 13th 5-year economic development plan also talked about investing in domestic solar capacity.

Another European market, Spain, with 20% unemployment and falling GDP, also embraced renewable energy subsidies to accelerate renewable energy targets and capacities. Eventually, caps on electricity pricing imposed to protect its electricity subscribers in the year 2000 while fossil fuel prices were climbing resulted in a huge debt obligation by 2008 for the Spanish government.[15] The additional and relatively smaller renewable energy FIT subsidies were the straw that broke the camel's back, and in July 2012, after significant growth of solar and wind capacity, the Spanish government curtailed its FIT subsidies, causing many existing projects to become uneconomic and in turn causing bankruptcies and failures, but most importantly making Spain a bad partner for private industry. With all this volatility, any investments in solar technology companies often caused agony for whipsawed portfolio managers.

Another odd phenomenon was the irrational link between solar or wind stock prices and the volatile oil price. In the developed world, very little oil is used for the generation of electrical power. The stock of a solar company like Sunpower in California therefore ought to be proof against the decline in the oil price, especially since electricity is arguably the least volatile of all the commodities. Nonetheless the market would go no further than apparently saying that oil, solar and wind companies were all energy stocks and they were traded in the same basket. It's true that many or most oil companies tried hard to explore renewable energy businesses. BP became "Beyond Petroleum," adding solar and biofuels activities. Shell had its own solar and hydrogen activity. Arco was one of the original solar investors. Exxon, as well as mastering the climate science in the 70s and 80s, put over

$600 million, as they say, a "rounding error" for such a large company, into an algae joint venture run by Craig Venter, the scientist behind the private effort to establish the human genome. Dr. Venter was bringing the sophistication of genetics to biofuels by scouring the world for organisms that would optimize production of lipids that could be made into biodiesel or jet fuel. Total, the large French oil company made a large investment in Sunpower and acquired SAFT, the lithium ion battery group in France. Chevron funded Chevron Energy Solutions to make an impact in the environment and improve energy efficiency and has remained a top solar installer in California. They also have an ongoing geothermal activity in Indonesia and partnerships with schools and municipalities for efficiency and solar power.

I think investors are still "once bitten and twice shy" about taking risks in early-stage technologies. Bloomberg New Energy Finance tells us that 2015 was a great year for renewable energy investment globally, but if stocks are not doing well, what is? An MIT paper[16] by Dr. Varun Sivaram, Dr. Bend Gaddy and Dr. Francis O'Sullivan called "Venture Capital and Cleantech: The Wrong Model for Clean Energy Innovation" examined the state of venture capital in cleantech startup companies for the period since the sector's boom in 2006 until 2011, a few years after the stock market crash. The paper compared the investment outcomes, the risks and rewards of two other industry areas, medical technology and software, for comparison. The conclusions were perhaps predictable. Cleantech companies require a lot of capital, take a long time to develop, rarely experience a satisfactory exit and have narrow margins. The paper said that across all three industries a similar amount had been invested because the venture capital firms had initially expected similar investment outcomes. The cleantech sector stood out as the worst performer with lower returns, and this was especially true for cleantech companies that were championing new materials and processes.

The paper concluded that several things are needed to encourage better cleantech performance. Patient money or long-term, strategic and institutional investors who are happy with a long-term return on their investments need to be involved. It went on to say that public policy needs to support smaller companies with grants. Great disruptive ideas are rare. They are the subject of initial deep suspicion, while simultaneously being sorely needed.

I am still supporting some disruptive initiatives that I believe could change the world. One is a cellulose-to-sugar (CTS) solution, which as far as I can see is extremely effective. It can convert any source of cellulose. Placing cellulose with a little clay as a non-sacrificial catalyst into a ball mill, a 150- year-old technology, for 15 minutes obtains an 80% conversion to lignin and sugar monomers without enzymes, heat, chemicals or even any liquids. There is also no waste, so the 20% unconverted material is the first to convert when put back into the machine. It stands a chance of completely disrupting the world's approach to biofuels, which are currently made from food (corn, sugar cane, soy) and consequently incur regulatory risk.

Other initiatives offer electrical storage solutions. One has developed a supercapacitor made with barium titanate as a dielectric, a dry electrolyte, with voltages as high as 3,400 volts, resulting in significant amounts of energy storage. These stories are industry disruptive and may not result in anything, but if they are properly managed they could massively improve life.

Socially Responsible Investing

Socially responsible investing (SRI) reaches back hundreds of years and reflects social issues of the time, whether they be slavery, religion or the protection of workers. "Sinful" companies in the drink, guns, sex or tobacco industries have also always been a negative target of SRI investors. There is a fund today whose goal is simply to sell the CONG public equities and go long the sustainable variety that, with the current evolution of coal stocks, has a 2015 return of over 30%.

Gender equality, civil rights or labor issues, economic development projects and avoiding unpopular wars are other issues that have been SRI foci. Unpopular regimes were targeted, sometimes already subject to economic embargos, such as Apartheid's racial segregation in South Africa. Apartheid ended at the same time the iron curtain that separated the communist USSR and its satellite countries from the West came down between 1989and 1991. The disinvestment by those who disagreed with the local political regime caused financial pressure for 75% of South African employers and finally led to a charter calling for the end of the regime. Later, this experience was also used successfully against Sudan to address the problems in the Darfur region in the south of the country.

Climate, environmental or renewable energy investment has only become a socially responsible investment target since the 1990s. The world is embracing the concept that climate change, among other impacts of fossil energy use, has become a significant and quantifiable investment risk. Investor conferences focusing on networking investors interested in environmental investment exposure now abound. SRI funds also have grown, and the amazing thing is that often they have been shown to have better returns as external costs disadvantage inefficient CONG technologies. The first SRI funds appeared in the 1970s. Today, SRI, also called environmental, social and governance (ESG) investing, has become a booming affair with $5.67 trillion in assets at the start of 2014, 76% more than 2012. In the United States, 18% of all investments out of $36.8 trillion under professional management are steered by SRI criteria. Activist or governmental or politically influenced funds often adopt SRI investment policies. The large Californian pension funds, CALPERs and CALSTRs, are both encouraged and eager to divest from fossil fuel stocks. Since 2012, there are now four times as many SRI mutual funds and more than 20 SRI- themed exchange-traded funds (ETFs).

One of the most important modern socially significant themes in the world is whether to support the burning of carbon fuels any more. Of course, one of the main ways to disincentivize managements of publicly quoted CONG companies like Exxon Mobil, BP and Chevron is by persuading long-term shareholders to jettison their positions. With this in mind, the founder of the Boston investment fund Grantham Mayo van Otterloo, Jeremy Grantham, authored a document in 2013 called "Unburnable Carbon"[17] about the importance of leaving carbon in the ground and making the transition from fossil fuels in time to save the climate from heating up by 3.6°F (2.0°C). Today, many large pension funds, university endowments and other socially responsible investors are contemplating the sale of their carbon-intensive energy stocks so that they both preserve capital and play a role in the shift away from CONG.

Grantham raises the possibility that the emergence of renewable energy is just another bubble, destined to do what Grantham says has happened to 34 other identified historical bubbles, where experience tells us we will see the index once more revert to its original level prior to the bubble forming. It is also true, however, that technological awakenings can also permanently change the world.

Currently, the Nobel Prize is awarded each year and prize money comes from the Nobel Foundation, which has an exposure of $400 million to public CONG energy stocks. Since Alfred Nobel was a man who in his later years decided to leave a legacy of positivity with the Nobel Prize for individuals who contribute significantly to moving humanity forward, it is now deemed to be a conflict that the foundation also holds CONG stocks. A group of world-changing physicists, chemists, journalists, lawyers, authors of the Intergovernmental Panel on Climate Change (IPCC) climate report and Peace Prize winners all argue this conflict can easily be corrected by divesting the CONG assets. Because this conflict is ironic, I think it is only a matter of time before it is resolved.

The last two hundred years have seen major, one-way, directional changes in our lives. It's a poignant reflection of the hope for a better future that sees human beings innovate away from dreadful average living conditions over most of the 200,000 years of our species' existence. Suddenly, things changed in the last two hundred years, building on the knowledge unearthed by organized, disciplined human curiosity that caused us to discover things about the universe and that answered some big questions.

Amazing progress is the result. In 1903, the Wright brothers flew the first powered aircraft, and in 2013—only 110 years later, a small fraction of the time of our species—the *Viking* space probe left the solar system. Science is a branch of human hope that empowers us to resist the pain of existence and become awestruck by our growing knowledge of the universe. Part of our yearning to explore results in a better quality of life for everyone. Renewable energy is a new phase in an ever-changing world, leading us to an improved economy without the externalities that are now running rampant from fossil fuels. We have become so inured to having a civilization based on CONG that a big change has not seemed possible. Up to this point, the question was, "Is it even possible for a community to be 100% sustainable?" Now the question is "How long do we have to dally before gifting ourselves these energy advantages?"

Google's RE < C

In 2007, Google announced an initiative to explore renewable energy. They came up with a simple formula RE < C, meaning you could obtain a unit of renewable energy at a lower price than a unit of coal energy. It embraced a widespread, optimistic intuition that the plummeting prices of renewable energy would fall far enough soon enough that coal would be more expensive. Google was financially in a position to back up its new formula and publicized its intention to invest millions of dollars into geothermal, solar and wind technologies in the search for a lower-cost, sustainable approach.

Google represented an optimistic technology giant opining about a part of the "new economy" that could easily result in a breakthrough for sustainability. GE saw the same idea, and many European energy companies such as Siemens AG had experienced significant impact from the growing renewable energy industry. Even utilities such as E.ON, Vattenfall, Duke Energy, NRG, Florida Power and Light (NextEra Energy), Pacific Gas & Electric and others all recognized that the world was changing. However, in 2011, after barely five years, Google suddenly let the initiative drop. Their global audience, including me, gasped in dismay and felt it had to be a mistake, a premature conclusion not based on reality. The hope was crushed. In subsequent years, however, Google has invested over $1 billion in various solar and wind projects anyway. They also announced in a blog that the thrill of renewable energy and its positive impact was still a guiding motivation for their teams.

Ross Koningstein and David Fork are the two Google engineers who revealed the thinking behind dropping the RE < C initiative in a November 2014 Greentech Media article.[18] They originally thought that existing renewable energy technologies that were close to commercialization could offset the worst of the climate change scenarios and invested in them for that reason. By 2011, the engineers realized that their investments were not achieving their goals, so they canceled the program to rethink it. They realized that even if Google had led the way to an immediate 100% general global adoption of renewable energy, the transition would not have resulted in significant reductions in carbon dioxide emissions.

After looking at different scenarios with consulting firm McKinsey and comparing them to NASA scientist James Hansen's 2008 model that showed a level of carbon dioxide in the atmosphere of 350 ppm was required to stabilize the climate, they were not encouraged. As of December 12, 2016, it was 404.48 ppm.[19] Even combining the Google best-case scenario, which offered a 55% emission cut by 2050, showed that 350 ppm was not attainable. Yes, even Google's best-case scenario resulted in severe climate change (ocean acidification and rising sea levels, eroding coastlines, freshwater shortages, shifting climate zones, drought, wildfires, spreading diseases, food shortages, climate refugees, and therefore wars, etc.). What was needed was a combination of both zero-emission energy sources and methods to extract CO_2 from the atmosphere. Even if Google had achieved what it wanted and found a method to replace all the coal plants in the world, it still would not have halted climate change. When you compound this with the fact that solar and wind were both intermittent and needed storage to balance demand, a perennial problem, the new realization finally put the lid on RE < C. Google instead fell back on an internal approach that encourages innovation.

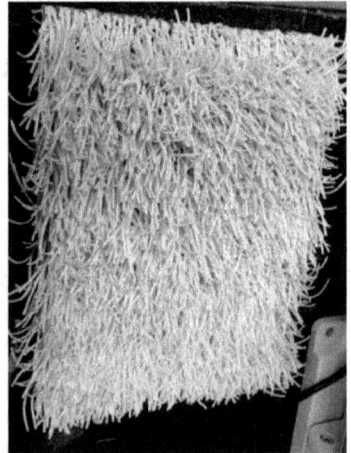

The CO_2 already in the atmosphere will take at least 100 years to be absorbed, and the warming momentum will continue all that time or more. If no more CO_2 were released, the warming would still continue for 100 years. This phenomenon is called *thermal inertia*. However, you only need to (I apologize) "Google" the phrase "technologies that absorb CO_2" and you currently obtain 14,700,000 results, most of which are discussion material on the subject. Among the references, however, are plenty of methods to extract CO_2 from

Figure 5: *CO2-absorbing plastic surface that resembles the high surface area of pine needles and can filter CO2 out of the atmosphere 1,000 times faster than real leaves. Source: Author's photograph.*

the atmosphere, not that they all work efficiently and economically, of course, but some are extremely feasible.

Google's policy did not fail because their formula didn't work, or renewable energy was not available at a lower cost than coal. It was dropped because the amount of CO_2 already in the atmosphere is what needs to be addressed most. Adopting clean energy on its own would not make the essential difference we all need. We also need to mop up all the "spilt" CO_2 already emitted before we can achieve the climate goals we have set ourselves. What greater driver do we all need than this?

On Monday, June 4, 2012, I went to visit Professor Klaus Lackner,[20] who at the time was working at the Earth Institute at Columbia University. Later he moved to Biosphere II, the large glass-and-steel building in Tucson, Arizona, also operated by Columbia University. His daughter Claire's science project involved bubbling air through a solution of sodium hydroxide, also known as lye or caustic soda. She used a fish tank pump and bubbled the air through the liquid all night to discover the next day that half of the acidic CO_2 had combined with the alkaline liquid sodium hydroxide ($NaOH$), turning it to sodium carbonate (Na_2CO_3). Although he knew that his daughter's science project used energy in the pump, he wondered if there was a low-energy method to extract CO_2 from the atmosphere and even a way to address the global CO_2 emissions challenge. He realized that tree leaves already absorb considerable amounts of CO_2 from the atmosphere by being a certain shape and they are held aloft in the wind to better accomplish this.

He discovered a polymer material, an anionic exchange resin, normally used to purify water, that absorbs CO_2 very quickly and that is covered with sodium carbonate. When dry, it is 1,000 times faster than trees at absorbing CO_2, a process called *engineered chemical sinkage*. When the material is wet, the CO_2 is conveniently washed away. This means you only need 1,000th of a forest to do the same job as the trees in that forest. Evolution formed leaves and needles as optimal gas exchange equipment for trees. One of the configurations of his polymer appears just like pine needles, as in figure 5. He formed a group called Global Research Technologies (GRT) to develop the product.

At Columbia University, he showed me an airtight bell jar in which you can monitor the CO_2 levels. When you put the plastic inside, the atmospheric level of 400 ppm declines. Then they dunk the plastic in water and the levels of CO_2 increase again. The plastic can cycle hundreds of times. A single plastic tree can absorb a ton of CO_2 a day; 60

million trees would absorb all the CO2 we emit. It's one of the few methods whereby the current atmospheric CO2 levels can be reduced. The trees can be located in deserts or right above underground CO2 storage sites so that the gas would not need to be transported in long, expensive pipelines. Each tree might cost $20,000, so 100,000 trees absorbing 3.6 billion tons of CO2 would cost $2 billion. A hundred million such trees would counteract the 40 billion metric tons of CO2 we add to the atmosphere every year. We already manufacture 80 million cars a year, so for an existential event, you would imagine it makes sense to consider this.

Lackner is now the director of the Center for Negative Carbon Emissions and professor at the School of Sustainable Engineering and the Built Environment of the Ira A. Fulton Schools of Engineering, Arizona State University. He has committed to an 80% reduction in CO2 emissions by 2050. As early as 1998, he realized that it was feasible to do this, especially if you allowed the wind to provide the energy to pass the huge volumes of atmosphere over the absorptive plastic surfaces. Synthetic trees were the idea that inspired a company called Global Thermostat, founded by CEO Dr. Graciela Chichilnisky in Stanford, California, with Seagram's heir Edgar Bronfman Jr. as chairman. They recognize that what we really need is a carbon-negative approach, not carbon-neutral approaches. In its latest assess- ment report, the IPCC said that since the lifetime of CO2 in the atmosphere is over a century, we need negative emissions to hold the line at 400 ppm. Paris COP (Conference of the Parties) 21 set this ambition in motion.

According to the UN IPCC Fifth Assessment Report & the 2015 Paris Agreement, Carbon Negative Technology™ is now the only way to avert catastrophic climate change.

The Center for Negative Carbon Emissions has synthetic trees on the roof of their building and need 10 to 20 million more to reach their goal. We hear a lot about reducing emissions, but today the narrative also needs to include carbon-negative solutions. A lot can be done with CO2, such as converting it back into fuels for transportation using renewable energy. Professor Lackner does not believe in the long-term capacity of carbon capture and sequestration methods like those used by oil companies, which send the CO2 down drill holes. For the professor, there is a solution in making the CO2 solid again as it is via natural processes that make many rocks, like limestone. Underneath the Alma Mater statue in the main quad of Columbia University there is a foundation rock made of serpentine, mostly made up of CO_2. Professor Lackner has figured out a way to speed up this formation.[21]

Another method might be to make actual hydrocarbon fuels from the CO_2 in the atmosphere. A company in the UK called Air Fuel Synthesis[22] was led by its original CEO, Peter Harrison, who was passionate and successful in being the first demonstration of the production of gasoline (or petrol to him and me) out of CO_2 and hydrogen from water. The hydrogen is added to the CO_2 and compressed, preheated and then distilled in a methanol reactor. The methanol is heated, vaporizing it into gas, which reacts with an aluminum and zeolite catalyst and finally is chilled to condense as gasoline. The pilot plant generated about a half liter per day. They use it to run a two-stroke engine and drive a motor scooter. The system uses energy to make atmospheric gases into fuel, but if the process uses all the excess renewable energy that exists, then it could easily be seen as a form of energy storage and at the very least another carbon-neutral option for liquid transportation fuels. Such initiatives are blossoming now.

Another professor, David Keith from Harvard University's School of Engineering, has a company called Carbon Engineering in which Bill Gates is an investor. They also develop technologies that take CO_2 out of the atmosphere. When air containing CO_2 is bubbled through a solution of calcium hydroxide, $Ca(OH)_2$, it forms a precipitate or powder of calcium carbonate, $CaCO_3$. Air enters an inlet of about 2 m^2 and 15 vehicles' worth of emissions are captured in a day. The CO_2-absorbing liquid flows by gravity across tightly bound PVC sheets. When CO_2 hits the liquid, it is converted to a carbonate solution. At the outlet, over 80% of the CO_2 has been removed. If combined with hydrogen, the CO_2 can be used to make new carbon-neutral hydrocarbon fuels.

These companies' solutions capture CO_2 that's already been emitted, while 60% of ongoing emissions come from cars, trucks and planes. Air capture of CO_2 can be installed on land that can't be cultivated and is 300 times less concentrated than in vehicle exhaust. The Kyoto Protocol carbon market will fund such carbon-negative initiatives in developing nations.

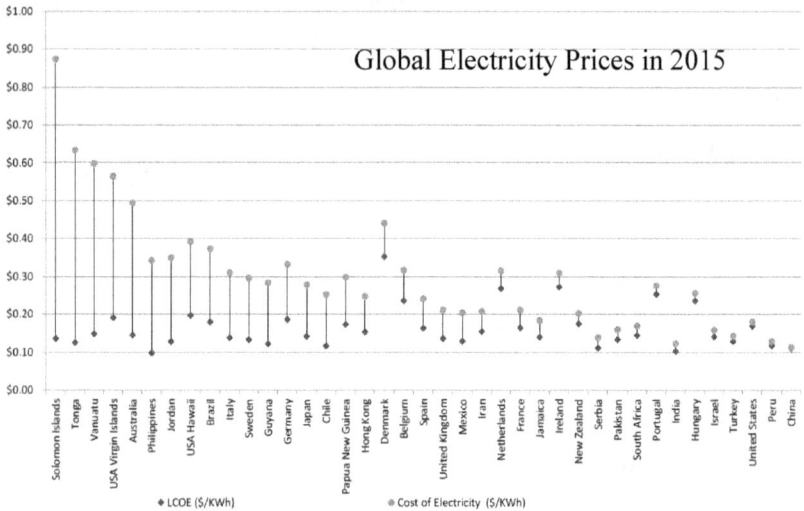

Figure 6: Cost of electricity in countries around the world. Source: Deutsche Bank Estimates.

The market traded $175 billion in 2012 and has fallen off since but is credited with successfully reducing CO_2 emissions levels by 30% in the carbon market trading nations.

It now appears that the original Google goal to obtain a unit of renewable energy at a lower cost than a unit of coal energy can finally be achieved, at least. It ought to be stated clearly here that a unit of renewable energy is already well below the huge cost of coal power if you include all the associated social costs, the externalities. As a function of the declining absolute costs of renewable energy equipment, recent power purchase agreements at less than 4 cents per kilowatt-hour in the solar space have considerably undercut the equivalent in coal.

A report by Deutsche Bank's solar analyst, Vishal Shah, in 2015 revealed expectations that 80% of the global electricity market will be at grid parity without subsidies, because grid electricity prices are one of the least volatile of all commodities, while solar equipment and financing costs keep falling and efficiency keeps improving.

Competitive impacts on balance-of-system costs will produce yet another 40% cost reduction over the next four to five years. Shah's 2015 figures for unsubsidized rooftop solar prices of $0.13–$0.23 per kilowatt-hour are already below the retail price of electricity in many markets and are headed toward $0.10 per kilowatt-hour, which already acts as a base line in figure **8**. The chart shows a relatively even lev-elized cost of energy (LCOE), but island subscription costs are huge. Unsubsidized solar power is a no-brainer, especially if you happen to live in the Solomon Islands, or any islands, for that matter.

In 2015, Saudi Arabian group Acwa Power International offered to build a solar park for $0.0585 per kilowatt-hour. Record-low PPAs have been recorded by Bloomberg New Energy Finance in Peru and Mexico. Another developer in Dubai was willing to risk a lot for the prestige of being the cheapest solar developer at just $0.0299 per kilowatt-hour. Though the 2016 bankruptcy of SunEdison cast a shadow over solar, renewable energy installations and companies that invest in projects to obtain cash flows, also called "yieldcos," the truth is that projects in sunny places, when funded at low debt rates, can indeed make money today at minimal power purchase agreement levels. Warren Buffet's NV Energy utility in Nevada agreed to pay an all-time low U.S. PPA of just $0.0387 per kilowatt-hour[23] in 2015 for electricity produced by First Solar's 100-megawatt project, sending a chilling signal that all the solar cost cuts and improvements in efficiency were being passed on to the utility and lost to the developer. Only the previous year, that same utility had paid $0.1377 per kilowatt-hour for other renewable energy capacity. By 2017, five years since it set the target in 2012, Google, like many large and small companies announced 100% of its power needs are met by renewable energy for all its data centers and 60,000- strong workforce.

Utilities Struggle to Adapt

The traditional way of obtaining electricity is via the utility. You can sub-scribe and pay a price per kilowatt-hour. We are familiar with the large, centralized powerplants that supply the grid, and new transmission, or power plants, are very large local projects. The current system often allows utilities to pay for upgrades by passing the cost on to the con-sumer in price increases. For many years, this has been unchallenged, and a state Public Utility Commission (PUC) provides some guidance for consumers and utilities. It would be normal, then, for utility manage-ment to be hostile to change, to have a risk-averse mind-set and to have an unwillingness to invest, preferring to leave that to the private sector. There is also a trend for utilities to simply distribute the energy that is connected into their grid systems by independent power producers (IPPs). Although it's clearly a difficulty for such a utility to make the necessary strategic changes now that renewable energy resources are beginning to be demanded in scale, there is also a lot of serious thinking going on about how to survive and keep the cash flows they have come to rely on.

Grids everywhere are bracing for the impact of more intermittent solar and wind and other renewables as well as electric vehicles and storage. Some traditional grid managers contemplate obstacles to solar because they interpret it as a reduction in demand and revenues and that it's not doing its part in supporting the cost of the grid. With an approach based on metered electricity, of course, this is difficult to shake off, and notably utilities in Nevada and Arizona and elsewhere have opposed retail solar penetration for this reason. A federal law called PURPA mandated that any utility would have to provide a grid connection for any power plant built by a private individual of a given size. In some grids, this electricity is paid for with subsidies, and in others, such as Hawaii presently, there is no payment to be expected from the utility for electricity they receive from you. Although it has not been transparently clear whether electric vehicles will grow fast enough, it is now becoming an object of strategic focus that EV demand and the savings from storage grid smoothing could easily offset any decline in revenues from the impact of distributed renewable resources.

The big picture is clear. Humanity is in the process of transitioning from fossil fuels to sustainable energy use. Many of the new energy resources come as distributed energy, even down to the individual rooftops of their residential subscribers. The signs, as you will see, are self-evident. Utilities often see solar as a way of losing money. What many utilities are not seeing, however, is that solar installations also provide significant support services such as remote grid support, lower the cost of matching peak demand, reduce supply strains and provide emissions-free capacity without the logistics or cost of fuel. They also reduce the necessary future investment in the grid infrastructure and make it easier to adapt to "smart grid" options. The drivers of this change are obvious when they are spelled out, and yet there are many who are still oblivious to this change, which is now well under way. We have only to switch the light on or turn the knob on our kitchen gas range to realize that we expect a guarantee that electricity and gas will flow 24 hours a day. Our modern age and developed economies depend on reliable, baseload and peak energy supplies. Essential activities, those that cannot be interrupted, including common activities in hospitals, factories, schools and residential buildings, all depend on being able to access energy at any time.

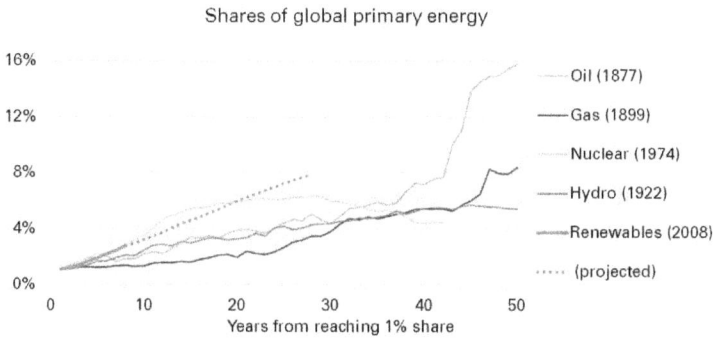

bp

Speed of transition

Shares of global primary energy

Figure 7: BP's 2016 Statistical Review was the first to study renewables in more depth, and this early growth chart shows that renewables are growing faster than any other form of capacity in history. Source: BP PLC. 2017.

The use of labor-saving electricity in our current civilization is amazing. Daily work that humans once had to do individually—farming, planting, building, sewing, weaving, making clothing, washing clothing, walking, harvesting, fermenting, burning, fighting—were "hats" or activities that everyone had to "wear" themselves.

Early energy sources, including wood, dung and then whale oil, were used to provide cooking and fuel for lighting. Its amazing how many hours are made available for other activities when you have a source of light after sunset. Along with whale oil, there were many other oil proxies, including animal fat, turpentine and coal oil. All these items were a bit expensive and under threat of discontinued supply. Whales were being hunted to extinction and the other forms of oil were often messy, smoky or smelly.

Coal was the main fuel of factories, ships and trains. Then it was used to generate electricity, along with hydroelectricity, bringing electric light to cities. Ships transferred from coal to oil by the First World War and most still run on diesel, although in many cases, the diesel is now used to generate power for electric motors.

Coal became the main way to light a city, and electricity was a major product of coal. Slowly over time, pollution accumulated from the huge volumes burned, and it became clear that coal use was also polluting our environment with many different chemicals, causing acid rain among other phenomena, but it took a long time before this information became as abundantly clear as it is today.

Offset markets for sulfur and nitrous oxide gas credits were established in America, which successfully caused scrubbers and other pollution control systems to be widely installed. In cars, as well, auto exhausts were found to be major sources of nitrogen oxides and sulfur dioxide, resulting in similar pollution control successes there too.

Fossil fuels, it must be pointed out, have been essential to the emergence of our current civilization. Nobody really doubts this point of view, but one author takes it a step further and wants us to double-down on fossil fuels, believing the devil you know is the best formula.

> *You've heard that our addiction to fossil fuels is destroying our planet and our lives. Yet by every measure of human well-being life has been getting better and better. This book explains why humanity's use of fossil fuels is actually a healthy, moral choice.*
> —Alex Epstein, *The Moral Case for Fossil Fuels.*

Alex Epstein (funded by the Koch brothers) is thoroughly accurate in the first part of his thesis that fossil fuels have brought us successfully to our current level of civilization. That's as far as his approach gets on a rational basis. Now, he continues, in a singular display of irrationality, we should stick with CONG as the world gets hotter since we can use it to cool ourselves! His reliance on fossil fuels going forward in the face of their accumulated negative impacts is misplaced. He is, though, correct in speaking against those well-meaning green or environmental advocates whose goals are purely to remove the human impact on the planet, leading us back into the past, to a simpler environment of self-sustainability, the hard undesirable, puritan world of cold showers and manual work.

In this book, I am making the claim that we can have our cake and eat it. We can not only get off the human footprint but also, thanks to using sustainable technology, actually use fewer units of energy for a given amount of work and have more energy available. You can drive an electric SUV recharged with renewable energy without any guilt feelings and get more mileage, more power output and more control. Epstein, at least, values human welfare as the basis of his conclusion that we ought not to stray from fossil fuels. His assessment, however, completely ignores humankind's unique capacity for innovation that will come up with even more solutions to directly combat the externalities of CONG and even continually improve on current renewable energy technologies and end up improving the quality of life. These innovations have already disrupted fossil fuels for the better.

Communities will do much better to adapt and change to accept sustainability. The *only* way we need CONG today is to help avoid an economic collapse as we transition to sustainable energy, but it is still undoubtedly the mainstay of our energy sources.

Along with these changes has come the development of the new, sustainable renewable energy technologies. We've used wind and flowing water for centuries to sail ships and turn mills. Wood is effectively a carbon- neutral energy source, and today we have a range of efficient, cheap technologies that can harvest energy from natural sources without any emissions or other impact on the planet.

We know that many significant transitions have happened in the past. One obvious one is the transition of mode of transportation from horseback to train that happened between 1800 and 1950.

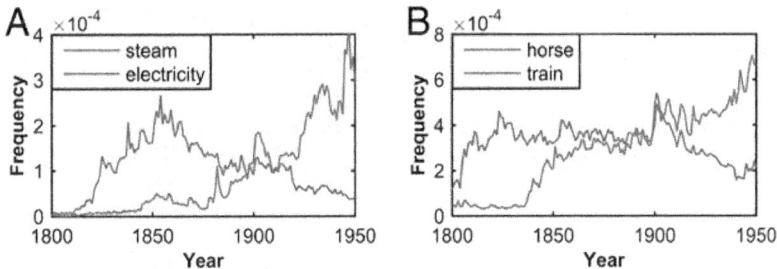

Figure 8. Two charts from the analysis of British newspapers from 1800 to 1950 showing the frequency of the terms "steam," "electricity," "horse," and "train." Source: Content Analysis of 150 years of British Periodicals. Universities of Bristol, Cardiff and FindMyPast Newspaper Archives. 2016.

The British Library keeps a collection of all the newspapers that have been published in the UK in every town going back in many cases to the start of 1800.

This provides a good picture of provincial life in the UK. In May 2010, a group called FindMyPast partnered with the British Library to enable a word analysis or an automated content analysis of all the printed newspapers in that 150-year period. They had digitized 17 million out of 40 million pages by 2016 and made them available to the public at www. britishnewspaperarchive. co.uk.

This means that you can enter terms like "electricity," "steam," "horse" and "train" and find the frequency that each term is mentioned throughout the period.[24] From these charts, it's very clear a common-sense linkage of the trends away from steam to electricity and away from horses to trains, it would be fair to suggest, paralleled the real trends. These transitions are caused by changes in technology and the efficiency and convenience of the new methods of using energy or transportation. In both cases, by 1900 the horse and steam were both eclipsed by trains and electricity. This same transition is currently happening for fossil energy and renewable energy.

The Economic Transformation

"I'm down here because Hillary sent me to tell you that if you really think you can get the economy back that you had 50 years ago, have at it, vote for whoever you want to, but if she wins, she is coming back for you to take you along on the ride to America's future." —President Bill Clinton, in his speech at the Democratic National Convention in Philadelphia, concerning the lost employment of West Virginia's coal miners. July 26, 2016, during that year's U.S. presidential election.

These two charts in figure 10, from Bloomberg New Energy Finance, show the billions of dollars invested quarterly in both of the main types of renewable energy technology along with the geographical spread on the lower chart for the period 1Q 2004 until 4Q 2015.

Each quarterly column is the same height for each chart and represents the overall dollars invested in that quarter.

Apart from the recognition that there is considerable growth over the period, these charts represent a total and accelerating investment of over $2.4 trillion, running at a current annual rate of about $290 billion annually. Other points discernible from these charts are the strong growth in Asia, especially in China and the stability of investment in the United States while Europe has declined significantly from its peak in 2011. In terms of technology, it's clear to see that wind is still growing overall but that the key to growth is the amount of solar installations worldwide.

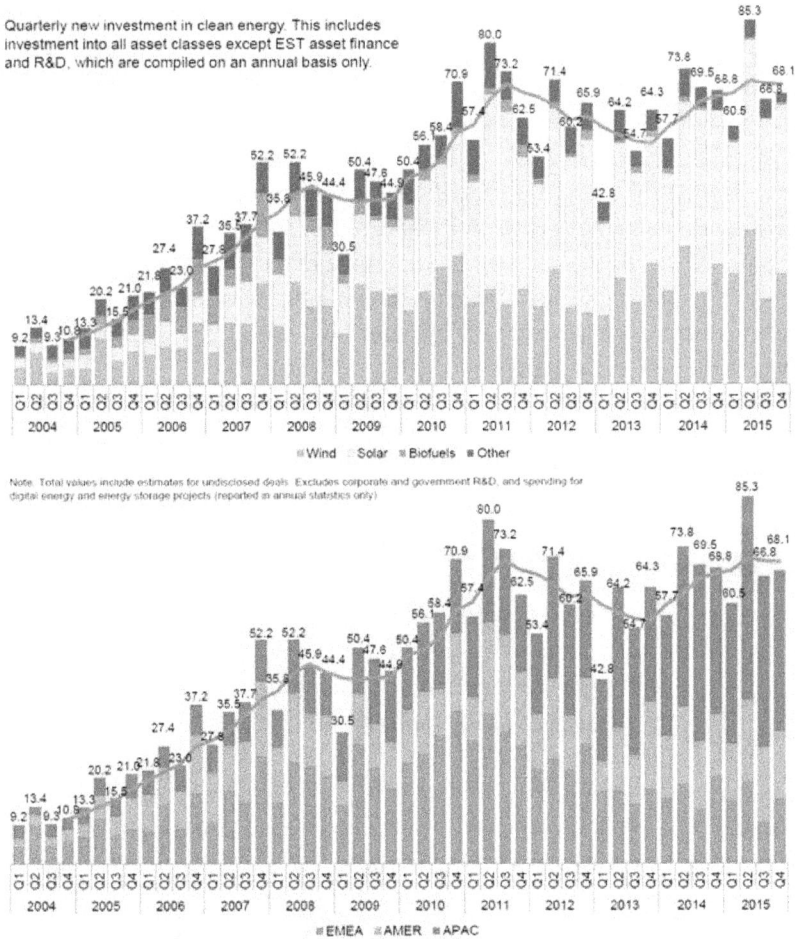

Figure 9: New investment in clean energy broken down by region and by technology.
Source: Bloomberg New Energy Finance.

Initiatives in storage and biofuels, which I believe are on the close horizon, will further transform the economics of energy. These numbers are backed up by the United Nations Environmental Program, which published their 10th report[25] called "Global Trends in Renewable Energy Investment 2016." The key takeaway from this report was that for the first time, in 2015, more than twice as much investment, $286 billion, went to renewable energy than to the global coal and gas-fired electricity segment of the market, which nonetheless attracted $130 billion, even though oil and gas prices were on a downtrend. Most of the renewable energy investment was made in developing countries, where $156 billion was spent versus $130 billion in developed countries. Distributed generation is a winning driver for growth in developing nations. During 2014, growth was 19%, with China leading the way with 17%, while developed nations' investments declined by 8%. Though China was still a large part, 36%, of the total, India, South Africa, Mexico and Chile put in significant growth too.

Structural change is clearly under way even though the huge pre-existing investment in conventional coal, oil and gas capacity is still just less than 90% of the total global energy capacity, meaning that, for the first time, renewable energy was just over 10% of the global total. Given the current trends, investment in renewable energy by 2020 will reach over $400 billion annually.

The Transition

"The power of population is indefinitely greater than the power in the earth to produce subsistence for man."—Thomas Robert Malthus, an alarmist viewpoint that thankfully did not come true because technology scaled food production. If anything, success in food production helped increase population.

Cassandra, a Trojan princess in ancient Greece, was granted the gift of prophecy by the god Apollo. She also had the temerity to reject his advances, so he cursed her: if she were to use her powers to make prophecy, she would never be believed. Cassandra-like voices of alarm have echoed through the ages when visionaries sensed challenges to the sustainability of their communities. Some examples are especially interesting. Thomas Malthus rang the alarm bells in the early 1800s when he realized that human populations required resources of energy and food and water to avoid famine. His name is associated with the demographics of resources and the horrible consequences when those resources are no longer available: starvation, disease epidemics and mortality.

Lester Brown, an American environmentalist admired for his detailed assessments of the state of the planet, has rung alarm bells from the perspective of dwindling global food supplies. He addresses the important task of quantifying the amount of food available and how many people are eating it and where. He concentrates on the geopolitics of food supply and illustrates a scary world to come, with untold hunger and mortality. Brown represents the "poster environmentalist," with a vision of a world that he sees is very likely to unfold. It can be terrible to think of the world starved of essential resources for the sake of the waves of humanity that exist and those expected to join us in the coming decades. In the context of global warming, the alarm becomes a scream because the effects have already started to exacerbate existing problems by generating climate refugees, enabling wars, spreading illnesses, causing wildfires, creating food and water shortages and causing chaos. The Syrian conflict can be traced to struggling farmers on land suffering a drought which affects the entire Middle East region and their government turning a blind eye to them when they needed help. The geopolitical consequences of energy and climate are taking us in the wrong direction.

Thomas Friedman, an American journalist and author, in his book *Hot, Flat, and Crowded*[26] declares fossil fuels and the current form of fission nuclear energy (CONG all together) as "hellish" energy sources to be distinguished from renewable energies (heavenly sources). Chapters 3 through 10 of this book investigate why this is undeniably a good distinction to make. The eight externalities cast CONG in the role of a significantly nastier protagonist than the current and generally accepted narrative allows. Not only does each externality affect us but also each CONG energy externality is additive, with a cost that can be added to the next until the real cost per unit of that energy source is derived. Various agencies have attempted these difficult calculations in the interest of understanding the scale of the damage and illustrating how much we already pay for our current tenuous living standards.

Today, we are much more sensitized to the risks of major changes that transitions like the adoption of renewable energy will give rise to. A new energy source has far more due diligence performed upon it than the earlier adoption of fossil fuels ever did. This raises the question of what would happen today if oil were to be introduced for the first time. It would likely stir negative reactions straightaway for its impacts and be rejected out of hand. It would be sad, though, to reject new sustainable energy technologies that appear green and clean and literally, as well as figuratively, represent a major breath of fresh air merely because we have become afraid of unintended (unidentified) consequences. A spirit of "once bitten, twice shy" combined with a willingness to overregulate will ironically intensify the harms of our current paradigm and delay our transition. We are at once bitten by the overall high cost of fossil fuels but shy to invest in improvements. The "better the devil you know" approach resists improvements to our energy paradigm that we nonetheless need to make. However, once we become aware of the cumulative nastiness of CONG, I hope the result will be a wish to rush away from CONG and embrace all the renewables we can get as quickly as possible.

Looking at the renewable resources we can enjoy in the future, we live in a world of abundance. Water is everywhere in the atmosphere and can be extracted with the use of renewable power; plant food can be grown rapidly and easily without much water using simple drip technologies. More energy than humans use annually arrives as sunlight within a couple of hours, and we know how to use it and how to obtain the rest from wind, falling water, biomass, efficiency, and the ground. We simply need the equipment to tap into this abundance, and for that we need a healthy political-economic situation that is not afraid of change and that protects innovative ideas and encourages the development of useful technologies. Making a transition from CONG to sustainable abundance can't happen overnight, but with human ingenuity it can happen in less than a generation. Renewable energy capacities, not including nuclear, are now at 10.4% of all global energy consumption, including hydro, according to BP's 2018 statistical review. The effects are starting to be visible, but we need to keep the pressure on.

If this transition was treated with the urgency with which the United States addressed World War II or the Russians evacuated their industrial base beyond the Ural Mountains in that same war, then we might have confidence that it was going to happen quickly and effectively. In reality, though, any sense of urgency is muffled by a so-called balanced media in which crazy arguments are given equal weight to science and false equivalence, resulting in confusion and a misinformed public. There are still forces arraigned against any change or adaptation, but in a free world those forces can't stop individuals, corporations and whole communities from making the necessary changes and enjoying the resulting economic improvements themselves. I sometimes get pessimistic about our ability to actually change, but pockets of innovation always seem to spring up and provide working examples that others can follow.

No doubt the democratic system will swing into action after the serious implications of the problem start to manifest. There appears to be a warming of the general U.S. public to the idea, for instance, that climate change is something to take seriously. This is possibly because recent history suggests that we are already witnessing an acceleration of the dark side of climate change. The spate of recent U.S. weather anomalies: Hurricanes Ivan, Katrina, Sandy and Matthew as well as an increased number of large and hugely destructive tornadoes and wildfires effecting intense localized damage, droughts, flooding such as the 2016 events in Ohio, and heat waves in India and the Middle East, where heat index temperatures reach humid highs in the region of 150°F (66°C) or more. In Russia, mysterious craters[27] where previously solid frozen methane had been present for thousands of years have literally evaporated dangerously in today's warmth. Methane is 21 times more dangerous for global warming than CO2.

All these events are indicated with high degrees of certainty by the scientific community to be the direct result of global warming. Billions of dollars of damage result from each event. There is a growing connection between the disaster damage and the climate problem. However, there is also a large population still unaffected by the issues. During Hurricane Sandy, if you lived on the coastline of New Jersey, the first and second floors of your home were flooded several times by the high tides and the surging seawater. Dirty water and mud ruined everything at ground level. The water was high enough to pour in over the windowsills on the second floor. If, on the other hand, you lived on the Upper West Side of Manhattan, you might have suffered through an hour or two of interrupted cable service, but you could continue to call out for food and watch your TV favorites. The soft belly of high-quality American living standards hardly stands a chance against the hardships of unabated climate change, which leaves many without electricity or clean water for weeks at a time. During Hurricane Sandy, the clash of experiences was like the surreal juxtaposition of an haute couture, center- spread advertisement in the *New York Times* opposite a news article on earthquake victims.

Yet we can turn back the clock on this climate damage. We can access abundant energy, food and water resources using simple technology that works. Of course, investment capital cost is an issue, but it becomes less so every day. Human technological optimism that has already shown incredible performance in the last two hundred years ought to be able to easily address such tasks as wringing water from the atmosphere, energy from the sun, the wind, water and earth, making new nuclear designs cheap and safe and obtaining food from new farming methods. It becomes clear that we can manage increasing populations, and not at the expense of the environment. Instead of grieving over the declining water table, we can use water- conserving farming methods that can multiply food production using drip irrigation, aeroponics and hydroponics. We need to keep in mind the volumes of water in the atmosphere that are condensable anywhere (albeit at a high energy cost) without pipes or plumbing infrastructure. Another tool is the miraculous parabola, a geometric curve that holds the answer to many of our energy problems and that has become ubiquitous in pursuit of some things, such as electronic signals, but not quite as much used yet for focusing beams of solar energy.

The Greatest Single Engineering Achievement of the 20th Century

"We live in a society exquisitely dependent on science and technology, in which hardly anyone knows anything about science and technology." —Carl Sagan

Thinking in terms of this Big History context, it's also amazing to think how electricity, which we have enjoyed only in the last 400 years, has become such an indispensable tool in our everyday lives. It is hard to underestimate how much we rely on it for so many services. In the last 25 years alone, the use of small electronic portable information/communication devices has revolutionized the world. In April 2012, the National Academies of Science in the United States named the "electrification of energy" as the greatest single engineering achievement of the 20th century.[28]

I can hardly imagine meeting somebody in a strange place without the facility and convenience of mobile communications, which offer maps and real-time location information with the Global Positioning System (GPS).

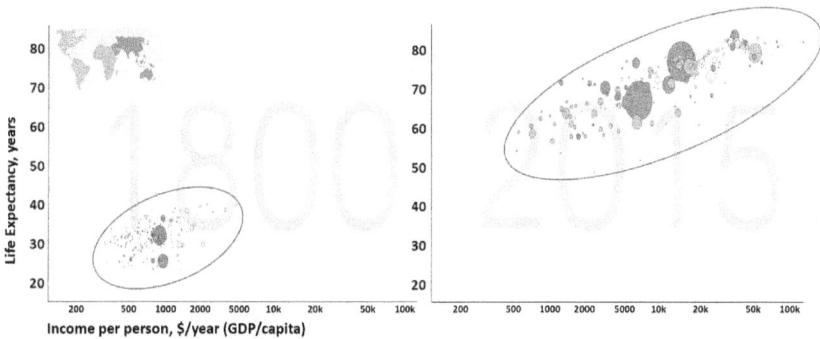

Figure 10. Hans Rosling's Gapminder charts, showing global longevity vs. income from 1800 to 2017. Source: Gapminder.com.

The arrival of email communications and video calling, building on the basics such as light and heat, have transformed civilization. Hans Rosling, a Swedish medical doctor, statistician and cofounder of the Gapminder Foundation, who passed away in 2017, has produced a remarkable video of the annual changes in income versus longevity of modern humans for each country. The video stretches from the dawn of the industrial revolution in 1800 all the way to 2017 for each country. Figure 11 shows the start and end charts side by side.

Nothing illustrates with as much force the impact of improving human conditions, food, shelter, clothing, income and psychological welfare as this video. During the period, most countries make large strides. Life spans start at an average of less than 40 years and, in the most advanced countries, end at double that, while incomes improve from $500 to $40,000 annually. Taken in the context of the 300,000 years of *Homo sapiens* existence, we can see that very suddenly and only in the last 200 years, for the first time, we have almost tripled our life span, effectively combated disease and largely rid ourselves of hunger and material needs. Human beings have released themselves from the bondage of long-endured ignorance.

Much of the reason for all this progress is improvements in our conditions due to technology. Humanity could well be about to go through another batch of changes. It is perfectly clear that the current environment is one of dynamic change, where competing companies innovate technological advances regularly, with a rush of new patent applications every year.

Innovation is a huge global resource. If we look at transportation, we know that air travel will be improved as aircraft evolve to be more efficient, comfortable and reliable, with greater ranges and possibly even leaving the friction of the atmosphere for even faster trips around the globe. Ground transportation is flirting with several new advances, including Elon Musk's driverless cars and rapid-transport Hyperloop train, which can't crash because it's in a tight vacuum tube and can go extremely fast between cities. Fast trains are appearing all over the world, as it is.

Maglev trains are highly expensive, so far, and consequently have been dragging their feet, but demonstration tracks in the United States, UK, China, Germany and Japan have been impressive examples of friction-free, wear-free rapid transportation. Richard Branson's offer of a few minutes of experience free of gravity just 68 miles above the atmosphere is imminent and presages a future of orbital hotels and, of course, the usual hackneyed sci-fi items, lunar and Martian colonies.

Electric vehicles also offer increasingly longer ranges with cheaper energy storage options. Musk, Bezos and Branson, among others, are offering access to unparalleled lifetime experiences at a reasonable cost. In medicine, it's clear that there is a super-hot series of developments always under way. Advances appear to be coming to fruition in cancer treatment, and ever better understanding of human genetics makes it highly likely that gene therapy will begin to provide reliable treatments for many ailments. Nutrition studies also clearly identify major illnesses that stem from overeating and processed foods, such as obesity, diabetes, heart disease and some cancers, and the importance of diet and exercise. Antibacterial solutions have been lacking, and dangerous diseases such as tuberculosis and smallpox have been evolving to survive the current batches of antibiotics. Few new versions of conventional antibiotics are available, and so new technologies are being brought to bear.

One interesting candidate is the peptide-conjugated phosphorodiamidate morpholino oligomer, or PPMO, synthetic sequences of DNA that can match with specific genes such as those of a bacterial or viral infection and beat it back. These are being developed at Oregon State University. They've not been tested on humans and represent a significant potential new armament against infections. Viruses frequently scare us with variants that promise to be virulent, flesh-eating and unstoppable, but, again, these are being addressed as we get a better grip of the DNA code. The advent of an efficient gene-editing technique called CRISPR means that these changes are all the more likely.

Another promising piece of research[29] by a team including Peter de Keizer from Erasmus Medical Centre in Rotterdam concerns elegant aging. Our current experience of aging is to have ever wrinklier bodies that increasingly groan and ache as we lose our mobility. How great would it be to retain our energy and health until we were very old? Some cells in our bodies turn senescent, which ages us. These cells don't die but accumulate and complicate the health of the body by releasing hundreds of proteins such as growth factors, interleukins, which are inflammatory, and proteases, which can break down other tissues. We were never meant to be living into our 80s, but as we do there are billions of these senescent cells poisoning us and enhancing our aging. De Keizer and his team have discovered that the protein that causes cell death, called P53, is restrained from doing so by another protein called FOX04. He isolated a protein that unshackles the P53 and lets it do its job. In mice the equivalent of 80 years of age, they discovered that getting rid of the garbage dump of senescent cells resulted in better fur quality, fitness and organ function. There were no side effects. The promise is to give back health to older cohorts of humans and improve the quality of life considerably, but the passage from mouse to man has always been a long one.

We have emerged from a long period of primitive communications technologies. Now we are going through a fascinating period when astonishing computer power has become available in our smartphones. Sensors and cameras and hundreds of thousands of software applications (apps) explore user-friendly solutions to all kinds of challenges, fast turning the old-fashioned, simple phone into a relic. The ability to use constantly improving communications technology is rapidly being augmented by miniaturization to wristwatch size and small enough to fit as a screen on a pair of glasses. Streaming high-definition video has transformed our viewing habits. Now we watch content we are interested in whenever we want, and the broadcast TV and cable industry are falling over themselves to try to win back viewers. People are binge-viewing whole TV series in a weekend that were previously used by TV channels to keep viewers watching advertising. Information on almost any subject is available anywhere there is a cell or Wi-Fi signal. Wi-Fi signals themselves get faster and better, with the possibility of a broader rollout of large WiMAX offerings providing total coverage of an entire town or city. The internet has made it possible to have free video telephone calls with anybody else who is hooked up, anywhere in the world at any time.

The Automobile Renaissance: EVs and AVs

The first electric production vehicle, I'm proud to say, came from London in 1884 and was built by Thomas Parker. It ran on his design of energy-dense secondary (rechargeable) batteries. Four years later, the Flocken Elektrowagen by Andreas Flocken evolved in Germany as a more practical vehicle. Electric motors were convenient and smooth running. About 30,000 electric vehicles were around by 1900. Then internal combustion engine (ICE) vehicles moved ahead because they had new sources of fuel with mineral oil, greater range and faster refueling. The electric starter motor took away the inconvenience of hand cranking the motor, made worse on cold mornings, and until the 1970s we never looked back.

In the 1990s, the California Air Resources Board (CARB) was one of the first public agencies to request more fuel efficiency and lower emissions and a track to a zero-emissions vehicle. A range of new electric vehicles was built by Chrysler, Ford, Toyota, Honda, Nissan and GM, which introduced the lead acid–powered EV1, about which the documentary "Who Killed the Electric Car" was made. All these vehicles were pretty quickly eradicated from the U.S. market, partly, it is assumed, by an anxious oil industry and its active lobbyists, unwilling to experience any risk of a cut in oil consumption. They had already killed public transport in many American towns.

Hybrid vehicles mix a powerful electric motor with an internal combustion engine. This means they have both a battery and fuel. Because the initial acceleration of an internal combustion engine vehicle is very inefficient, electric and ICE hybrids combined their relative advantages and used the electric motor, which has a constant torque, even at 1 revolution per minute (rpm), to accelerate a car to, say, 25 miles per hour, at which point the ICE would kick in. Such hybrids are still commonplace and have more than doubled fuel efficiency to about 60 miles per gallon. A favorite model is the Toyota Prius.

Interestingly, this effect was not lost on the very first automobile engineers. William H. Patton filed a patent for a gasoline–electric hybrid in 1889, and eleven years later, Henri Pieper and Ferdinand Porsche developed cars with electric motors in wheel hubs. Vehicles like the Chevy Volt use an optimally adjusted internal combustion engine to generate electricity and keep a battery charged, a combination that gets you many more miles per gallon of gasoline. A fuel cell vehicle is also an electric vehicle, and it extracts the electrons from the hydrogen fuel gas using a fuel cell, which generates water as an "exhaust." Another advantage of electric vehicles is that deceleration is used to efficiently generate more power and recharge a battery. David Arthurs, an electrical engineer from Arkansas, invented a regenerative braking hybrid Opal GT in 1979 and achieved efficiency of 75 miles per gallon. Regenerative braking is now commonplace in electric vehicles. When the driver needs to brake, foot pressure on the brake pedal engages the vehicle's electric motor, but this time the wheels turn the motor and it becomes a generator. Motors and generators are reversible. Resistance from the motor slows the wheels down, but that resistance is the generation of new power, which can be held immediately in supercapacitors and then charged at the battery's pace. Besides the greater range, electric hybrids are conveniently charged at home, overnight. Goodbye to the gas station!

After a brief hiatus, almost every auto manufacturer in the world now has a new electric vehicle (EV) model or two, starting with expensive, stylish sports vehicles that appeal to the well-heeled consumer such as the Tesla Roadster sports model, which was introduced in 2008 in the $100,000 range.

The main reason for the high cost of EVs is the lithium ion battery. Up to this point, batteries have been expensive, but the cost of EVs dropped by 35% in 2016, and the trend is on par to make the EV as affordable as a regular gasoline counterpart by 2022. A 2011 Harvard study[30] by the Belfer Center found that the high price of electric cars was not offset over their lifetimes by efficiency and higher miles per gallon equivalent that owners experienced. McKinsey & Co.[31] showed that battery advances have helped prices drop from $1,000 per kWh in 2010 to about $227 per kWh today, a trend that shows no sign of slowing down and that is likely to reach at least $73 per kWh.

The electric car has been getting steadily cheaper. Tesla's model 3, unveiled in March 2016, is aiming for a low price point of $30,000. It immediately garnered 325,000 reservations. Global sales of EVs are blossoming. Electric motor efficiency means that even if the electricity comes from coal, the extra mileage gained dilutes the pollution per mile. If renewable energy that can come, literally, from your back yard is the source of the electricity recharging the EV's battery, as is inexorably, increasingly the case, then every EV trends toward being a zero-emissions vehicle.

Falling prices, better torque and acceleration, the possibility of home charging (and charging your home!) with home solar or wind power, improving ranges, beautiful designs and potential breakthroughs in all these bottlenecks are increasing demand for EVs, and the prediction is that by 2023 enough EVs will be on the road to reduce oil consumption by 2 million barrels per day, from 96 million. Rising volumes also lead to economies of scale.

Today EVs are such a tiny fraction of the global auto market that oil companies may be forgiven for missing the threat, but affordability means that all the other selling points will lead to significant market share gains. In 2017, U.S. EV sales grew by 25% to 200,000 vehicles, with over 30% growth annually since 2012 at a time when sales of fossil fuel cars are falling. Other new technologies in history have experienced large growth rates like this. Solar panels and LED revenues emulate this growth rate today.

Bloomberg New Energy Finance say they believe this will happen by 2023. This will create an oil glut similar to the one that happened in 2014 that saw oil prices decline from $110 a barrel to $45 a barrel. They go on to say that by 2040, an electric car will cost less than $22,000 and 35% of new cars worldwide will be plug-in electric. With their capacity to be refueled at home with solar, wind or grid renewable energy equipment, they stand a huge chance of reducing the carbon intensity of transportation.

Green cars are increasingly benefiting from subsidies, while ICE vehicles face mounting penalties in location and parking restrictions. Cheap fuel drives ICE purchases in the United States while electric vehicles remain expensive, but there is still a $7,500 federal tax subsidy for the latter. In China, the effort to curb pollution emphasizes the importance of electric cars. Growth there is significant at 55% in 2016. While I think that electricity will replace gasoline progressively in the 2020s, another major impact on oil consumption will be the replacement of fossil transportation fuels with cellulosic ethanol, butanol and biodiesel. I'm aware that an industry insider will see this as naïve in the extreme because despite EPA mandates and a super RIN subsidy, it's not yet been done at any scale outside Brazil, but this is the time we will see this change. The huge resource of global cellulose is easily able to replace new oil reserves in the ground without any exploration premium and now can do it at a much lower cost. This means that the life spans of internal combustion engines can also economically be extended in addition to cutting their carbon intensity completely.

What happens if all vehicles were electric? There would be an extra draw of power from the grid, but how much? Peter O'Connor, a scientist researching electric vehicles and the grid for the Union of Concerned Scientists (UCS), suggests[32] that there would be an increase of 25% in electricity demand. Light vehicles travel 3 trillion miles a year, he points out, which at 3 miles per kilowatt-hour means an extra trillion kilowatt-hours on top of existing U.S. electricity consumption of 4 trillion kWh. The CO_2 released by all the light vehicles in 2014 was 1 billion tons, or 20% of all U.S. emissions, compared to electrical powerplants, which released 1.9 billion tons producing 3,931 billion kWh. The 25% increase in power output for electric vehicles would release 0.5 billion tons of emissions, which is half the current fossil fuel emissions of autos today. If renewable energy was any part of this production, emissions would be lower. Since I believe that by 2050 we may have much more renewable energy, emissions will be a lot lower and we will be well on the way to a sustainable world. EVs could be 40% of all new car sales by 2030. UCS has a program called "Half the Oil" making recommendations that result in cutting half of the U.S. oil demand by 2035.[33] I am sure they have no idea yet how large a contribution may come from cellulosic biofuels.

In 2016, there was the start of a peculiar government and car company behavior built on all this positivity in EVs. Volvo said that by 2020 they would no longer produce an internal combustion engine vehicle. Governments, mainly in Europe, set goals for new EV infrastructure and the end of oil consumption

Along with the arrival of EVs, however, is a coincident revolution that may change everything. Business as usual is to expect everyone to buy a family EV instead of the traditional internal combustion engine vehicle. Looking out of my January window onto West End Avenue on the Upper West Side of Manhattan, every street and even the main drag, Broadway at 81st Street, is lined with parked vehicles. These are mostly passenger vehicles, family vehicles that are safely parked for the residents of the large number of local apartment buildings. The cars are all relatively high-value items, but they lie dormant, unused for most of the day. They have a very low-capacity utilization for something quite so valuable. Globally, this represents a massive amount of machinery that is also often used only for the purposes of taking a single person from A to B. If the average daily use of a vehicle in the United States is 21 miles, in the city it is far less. That 21 miles takes a half hour to drive, so that means for 23 hours a day or more, most cars are sitting idle.

Uber, Lyft, Juno and Via are just four of many internet-streaming, smartphone-friendly, computerized map, software ride companies that have openly challenged the taxi and the ownership of personal vehicles, especially in a city. Now that the thousands of "smart ride" vehicles are already out in the city and can come pick you up any time of the day to take you anywhere, it's very difficult to see why one would require a personal vehicle.

To be clear, the costs are the first barrier, but they are coming down and a group like Via offers a ride almost anywhere in the city for only double the subway fare, taking you from your start block to your target block. There is plenty of competition to keep ride costs in check. To cut costs even further, Uber and others are experimenting with autonomous vehicles (AVs) that you can summon in exactly the same manner but that have no driver to pay, or get tired, or make human driving errors, or strike a bad tone with passengers. The competitive environment means that much of the driver expense will likely be credited to the passenger, who will experience better value over time.

Research has shown that many people might not acquire a personal or family vehicle if there is a ride, autonomous or not, conveniently and reliably available. Autonomous vehicles can be either internal combustion or electric.

They can communicate with each other and know the traffic conditions. Current highways are more than sufficient to much more quickly get all those who still have to commute to their workplaces in AVs if they space themselves at only 2 seconds apart, for example, which they can do. An amazing miracle of modern transportation can come about. Human errors kill over 30,000 Americans annually, and 1.2 million people a year die in driving accidents around the world, 40% of which involve drugs, alcohol or fatigue and will be the first to go. These numbers will plummet along with insurance premiums for accidents.

Optimization of the system will involve a little tweaking from regulatory authorities, and municipalities will experience a drop-off in incomes from parking and speeding tickets but a huge savings in mass transportation expense and new uses for the large amounts of land released in a community that is no longer required for parking, or employing traffic wardens, or traffic police. I often use Via and ride share with nice strangers, reading emails, taking in a documentary or typing out texts until I arrive at my destination. The Via vehicle has managed to keep more passengers on seats than any other vehicle on the road during a whole day. This is exceptional capacity utilization for automobiles. You have your time back. Consumers express considerable excitement about these developments. Regulatory authorities are already examining available data, such as the 2 billion rides already given by Uber and are coming up with rules to authorize use of AVs in the most optimal manner.

Google, Tesla, Apple and many of the auto manufacturers are well advanced in their production of autonomous vehicles. AVs will be available for consumers to buy by 2018. You can call the AV to drop your kids at school and then come to pick you up and take you out to dinner in town. Instead of a valet or spending 15 minutes looking for parking, you merely instruct the vehicle to find somewhere to lurk and, when you are ready, you can send it a signal and it will pick you up, just without the valet's smile.

By the 2020s, when EVs cut oil consumption, the arrival of AVs could also cut the number of vehicles required to transport us to a fraction of what we currently have. This implies a huge drop-off in revenues for oil and car companies, not to mention that there won't be a large increase in energy consumption either. National Renewable Energy Laboratory (NREL) and the University of Maryland have said that the use of AVs in normal traffic will result in a 15% fuel savings. Dropping the pursuit of high performance and sizing a vehicle to its passenger load can also effect additional, significant savings. If the numbers of vehicles also drop, this will lead to a large drop in greenhouse gas emissions. Fewer people will require a driver's license.

It's not clear how many people will make the switch to the equivalent of just owning a cell phone instead of having a home fixed-line telephone too. In cars, the equivalent to just having a cell phone is to rely on the sharing economy and call a vehicle with your phone when you need it. There will be no need for registration, insurance, parking fees, servicing, new tires or running the risk that the city will take your car to the pound when you don't park it in the correct location. Cities will find new space for dwellings and parks. Longer commutes will not be inconvenient because you can be productive, meaning more families will live in attractive, countryside neighborhoods.

If autonomous vehicles can pick up someone from anywhere at a competitive rate, then mass transportation systems will be disrupted and, along with them, the premiums that real estate commanded for being conveniently located near light rail or bus services or main roads. Living in the burbs or shopping at the mall has advantages and disadvantages, but autonomous vehicles bring the city center or cousins living on the other side of town much closer together.

A group called 99mph examined the impact of AVs on real estate. They said that $1 trillion of property value will shift across 13 major U.S. cities, led predictably by Los Angeles, where an increase of $277 billion of convenience value will accrue to suburban property prices, followed by New York at $189 billion, with both cities experiencing an 8–13% shift in values away from properties currently experiencing the transportation benefit to all those which are not. Millions of people around the world will find their commuting experience improving by being able to sleep, eat, read, discuss and so forth as they ride. The commutes may also become literally shorter as vehicles drive closer to each other and use optimal routes, meaning that remote spots are less remote by AV.

Of the $30 trillion value of real estate in the United States, as established by Zillow, $1 trillion is not an excessive change but still a very significant second-order impact of AVs. Already urban public transport has been suffering from the impact of Uber and its competition. A lot of commuters and employers will reconsider commuting in terms of distance and the opportunity that AVs bring to the table. Property prices anticipate changes ahead of time, so although it may be the case that it might take two decades before we are all habitually summoning a vehicle to arrive in the next minute, prices will have preempted this well beforehand.

This is another case of massive current inefficiency being solved by the internet's sharing economy, resulting in more service or GDP per unit of energy. The global transportation system is about to be changed out of all recognition in the next decade by the internet and smart technology!

Energy Storage and Intermittency

"If necessity is the mother of all inventions, we have the biggest necessity and it's got to be the mother of many inventions." — Steven Chu, Secretary of Energy of the Obama administration.

The increasing interest in energy storage intended for off- and on-grid applications can be attributed to multiple factors, including the capital costs of managing peak demands, the investments needed for grid reliability, and the integration of renewable energy sources. Although existing energy storage is dominated by pumped hydroelectric, there is the recognition that battery systems can offer a number of high-value opportunities, provided that lower costs can be obtained.

There are many chemical and mechanical storage options. Many systems are inefficient or slow or have limited cycles or cost too much. Today there is a burning desire to overcome wind and solar intermittency and the waste of energy caused by burning too much CONG fuel (to ensure continuity of electrical supply) and sufficient supply to match demand. These are huge drivers to innovation. Energy storage can keep the supply as small as necessary.

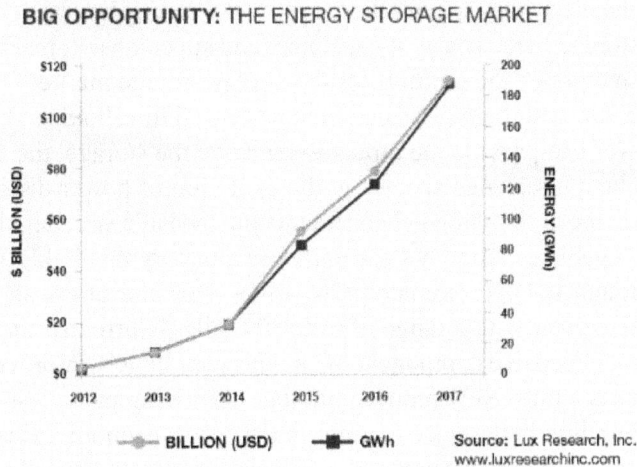

Figure 12: Growth of the electrical storage market in billions of U.S. dollars and gigawatt

The essential overview is that there are currently four types of storage available. Mechanical types that use potential energy such as pumped hydro, compressed gas, or gravity systems. There are the chemical batteries we are familiar with, found in our smartphones and devices and cars. Also, electrochemical systems store excess electricity by using it to generate hydrogen that can be stored under pressure for later use. There is also thermal storage typically in the form of ice or hot water.

Since the average lifetime of a vehicle is 21 years through several owners, we are in a generational transition to a new electrical paradigm given some substance by the calendar of car ownership.

GASOLINE CAR		SOLAR ARRAY		ELECTRIC CAR	
$ 2.00	per gallon	8 kW array		10.5 kWh required to drive 21 miles	
25 Miles per gallon average		5 Hours daily irradiation		0.5 kWh per mile	
21 Miles average range driven daily		40 kWh generated daily		26.25% of the solar array power for the car	
$ 1.68	Daily cost for 21 miles	$ 3.40	Cost per Watt of solar installation	$ 1.11	Cost for the daily 21 miles
$ 50.40	Monthly gasoline cost for vehicle	$ 27,200	Total installed cost of array	$ 33.32	Monthly solar cost for vehicle energy
OLD HOUSE		30% ITC Credit for solar installations		NEW HOUSE	
50 kWh daily		$ 19,040	Net after subsidy	29.5 kWh Remaining electricity for house	
$ 0.15	Utility cost per kWh	8% Interest rate		73.75% of the power required for the house	
$ 225.00	Monthly subscription to utility	$ 126.93	Initial monthly interest	$ 236.46	Monthly household solar electricity cost
$ 275.40	Total monthly car and house cost	$ 4.23	Cost per day for total solar array	$ 269.78	Total monthly car and house energy cost
		$236.46	Monthly principal & interest	97.96% of old house energy costs	

Figure 13: Comparison of the energy consumption of a traditional car/house pairing on the left with a solar-powered car/house on the right. Source: NEF Advisors, LLC.

In the electricity sphere, there is now the emergence of a fifth, solid-state storage in the shape of capacitors and supercapacitors. Figure 12 shows a picture of the outlook for the energy storage market from Lux Research. Once storage costs decline to a certain threshold, and solar or wind power can provide the inputs to recharge the storage, the utility is likely to be a threatened species in the evolution of power distribution, only made more so by the enhancements inexpensive storage will make to electric vehicles. Top EVs currently use a battery that holds about 85 kWh, enough for approximately 265 miles. Porsche offers an electric sports car variant with a range of over 300 miles. Currently, the battery is the most expensive component of an otherwise hugely improved vehicle. There is a massively reduced number of moving parts, less weight, more reliability, better acceleration, much better economy. Everything is made possible only by expensive batteries that have limited capacity, limited charge/discharge cycles, and long charge times.

A house uses between 30 and 50 kWh a day for all its services. A reasonably sized solar installation can provide all the power a house and car need every day. If the average distance traveled by a car is 21 miles (shops, school, and friends) and 0.33 kWh is required to drive a single mile, then only 7 kWh of electrical energy is required to cover that distance. An 8-kW solar array will generate this in an hour and power for the house in the remainder of the day on a sunny day.

A company in the UK called Edison Power[34] has a capacitor the size of a ream of paper and made of paper and graphene that holds a kilowatt-hour of power. Currently used in an electric bike, it has a range of 73 miles and is light, compostable, nontoxic, and cheap. The unproven but likely case is that it can charge in 15 minutes or less through thousands of cycles. Similar advances are being made with dry electrolyte capacitors that hold more energy by having very high voltages.

Low-cost energy storage has the potential to foster widespread use of renewable energy, such as intermittent solar and wind power. Other renewable energy such as run-of-the-river hydro, geothermal, anaerobic digestion, biomass-to-energy, waste-to-energy, ocean currents, and tidal energy are effectively base load with high capacities that could possibly exceed even the capacity of nuclear power. To date, wind and solar energy sources have been unreliable: Winds can be capricious, and cloudless days are never guaranteed. With cheap energy-storage technologies, renewable energy can be stored and then distributed via the electric grid at times of peak power demand. Already the vision of a house with an electric car supplied with power from the house's own wind and solar arrays is starting to be an economic reality.

Innovation

> *"If you believe you are too small to change something, imagine sleeping with a mosquito…and you will see which of the two prevents the other from sleeping."* —The Dalai Lama

Many times, I have visited small companies busy developing new technologies. To my persistent questioning the chief technology officer or chief executive officer explains the company's concept often only to finally joke, "If I tell you, I'll have to kill you" as a nuanced and gentle persuasion away from the act of telegraphing the details of their lifesaving technology to the world. It is precisely this part of human activity—

solving challenges, innovation— that is an undervalued human asset. If change is the only constant, then adapting to change is something more valuable than anything else. We have seen human ingenuity, creativity, and planning capability work miracles in war and peace. There is no reason for this human capacity for innovation to be absent now from the next big challenge we face as a civilization, and it clearly is not. Motivated humans are very powerful and creative. In the animal kingdom, we don't run the fastest or have the ability to fly or see well in the dark. Our minds are our adaptive organ and have already conferred upon us the ability to survive in many different environments, even in space. Our minds now provide us with the path to sustainability. We have all the means we need to fix our CONG issue now, and every month that passes, those forces get more powerful.

During wartime or emergencies, human beings do amazing things. They adapt to circumstances and prevail over difficulties by developing solutions. This is often an underestimated capability. It's always been something that you can count on—but nobody does count on it happening because where and when humans get inspired are quite unpredictable. I am a nut for new technology. I love to play with new things as they emerge. I always want the most recent PC or smartphone. When brought to bear on renewable energy, this love for technology turned into almost an obsession about what was possible. I have a long list of over 400 different renewable configurations that have been tried. The question is, given current knowledge and developments, is it possible that human beings could live on this planet in huge numbers and not have any footprint at all, or at the very least be described as operating a sustainable economy? It sounds implausible. How could all of us find enough food, energy, and water from the environment without creating pollution, agricultural eco-damage, climate change, and resource wars?

It turns out that not only can we but also, we've been able to do so for a long time. It's a matter of economics. A free-for-all approach allows the externalities of current energy resources to damage the commons. If this damage is reversed by using available sustainable technology, then we can obtain all the food, water, shelter and energy we need without damaging the ecosystem. I'm excited to be juggling several technologies that have sprung from cleantech that arguably can together solve every resource challenge that human beings are experiencing. I have recently been introduced to a new technology to make cellulose into sugar. The U.S. Department of Energy is putting big incentives and grants into the market, motivated in part by reducing dependence on foreign oil. They have a biofuel credit called the RIN, or Renewable Identification Number, which is a number attached to each gallon of biofuel produced. Different classes of biofuels have a multiple of that credit, depending on how they affect greenhouse gas emissions. The cellulosic D3 RIN credit was selling for over $2.50 in February 2017, so if you manufacture fuel made from cellulose, you have a nice incentive to do so.

For about 40 years, the pressure to increase yields of sugar from cellulose has been huge, and today several methods are being worked on to accomplish this. Breaking down large molecules into smaller ones is called hydrolysis. Enzymes break up cellulose molecules into glucose, and supercritical water (extremely hot) can also be used. Cellulose biofuel plants that use the enzyme method resemble drug companies, with stainless steel fermentation equipment and white-coated PhDs, all of which represents a high cost. These methods are wasteful and inefficient. But there has been a major change in this picture. The Florida company Alliance BioEnergy Plus employs a mechanical method that reacts 100% of any cellulose in minutes, producing pure monomer sugars and lignin with no waste, a minor miracle.

This is the era of robotics. So many companies are replacing their workforce with robots. We are heading for the singularity, the time when artificial intelligence will start thinking for itself in a nonbiological matrix and then rapidly excel compared to the intellectual performance of mere human beings. Human advances have always been accompanied by a well- deserved hubris. These quotes come from a few of the historical moments when human beings thought they had reached a pinnacle of achievement:

> *"The advancement of the arts, from year to year, taxes our credulity and seems to presage the arrival of that period when human improvement must end."*—From a report to Congress in 1843 by Patent Office Commissioner Henry Ellsworth.

> *"This 'telephone' has too many shortcomings to be seriously considered as a means of communication. The device is inherently of no value to us."*— William Orton, the president of Western Union in an internal memo in 1876.

"The Americans have need of the telephone, but we do not. We have plenty of messenger boys."—Sir William Preece, chief engineer, British Post Office in 1876.

"Fooling around with alternating current (AC) is just a waste of time. Nobody will use it, ever."—Thomas Alva Edison, 1889.

"The ordinary 'horseless carriage' is at present a luxury for the wealthy, and although its price will probably fall in the future, it will never, of course, come into as common use as the bicycle." — *Literary Digest, 1899.*

"Everything that can be invented has been invented."—Commissioner of U.S. Patent Office Charles H. Duell, 1899.

"The actual building of roads devoted to motor cars is not for the near future, in spite of many rumors to that effect." —Harpers Weekly, 1902.

"The horse is here to stay, but the automobile is only a novelty—a fad." —President of the Michigan Savings Bank advising Henry Ford's lawyer not to invest in the Ford Motor Co., 1903.

"That the automobile has practically reached the limit of its development is suggested by the fact that during the past year no improvements of a radical nature have been introduced." — *Scientific American, 1909.*

"While theoretically and technically television may be feasible, commercially and financially it is an impossibility."—Lee DeForest, "Father of Radio" and a pioneer in the development of sound-on-film recording used for motion pictures, in 1926.

"There is not the slightest indication that nuclear energy will ever be obtainable. It would mean that the atom would have to be shattered at will." — Albert Einstein, 1932.

"A rocket will never be able to leave the Earth's atmosphere." —New York Times, 1936.

"Television won't be able to hold on to any market it captures after the first six months. People will soon get tired of staring at a plywood box every night."—Darryl Zanuck, film producer, cofounder of 20th Century Fox, 1946.

"Where a calculator on the ENIAC is equipped with 18,000 vacuum tubes and weighs 30 tons, computers of the future may have only 1,000 vacuum tubes and perhaps weigh one and a half tons." —Popular Mechanics, 1949.

"I have traveled the length and breadth of this country and talked with the best people, and I can assure you that data processing is a fad that won't last out the year." —Editor of Prentice Hall business books, 1957.

"The world potential for copying machines was 5,000 at most."—IBM management told the founders of Xerox, 1959.

"There is practically no chance communications space satellites will be used to provide better telephone, telegraph, television or radio service inside the United States."—T.A.M. Craven, Federal Communications Commission (FCC) commissioner, 1961.

"There is no reason for any individual to have a computer in his home." —Ken Olsen, founder of Digital Equipment Corp., 1977.

"No one will need more than 637KB of memory for a personal computer. 640KB ought to be enough for anybody." —Bill Gates, cofounder and chairman of Microsoft, 1981.

"...the idea of a wireless personal communicator in every pocket is a pipe dream driven by greed." —CEO of Intel Andrew Grove, 1992.

"There's no chance that the iPhone is going to get any significant market share. No chance." —CEO of Microsoft Steve Balmer, 2007.

Islands everywhere use fossil fuels to provide power but also have plentiful natural resources. I visited Bora Bora, in the South Pacific, to see Intercontinental Hotel's adaptation of a technology that is halfway to ocean thermal electrical conversion (OTEC) but that leaves out the electricity generation part. Okay, I admit, I was also happy to escape for 10 days to celebrate my 50th birthday, but my predilection for technology took me to the hotel basement where I was guided by the French manager to see the equipment. He walked me out to the edge of the "motu" (island on which the hotel was situated) to see where the incoming freezing abyssal ocean water originated, and we traced the pipe to the hotel cel-

Anaerobic Digestion, using manure, human waste or waste food
Biomass to Energy, Corn or Sugarcane Biofuels
Biomass to Energy, Oil Crops to Biodiesel
Biomass to Energy, Cellulosic Biofuels
Biomass to Energy, Wood Chip to Thermal
Biomass to Energy, Biomass Briquettes
Efficiency, Use of Low Electrical Demand Appliances, LED, HVAC
Efficiency, Combined Heat & Power (CHP)
Efficiency, Demand Response
Geothermal, Injection and Production wells
Geothermal Local Loop
Hydroelectric Dams
Hydroelectric Run of the River
Ocean Wave
Ocean Tides
Ocean Thermal (OTEC)
Ocean Current
Ocean Chemical
Solar
Storage, Mechanical
Storage, Chemical
Transmission, Low Energy Loss
Transmission, Through the Air
Waste to Energy, Use of MSW
Waste to Energy, Use of Cellulose
Waste to Energy, Use of Waste Thermal Energy
Wind

Figure 14: 27 basic types of renewable energy technology. Only the blue ones are intermittent. Source: NEF Advisors, LLC.

lar, where a heat exchanger moved the coolness over to a more benign working fluid that was then circulated throughout the hotel's "rooms," the small, individual thatched cottages built on stilts in the impossibly blue shoreline waters. Blowers in each room pass air across heat exchangers and impart the coolness to each room. The system succeeds in reducing costs significantly.

On Marlon Brando's nearby island of Tetioroa, the same system has been adopted to cool rooms. In this case, solar panels and coconut oil form a hybrid power generation system that generates power that is then stored in zinc bromide flow batteries made by a company called ZBB. This stored electricity provides economic power for the cooling pumps and lighting in the individual visitor dwellings. Increasingly, islands are getting this message, especially as it becomes more economic to do so.

But there is also resistance to such advances. Alex Epstein is a young and energetic self-described philosopher who founded the Center for Industrial Progress, a for-profit think tank that believes that technology can lead to better lives for everybody. I was intrigued, because this is also what I believe, until I saw that his route to a better world was a continuation of the use of fossil fuels. He is prepared to embrace the CONG devil we know but almost entirely ignore evidence of its damaging externalities or the fact that global energy capacity is actually transitioning away from fossil fuels at a rate of knots to energy sources that have many advantages over fossil fuels.

In his book *The Moral Case for Fossil Fuels*,[35] Epstein states clearly that the reason for the high standard of living we enjoy is precisely because of fossil fuels. On this point, we can certainly celebrate his conclusions and cannot take anything away from the last century of growth that has led to our current civilization, but I wish he'd stopped right there. He goes much further and quotes climate-denier scientists such as MIT's Dr. Richard Lindzen, who did good work a long time ago but whose theories about clouds and the equator have since been discredited. He is one of the marginal, non-consensus climate scientists who still publishes. To say that none of the evidence for climate change has been settled—a very familiar criticism—is to simply turn a blind eye to the mountains of data that are available via very high-level, trustworthy sources such as the NOAA and NASA.

Epstein argues that going green, an environmental goal that helps us escape from oil and CO_2 pollution, confers no real advantages on human beings. His view of environmentalists' vision of the quality of life in some ways echoes my own: It may not be as good as the vision of a life where technology is powered by lots of energy. In this point, we disagree because I think we can simply replace lots of reliable energy from fossil fuels with lots of reliable renewable energy. Solar, wind, hydro, storage and fourth-generation nuclear can put all of this in place for us and do it magnificently.

In a talk with Bill McKibben,[36] a detail-oriented environmentalist, Epstein failed absolutely to realize that the CONG externalities are damaging us. He claims that the impact of climate change is still to come and is not here now. He cleverly but naively says, "Nature doesn't give us a safe climate that we make dangerous [with greenhouse gas (GHG) emissions], it gives us a dangerous climate that we make safe with fossil fuels." He fails to see that in fact Earth's climate is currently relatively safe and that we are progressively making it more dangerous. Renewable energy promises to replace the reliability of fossil fuels. We can defend ourselves much better by getting rid of the GHG emissions and increasing the amount of power available without the large amount of damage from externalities.

Epstein uses a puzzling example. He mentions that a premature baby held in an incubator that needs a constant energy supply in the Gambia (!) is almost certain to die because it cannot rely on 24/7 power required for an incubator to run. He is making the point that intermittent power will fail such a baby. He fails to mention that this baby's country is not relying on intermittent solar or wind but on fossil fuels that often behave in an even more intermittent manner than solar or wind. I guess he thinks nobody will think his example through. I well remember being in Nigeria, where electricity was generated by the Nigerian Electricity Power Authority (NEPA), which also conveniently stood for "No Electrical Power Available." Developing nations often experience interruptions, brownouts and blackouts in their fossil fuel energy supply. His naïve illustration would only have worked if the premature Gambian baby's incubator was using solar power without battery storage. Actually, a premature baby in the Gambia will likely benefit from a hospital generator that supplies 24/7 power anyway. I don't want to downplay developing nations neonatal challenges.

Epstein maintains that the high quality of human life is only on offer if there is a 24/7 supply of energy, good food and water and that fossil fuels are about the only way this can be maintained. The truth is that solar or wind combined with an effective grid storage system or good nuclear, hydro, geothermal, tidal or wave power can allow the baby's incubator to work 24/7 and can keep the baby alive without issues. Epstein calls renewable energy sources the "unreliables," but—and this is elementary—with storage, they become clean sources of reliable, 24/7 energy and without any of the externalities we study in this book, making renewables the better choice by far. Also, as shown in figure 14, we have 27 basic types of renewable energy at our disposal, only 3 of which are seriously intermittent without storage.

Epstein doesn't care that we are overcoming renewable energy challenges, and he's blind to the externalities of CONG. The fact that renewable energy can be baseload doesn't figure in his mind-set. He depends on having an uneducated audience. If he recognized the scary tipping points that we face because of methane and CO_2 emissions, he would agree that we need fossil fuels simply to guarantee a transitional source of energy until the sustainable alternatives are present in sufficient capacity and quality to replace them, and then our transition will be complete.

Adoption of clean energy is accelerating as a whole list of factors support it: falling equipment prices, nonvolatile electricity prices, grid parity, Obama's Clean Energy Plan (as was), the unison of international support with COP 21 in Paris in November 2015 and COP 22 in Marrakech in 2016, extended global subsidies for renewable energy, the 13th Chinese 5-year plan with its 1-year plan amendment to increase solar and wind installations and the decision by the Irish to disinvest in coal and oil, among many others. These all accelerate the momentum so that a complete transition looks as though it could happen within as few as 25 years. If the human race manages to do this, it will have been just in time, but I am afraid that we will be left with the worst effects of climate change and will not be able to reverse them easily, potentially for an extended period, without a method to absorb or store the excess CO_2 already in the atmosphere.

Environmental Goals Really Add Value

After the COP 21 meeting in Paris in November 2015, 195 nations agreed to combat climate change by limiting global emissions and warming to 3.6°F (2.0°C) in a program that comes into force in 2020. All parties are required by best efforts to find the investments required for a sustainable low-carbon future. These signatories, including the United States, have agreed to cut greenhouse gas emissions by 80% from 2005 levels by 2050. Professor Geoffrey Heal of Columbia University in New York City wrote a National Bureau of Economic Research working paper[37] asking what it would take over the coming decades to cut emissions of GHGs to this level by 2050. This is also a plan that the Clinton administration had already indicated it would have adopted. Professor Heal suggests in his report that this can be accomplished for between $42 billion and $176 billion per year and by moving the United States, the world's largest economy and second-largest emitter of GHGs, away from the use of CONG energy by 2050. He says that we need to replace 66% of the CONG generating capacity and combine solar and wind with the advantages of storage as well as an increase the use of nuclear power.

Needless to say, I make the point that every kilowatt of capacity that moves from CONG to renewables has a disproportionate effect in reducing overall external costs, making it a no-brainer move. Heal concludes that the 80% GHG reduction is possible at a cost of between $3.3 trillion and $7.3 trillion. We know that there are also new technologies being perfected for commercial launch. It's also quite likely that there will soon be a trigger that will help to significantly improve lives.

There are signs that initiatives across the world are having huge beneficial economic impacts. In Europe, emissions since 1990 have dropped by almost 25% while the economy has grown by almost 50%.[38] Recent analysis in California shows that the adoption of state laws mandating installation of a certain capacity of renewable energy resulted in adding $48 billion to the economy and produced 500,000 jobs. In the Northeast, the Regional Greenhouse Gas Initiative (RGGI), a carbon market in which various states participate, is just celebrating its first decade of service. Billions of dollars have been saved and thousands of jobs have been created. The auction market is overwhelmingly popular even among local Republicans, with 8 out of 10 people believing that continuing the program is the right thing to do. An analysis by Synapse Energy Economics discovered that the market was the most efficient way for these states to achieve climate goals and boost employment, air quality and public health. Clean energy means more jobs and less pollution. When you remember that we can quote all the CONG externalities and not just one or two, it is expected that pollution will be cut by 5% per year and that over 60,000 jobs a year will be added from 2020 to 2030.

The Chinese have taken over the lead in installation of renewable energy by specifying it in their 5-year plans. Although they sit on mountains of coal, the population's alarm about pollution has alerted policymakers to not just the need to avoid coal but also the business attraction of renewable energy. The Chinese National Energy Administration intends to spend $360 billion by 2020 to become a world leader in one of the world's leading growth industries. They expect it will generate about 13 million jobs in the period and have a massive effect on reducing the externalities which they actively suffer from in the large cities. China has been on a ramp-up of coal-fired power stations over the last twenty years and emits twice as much greenhouse gases as the United States. Chinese coal companies also are very voluble about these changes, and during this strong growth phase, there certainly has not been total connection of new capacity to the various Chinese grids. They have even taken into account that much of their large population lives on the ocean, and sea level rise is a big concern. Even well before COP 22 in Marrakech, the Chinese were aware that the economic promise of renewable energy was the real win-win situation. In 2015, the Chinese installed one wind turbine per hour. Its highly likely that China will reach its goals by 2018.

The Trump election has all the hallmarks of taking the United States back into the dark ages of energy when a really great vision of a new age is what's actually on offer. On the private and state scale, falling costs have made a huge impact in the United States. Variable stop and start subsidies, which don't measure up to those given to fossil fuels, have made planning very difficult, although the new 5-year ITC subsidy is a breath of fresh air and will see much growth. However, those subsidies are under threat, and if coal is brought back, it will literally bring us to several tipping points that make climate change reversal very difficult. Its highly likely that the market share that China will take from this investment will rob the West of renewable energy jobs too.

Benefits of a World Without the CONG Externalities

"There will come a time when the Earth grows sick, and when it does, a tribe will gather from all cultures of the World, who believe in deeds and not words. They will work to heal it. They will be known as 'Warriors of the Rainbow.'" —Cree prophecy.

As individuals, we sometimes spoil ourselves with a new pair of expensive shoes, a fancy dress, a new suit, a dinner out, a house, a car, etc. Often these items are much more expensive than we really need to pay to obtain a similar level of utility. The vision of a global energy supply that is secure and off the human footprint is enough of a high-quality item that humanity would clearly be justified in spoiling itself, even if it means paying a bit more. We should not wait for pricing signals but should instead embrace sustainability as a core principle. If there were no externalities such as climate or geopolitical issues, it might be okay to muddle by with the CONG formula, but we have some frightening drivers pressing us toward change. We need an infrastructure change that will bring us better conditions immediately. Luckily, those expensive shoes are much better value today anyway! The cost of renewable energy equipment has been tumbling and there is further to go.

The world's fossil fuel habit entrains enough costly externalities that, in fact, sustainable energy solutions are already demonstrably cheaper. To me, it seemed somehow incongruous to have all the military training I went through, with screaming sergeants, difficult drills, weapons training with lives on the line and a small war, only to suddenly shed the experience on being demobilized and find a 9-to-5 job. If it was important, why wouldn't I identify with the same high risks and embrace a goal like a sustainable world with the same commitment? I felt like there was a clear mission to be found here in the wider scope of life. It became clear to me that humanity must become more sustainable in as short a time as possible to benefit from the improved quality of life for all promised by such a transition and to put an end to the long list of ills, from pollution, species extinction, hunger, agricultural stress, monocultures, ocean distress, wildfires, droughts, bleaching coral, rising oceans, geopolitical issues, dead people and many others nasty things we endure that are all part and parcel of CONG externalities. We must put all these behind us so that we can focus on the alternative, tantalizing vision of humanity that can demonstrably live better and aim for as high a collective quality of life as possible. On checking out the existing technologies, I came to understand that we already have everything we need to accomplish this.

The transition, an accelerating global transition from fossil fuel energy to renewable energy, is well under way. We are moving from fossil fuel energy such as oil, gas, coal and other associated things like condensates and peat, which have become inconvenient and expensive energy resources, to clean and logistically easy energy resources. Renewable energy includes a long list of technologies starting with intermittent solar and wind power and going on to include many other formats such as anaerobic digestion, waste-to-energy, combined heat and power (CHP), geothermal, biomass-to-power, biofuels, biogas, ocean tidal, current, thermal and wave power plus other natural thermal resources, run-of-the-river hydro and other water-based resources in rivers and the ocean.

One message that emerges is that the environmentalist stereotype embraces a return to the past: cold showers and a lower energy intensity paradigm that uses an insufficient amount of energy for us to travel, pump water, drive vehicles, communicate and stay warm and cool. It is, in fact, not the vision that's needed but an outmoded "panic" vision based on fear of the effects of dwindling conventional fossil fuel energy, water and food and ignorant of the possibilities of changing resources to effectively replace CONG with plentiful energy.

Instead, we are happily faced with the combination of accelerating adoption of clean technologies and increasing energy efficiency, where the quality of life improves while the energy required to maintain that quality of life declines. More efficient, lower-demand appliances such as light-emitting diodes (LEDs), air conditioners, cookers and electric vehicles (EVs) will all combine to offer us a maintained or significantly improved standard of living and with lower energy consumption. This is actually a new world, a more productive economy, where our ability to accomplish work and enjoy leisure is enhanced.

We can continue to drive big (electric- or carbon-neutral-fueled) cars, travel in planes around the world (also with carbon-neutral fuels and later electric), communicate, eat exotic foods grown in and flown in from faraway countries and have fully lit, heated and cooled homes—all using much reduced quantities of energy from renewable resources.

Annual Vehicle Miles Traveled Per-Capita, 1946-2012

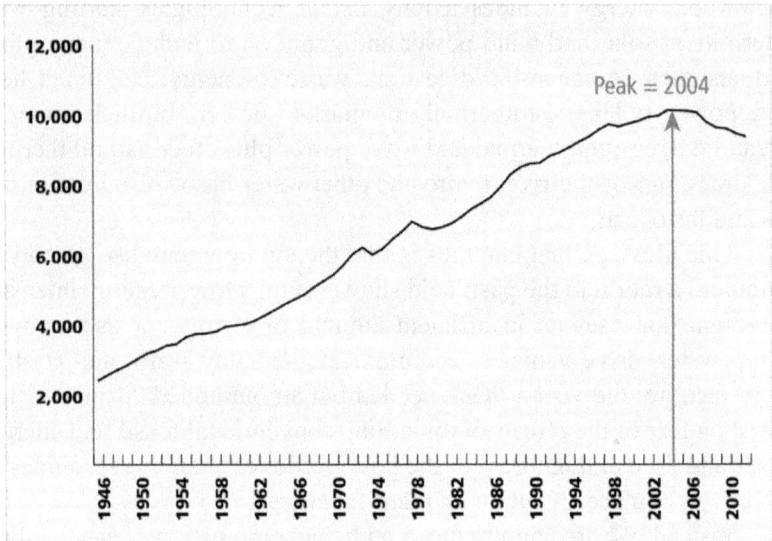

Figure 15: Chart of the increase and peak of vehicle miles driven since 1946 in the U.S. The oil shock of the 1970s is only a temporary interruption in the progress. The distances continued to increase even with major increases in the price of gasoline. Source: FHWA Highway Statistics and Traffic Volume Trends. Population data from the U.S. Census Bureau.

In short, we can have our cake and eat it. In retrospect, in the 19th and 20th centuries there were plenty of warning signs to suggest that a human energy paradigm shift was going to happen. Per capita use of energy exploded. After the invention of the factory and trains in the 1800s, which were energy intense and for which coal was a principal resource, there was the invention of the automobile. Industrialization made it possible for widespread and accelerating adoption of millions of automobiles. Roads were built all over the world.

A 2013 report included the chart in figure 15 of the increase in vehicle miles traveled per person since 1946 to illustrate just the transportation energy that was beginning to be used. The peak in 2004 suggests a major maturity in the U.S. distance driven annually, resulting in a maximum energy demand or peak oil demand. This point also ties in with improvements in energy efficiency and GDP per unit of energy figures.

The use of more energy called attention to the increase in emitted CO_2. Joseph Fourier, a French scientist who accompanied Napoleon on his sojourn to Egypt, recognized that a planet the size of Earth ought to be much cooler than it was at his time and that it must be the effect of the atmosphere that retains some of the excess heat. Irish scientist and president of the Royal Society in London, John Tyndal, in the 1850s discovered that water vapor absorbed heat and contributed to the greenhouse effect. Swedish scientist Svante Arhenius, who won the 1903 Nobel Prize in Chemistry, warned that human CO_2 emissions could warm the atmosphere and prevent a new ice age. More recently, the increase of CO_2 has been directly observed and warnings issued by such as Amory Lovins in his 1972 testimony to the U.S. Congress that rising industrial CO_2 volumes were going to affect the climate. The big oil companies were among the first to recognize the growing impact of CO_2.[39] As early as 1957, Humble Oil, the Exxon precursor, published a paper on absorption of CO_2 in the atmosphere and oceans, making the point that the extra CO_2 was coming from burning fossil fuels.[40] In 1968, Exxon's own scientists acknowledged that fossil fuels were the source of rising CO_2 that best fit observations. The dirge of climate impacts starts from a realization that CO_2 rise will raise sea levels, melt ice caps, change fish distributions and increase plant photosynthesis.[41] In 1977, Jack Black, a research scientist, presented a paper called "The Greenhouse Effect" to Exxon's management committee warning that humankind has "five to ten years before he needs to make hard decisions regarding changes in energy strategies." In 1983, the company cut its annual climate research spending from $900,000 to $150,000, despite a total research budget of $600 million.

Internal Exxon records released on September 16, 2015, by Pulitzer Prize– winning group Inside Climate News[42] revealed that long before global warming emerged as a global issue, Exxon had gone to the lengths of forming an internal brain trust to better understand the impact of rising CO_2 levels. They launched a supertanker, the *Esso Atlantic*, equipped with instrumentation for sampling air and ocean water to research and understand the ability of both to absorb CO_2. They determined that twice the CO_2 in the atmosphere would increase global temperatures by 2°C to 3°C. We have gone from 280 parts per million of CO_2 in the atmosphere to 410 ppm and are on the way to 500 ppm now.

After cogitating this information for a few years, Exxon completely switched sides, denied this was the reason for the studies, and stopped funding the project in 1982. Any kind of private or government action against carbon emissions, they knew, would result in damaging cuts in fossil fuel consumption. Instead of becoming leaders in addressing the issue, global oil companies founded the Global Climate Coalition to control government activities in reducing fossil fuel consumption.

Exxon went on to fund thinktanks and lobby government with an alternative narrative that suggested that climate science was unresolved and that government expenditure on the problem was potentially disastrous. It is most likely that they could have continued to do the science, acknowledged the difficulties and also continued to sell fossil fuels. We will never know how much further ahead we would be had they not consciously decided to deny the issue, especially because the changes actually present an economic opportunity, a point of view that must have been very difficult to sell internally.

Their change of heart was noticed. In 2006, Bob Ward, Britain's Royal Society senior manager for policy communication, sent a letter to Exxon that accused them of "inaccurate and misleading" information about climate and demanded that they stop giving money to organizations that distorted the science. Exxon spent more money on these institutes than on the climate science budget itself, although all of it was small change in relation to the company's balance sheet. In 2008, under mounting pressure from activist shareholders, the company announced it would end support for some prominent groups such as those Ward had identified. In 2015, New York Attorney General Eric Schneiderman opened an investigation into whether, like the tobacco companies before it, Exxon knew something about the climate that it then was lying about to protect its business. He issued a subpoena demanding in-house documents. In 2016, Representative Lamar Smith (R-TX) issued congressional subpoenas to the attorneys general of New York and Massachusetts for trying to hold Exxon accountable for its decades of climate denial.

The situation was clarified in the summer of 2017 when a peer-reviewed study written by Geoffrey Supran and Naomi Oreskes from Harvard compared the messages promulgated within Exxon to their own people to the public messages coming from the company on the subject of climate change. Supran and Oreskes demonstrated that Exxon had a very clear understanding of the situation and did not tell the public any of this. Of Exxon's own internal documents, their scientists' academic publications, 80% agreed that climate change is real and caused by humans. Completely at odds with this, fully 81% of advertorials placed by Exxon in newspapers such as the *New York Times* promoted doubt, implying that the science was in fact "unsettled." The authors say this shows clearly that Exxon intentionally deceived the public.

The situation cannot go away because Exxon's oil is necessary to avoid an economic decline while alternative energy is still growing, but as a regent passes sovereignty to the child king, Exxon and governments must arrange a peaceful way to transition that keeps everyone happy, and how that will happen is still a mystery.

Peak oil indicator events also remind us that our supply of fossil fuel is limited. King Hubbert, a Shell Oil geophysicist, explained to an incredulous audience in 1957 that U.S. oil output would peak in the 1970s, and it did. The adoption of fracking technology enabled the U.S. production of oil to increase again in the 2000s, and today the United States is suddenly ahead of Saudi Arabia as the world's largest oil producer. Normally, this would be something to be proud of, but with carbon emissions increasingly under attack, it's also making America one of the largest contributors to global warming. North Sea oil reserves hit their maximum volumes of output in the 1980s. Much is written about the huge scale of the Saudi Arabian oil wells and their ability to maintain leading global production, but they, too, are finite. One can only hope that the wealth these oil-producing nations have produced is sufficient for them to invest in renewable resources in time to neatly switch over their own dependence from fossil energy to renewable energy. There are signs that this is happening. There are huge gains in renewable energy capacity around the world and even plans to address atmospheric CO_2.

In 1992, the United Nations held the first climate conference in Kyoto, Japan. In 1999, the German parliament enacted a generous feed-in-tariff for renewable energy introduced by parliamentarian Herman Scheer. This catalyzed a sudden increase in wind and solar installations in Germany and kicked off an industry that all too suddenly became popular in China as well. Chinese production was heavily subsidized and used European and American technology to manufacture solar panels that were then exported cheaply to the German and U.S. markets. The European Union was, in general, politically much more favorably inclined than the United States to respond to the climate change challenge and other signals. One reason was that natural gas and gasoline prices in Europe were significantly higher than those in the United States, making the then high cost of renewables easier to bear.

On May 10, 2005, General Electric chairman and CEO Jeffrey Immelt launched a program called Ecomagination, designed to boost R&D in clean technologies and reduce GE's environmental footprint. The message was clear: A Dow 30 company was saying that climate change could be good for business. Adweek said that GE paid $90 million for the publicity campaign of Ecomagination advertisements and inserts in the *Wall Street Journal, Washington Post* and *New York Times*.

In a sign that adoption of policies to improve climate goals could have a big, positive impact on an economy, California's embrace of climate policy has been a tremendous asset to the state. They embraced a new law, Assembly Bill 32 (AB32), called the Global Warming Solutions Act of 2006, that required California by the year 2020 to reduce emissions of greenhouse gases to the levels it experienced in 1990 or to cut about 25% from the 2006 levels, when the law was proposed. There has been an intense struggle about implementing the law between the progressive renewable energy adopter group and the fossil fuel industry, which claimed it would cost California heavily to adopt such a measure. The conservative fossil fuel industry believed that the CO_2 reductions would go hand in hand with job losses and economic contraction. The measure was adopted and signed into law by Governor Arnold Schwarzenegger.

An independent and nonpartisan group of business leaders, investors and advocates for smart economic policies called E2 carried out a study[43] on the impacts of the Californian law and discovered that the climate policy has generated over $48 billion in the last decade and added half a million jobs. E2 discovered that every single one of the 80 assembly districts has benefited from the state's adoption of the law. This initiative also places California in the lead in U.S. states as the nation looks for economic revival. The actual emissions reductions are on target to be achieved, and now the state is considering increasing the emissions reductions to 40% below 1990 levels with another measure, Senate Bill 32 (SB32),[44] which was just approved. Governor Jerry Brown is supported in this by a large group of California mayors and business leaders, all very aware of the benefits of the first law. "Our cities continue to bear witness to the consequences of a changing climate," the mayors wrote. "From record heat and fire to the continued water quality and availability challenges of the drought, we are increasingly challenged by the consequences of climate change."

The Resiliocene

> *"In my opinion, all previous advances in the various lines of invention will appear totally insignificant when compared with those which the present century will witness. I almost wish that I might live my life over again to see the wonders which are at the threshold."* —Charles Holland Duell, commissioner of the United States Patent and Trademark Office, 1902.

The great energy transition has begun. Solar and wind have been the pillars of renewable energy technology. Transportation fuels are following as we will shortly be using cellulose to generate carbon-neutral biobutanol and ethanol. A 150-year-old technology that currently gives us talcum powder and cosmetics and that is also used in the chemical and mining industries, the ball mill, is also capable of splitting the "recalcitrant" cellulose molecule into its constituent parts: pure lignin and monomer sugars like glucose. The sugars are platform chemicals and can be fermented into hundreds of products such as plastics, foods, drugs and fuels such as ethanol, butanol, biodiesel and aviation fuels. The reason it appeals so much is the massive amounts—the billions of tons—of cellulose available in nature that can literally replace the use of gasoline and at a lower price point. As I type, this technology is going commercial for the first time, way past an idea in a smart mind.

Additionally, home and business owners will soon discover that air- sourced heat pumps can efficiently heat and cool buildings over freezing winters and humid, hot summers using just a fraction of the power: 25% today, but less than 16% in the near future. Yes, heat your entire house and only need to supply 16 of the hundred units of heat you generate for the entire winter. This will likely result in another push for small wind and solar, since it's these sources that can now provide the power required and replace oil and natural gas in a massive new wave of renewable energy adoption.

100% Resilient
The impact of sustainable energy is rapidly making itself felt in all corners of the world, with China as one of the leading nations. Discussions about 100% sustainable energy systems are now common-

place with examples all over the world of economies committed to the task. For example, the Portuguese island El Hierro in the Azores of western Africa employs a group of wind turbines that generate enough power to keep a pumped hydro storage system full so that baseload power can be supplied to the entire island. Cape Verde is another example of an Atlantic island going 100% renewable. Towns in developed countries, some of which are listed here, are doing the same. The countries which have committed to 100% renewable energy include Bangladesh, Paraguay, South Korea, Scotland and Iceland. Distributed energy and off-grid systems are growing fast in developing regions. G7 nations have committed to phasing out fossil fuels by the end of the 21st century, and even Pope Francis's encyclical letter[45] "Laudato Si" talked about replacing fossil fuels.

Forty-six countries in the world, mostly small mountainous ones, already have a hydroelectric energy culture that provides at least 60% of their generating capacity. Some larger countries in Europe, such as Portugal, Denmark and Holland, all famous for embracing large capacities of wind, enjoy days when 100% of their energy comes from installations of biomass, hydroelectric, wind and solar.

Even a very large country, Germany, the world's fourth-largest economy, already enjoys over 30% of its energy capacity from renewables, or almost 40 gigawatts of wind and about the same from solar panels and less from biomass and hydro. On May 16, 2016, almost 100% of Germany's total electricity demand was supplied by renewable resources.[46] This is a clear statement that we can all go to much higher levels of transitioning to renewables. All over the world communities are embracing a goal of 100% renewable energy because it's economic and they can do it. Many cities are committing to renewable energy to solve one or more of the problems imposed on us by the externalities of CONG. They are also motivated by the new jobs, the lower costs, the extra efficiencies and the economic gains that other towns are experiencing. Some of those cities in Europe include Malmo in Sweden and Copenhagen in Denmark. In this book, I mention all these places committed to renewable energy in full awareness that their status will change rapidly; a good website called Go100% keeps up-to-date with such changes worldwide.[47]

The Sierra Club has initiated a campaign led by Jodie van Horn called "Ready for 100" to help 100 American cities have the chance to understand that the transition to resilience is far from a cost but instead a major jobs creator, a major cost cutter, a way of putting incentives to work and a major efficiency resource that results in economic benefit. The campaign helps cities formalize their commitment to clean energy resources.[48] Citizens themselves have often experienced firsthand blackouts or polluted air, which has supported a belief in climate change and stimulated a need to do something about it. Several cities have simply taken matters into their own hands, revealing a level of commitment to the transition characterized by a desire for better conditions enabled by clear pathways to obtain them. Clean energy is an opportunity, not an obligation or cost.

The Carbon Disclosure Project recently completed a study[49] of global cities to examine their carbon intensity and discovered that of 570 cities that responded to the questionnaire, over 100 obtain at least 70% of their electricity from renewable resources and 42 of these were already at 100%. Cities such as Reykjavik in Iceland and Basel in Switzerland and Winnipeg and Prince George in Canada are on the list. Many of these cities are on hydro power, but increasingly a number of cities exploit wind, solar and geothermal as well.

Following is a list of U.S. cities that already use 100% renewable energy and those that have committed to that way. Renewable energy has traditionally done worse when the oil price is low, but the expectation is that a decoupling will gradually occur as equipment costs come down and the value of carbon neutrality and resilience becomes increasingly evident. Many cities are already working toward a percentage of capacity such as 20%, 30% or even 50% by a particular year. Following are examples that saw the economic advantages and dived right in:

U.S. Cities at 100% Renewable Energy as of End of 2017:

Rock Port, Missouri, was the first city in the United States to reach full sustainability. It has four wind turbines on farms in Atchison County that generate 13,000 megawatts a year and whose capacity is likely to go up to 16,000 megawatts. They have contracted with the Missouri Joint Municipal Utilities to take excess wind power into the local grid for use elsewhere.

Aspen, Colorado, with a population of just 6,658, by 2015 was getting 100% of its energy from wind, geothermal and hydro with a bit of landfill gas and solar. It was the third U.S. city to reach its goal. The city signed a contract with its utility, Municipal Energy Agency of Nebraska, to reach these goals. To be fair, Aspen was already 75–80% renewable energy initially, but they eliminated their coal and added the wind to replace the last 20–25%.

Burlington, Vermont, the capital of Vermont, with a population of 42,282, many of whom are environmentally aware, was the first city in the U.S. to go completely on renewable energy. From 2004 to 2008, they had a transition of thinking from a general desire to become resilient to a realization that the city was better off economically if they did it. The Burlington Electric Department acquired a 7.4-megawatt hydropower facility on the Winooski River that also blessed the city with tax benefits. They now use hydropower, landfill methane, wind, solar and biomass to completely replace their dependence on coal and oil and completed the process in 2014.

Georgetown, Texas, might not appear to be the epicenter of energy economics, but Mayor Dale Ross turned the city to 100% renewables even though the state itself is better known for oil. They are enjoying their moment in the sun, as it were, because TV crews from worldwide news organizations regularly show up to tape a segment on the town's sustainable credentials. They use wind power from west of Amarillo, and a new 150- megawatt solar power installation will come online in 2018.

Greensburg, Kansas, with a population of just 785, was literally razed by the largest tornado there can be in May 2007. The city had the choice to rebuild using the technology of the past, what they already knew, or to embrace what's possible today. As farmers with good knowledge of the land, wind, solar and water, and with incredible foresight, they chose the latter and teamed up with NREL experts to rebuild. There was a lot of interest in this case about demonstrating the recovery of a community with green energy. This gave the city the impetus to turn around, employing the motto "Rebuilding…Stronger. Better. Greener," and now the city is mostly powered by wind since 2013. The city installed a 12.5-megawatt wind farm that covers every residence, farm or municipal building's needs. The city's economy has improved remarkably since the rebuilding finished.

Kodiak Island, Alaska, was working on a renewable plan since 2002, when the Kodiak Electric Association planned to reduce dependency on diesel and install wind power and made it to 100% by 2012.

The 15,000 residents obtain their power from the wind and also some hydropower, which had by 2014 already saved the community $22 million versus their cost in diesel power, or 2.8 million gallons per year that released emissions of 62 million pounds of CO_2 annually.

Cities Committed to 100% Renewable Energy as of End of 2017:

Columbia, Maryland, announced in September 2015 that it would offset 100% of its energy use with renewable resources and signed a 20-year PPA with SunEdison. The company built the Nixon Farm Solar plant, which supplies enough power to replace 100% of the city's requirements.

East Hampton, New York, with 20,500 residents, is on the south side of Long Island, and when Hurricane Sandy hit in October 2012, the entire community lost electricity for several weeks. They knew they needed more resilience. This wakeup call caused them to act fast. The town board passed a unanimous resolution in May 2014 to adopt the goal of converting to 100% renewable energy by 2020 using solar and wind and the adoption of a microgrid and backup power source. The changes rub up against political challenges and need the support of the state of New York and the Long Island Power Authority (LIPA). The town is depending on LIPA to approve a large offshore wind project at Montauk on the tip of Long Island to meet the town's 100% goal, but the current sentiment is one of group determination.

Grand Rapids, Michigan, with a population of 192,000, the second- largest city in Michigan after Detroit, wants to go 100% renewable by 2020. Its office of Energy and Sustainability made the clean energy plan. It plans to cut its demand, install solar and geothermal power, and purchase renewable energy certificates.

Honolulu, Hawaii, wants to be fully renewable by 2045.

Ithaca, New York, committed to 100% of the power demand from its municipal buildings, street lights and traffic lights coming from renewable energy as soon as January 2012. They bought renewable energy certificates to equal 100% of their existing energy consumption via their Municipal Electric and Gas Alliance Inc. (MEGA), a nonprofit aggregator of gas and electricity. The city also has a 2-year power purchase agreement with Integrys Energy Systems of New York, Inc. By

2016, they reduced their carbon emissions to 20% below the levels of 2001.

Lancaster, California. In 2013, R. Rex Parris declared that he wanted his city to be the solar capital of the world and the first city to produce more electricity from solar than the city consumed. He placed solar panels on city buildings and the performing arts center and the stadium in one of the most expansive solar community programs ever undertaken.

Los Angeles, California, while not officially on the Sierra Club's "Ready for 100%" list of American cities searching for 100% renewable energy, they took a decision in September 2016 to commit to study the pathway to 100% clean energy. They have almost reduced coal power to nothing, already have a sizeable wind fraction and know that the benefits are spread to every city dweller. The Los Angeles Departments of Water and Power are doing the work to assess the moves required to make the transition to 100% clean energy.

One LA neighborhood, Mar Vista, has already committed to 100% clean energy by 2018. By July 15, 2010, they had already achieved 4.7% renewable energy, and their progress is inspiring other LA neighborhoods to do the same and create more Green Power Communities.

Nassau, New York. In December 2015, the town's board voted to obtain 100% of its power from renewable energy by 2020. This town is located just outside Albany and pledged to have all six of its buildings disconnected from the grid.

Palo Alto, California. In 2013, the city's council voted to commit to using 100% renewable electricity. They've had green power already for decades, but the decision to go 100% was possible because the town owns 100% of its own utility.

Rochester, Minnesota, is a small city of 100,000 souls that wants to be 100% by 2031. It lies within range of the excellent wind resources of the American Midwest and most likely will choose this among other methods to obtain its goals.

San Diego, California, is a large city with a population of 1.37 million. In December 2015, the city council voted unanimously to transition to 100% solar, wind and hydro renewable energy by 2035, including 90% of its transportation fleet. It was not a partisan vote.

San Francisco, California, with a population of 834,000, is an example of a large city that already benefits from a high capacity of renewable power. In 2002, they started to work to reduce greenhouse gas emissions with an electricity resource plan that has closed two CONG power stations and expanded the use of hydro power for 17% of the city's needs. The city is currently 23% renewable energy from solar, wind, biogas and geothermal. Another 21% is hydro, making a total of 44% and growing. In 2011, they decided they wanted to be 100% renewable by 2030. Mayor Ed Lee wants the city to do its share to combat climate change and make a positive difference and be a role model for smaller communities. He asked a mayoral task force to advise the city on how to get to 100%.

The San Francisco Environmental Department spearheaded the plan with six strategies to obtain 100% renewable energy by 2020. Solar installations, such as the Sunset Reservoir system, the size of 11 football fields with 24,000 panels generating 5 megawatts of power, have been placed within the city. Even the historic City Hall roof was made into a solar roof. Every city municipal building now is 100% greenhouse gas–free. A new residential development called Hunter's Point will be the first to be 100% greenhouse gas–free. A plan called CleanPowerSF means that residents can get some clean electricity at the same price as from fossil fuels, but they can also opt to have 100% and pay a slight premium. This is in addition to the growing adoption of solar power on residential roofs that take advantage of lower costs and state and federal credits. All this is being done in a cost-neutral basis as a result of the 10,000 extra jobs expected.

San Jose, California, has a 960,000 population. Chuck Reed, the mayor, pledged in 2007 to get to 100% renewable energy by 2022 using all the facilities available in the Bay Area and from the state of California. This will be done as part of a 10-point plan that includes a reducing energy use by 50%, planting 100,000 new trees, recycling wastewater 100%—currently a California "must-do"—and installing 100% renewable electric power.

Santa Monica, California. In January 2016, the city announced it would install wind and solar to run its municipality and obtain its installations from 3 Phases Renewables, a renewable energy power solutions provider.

Australians have also been yearning to decarbonize, decentralize and democratize the local energy supply. They have been employing a lot more storage and the unlimited supplies of solar irradiation that Australia is famous for. Towns in New South Wales that have embraced a 100% renewable energy regime include Tyalgum, Uralla, Huntlee,

Coffs Harbour, Byron Bay, Lismore and Mullumbimby. In Victoria, there is also Newstead and Yackandandah.

Even China and Latin America are working on the challenge. One notable project is in the Middle East and it's not Masdar, which was an excellent start, but which is currently bogged down and has its official opening as a city of 40,000 residents set for 2030 now. A new plan for a Saudi Arabian city called Neo-Mostaqbal, or Neom, which means "new future," was announced on 24 October 2017. It will be located right up in the crystal-blue waters and beige mountains of northwestern Saudi Arabia in the northeast of the Gulf of Aqaba. The vision of a post-oil future is inspiring designers. The next Saudi Arabian king, Crown Prince Mohhamed bin Salman, announced with his conservative Wahabi clerics in mind, *"We are returning to what we were before, a country of moderate Islam that is open to all religions and to the world. We will not waste 30 years of our lives, wasting time dealing with extremist ideas. We will destroy them today."*

The $500 billion city is unusual, being built to straddle the borders of Jordan, Saudi Arabia and Egypt. It's worth noting that if it includes Egypt, then it must also include a tiny slice of Israel that touches the Gulf at Eilat. The architectural blank slate will, of course, benefit from all the latest improvements in renewable energy as a top priority, with Gulf desalination and storage solutions on the front burner. The vision will be completed with magnificent civil engineering and vertical farms to create a fertile oasis to feed its population actively employed in bio-tech, future transportation, additive manufacturing, robotics, leisure, gaming, media, sports and recreation and global events. It will benefit from free high-speed internet communications. The website buzzes with excitement and passenger drones and almost utopian optimism, a breath of fresh air in a region that has known so much struggle.

This raises the question: What kind of world are we looking at after the transition? The obvious answer is one in which all the CONG externalities have gone. It would be, at last, sustainable, where the air really is fresh, and the ecosystem is un-poisoned and has been returned to a natural cycle, once again rich with variety. It is one in which fish populations and coral reefs and endangered species have recovered. No smog or health concerns arise from human energy activities. Human health and geopolitics could be vastly improved.

Can we even visualize a world where basic sustenance is plentiful and nutritious, and water is likewise abundant? Since the externalities of CONG are not currently paid for in our energy bills but must be paid eventually by our taxes, the promise of a sustainably resourced community also includes the potential of a seriously reduced tax load, improved energy security and better international relations. The climate issue could finally be resolving itself, and pollution could be a thing of the past. Hospitals would no longer have to treat as many patients suffering from the effects of exposure to toxins, carcinogens, polluted air or coal mining or even driving accidents.

Today, when the oil price climbs, it acts as a tax or a governor on the economy. When the oil price falls, it's a stimulant. In 2008–9, the oil price climbed to $147 per barrel, helping to set off a recession, which in turn reduced demand for oil, which in turn sent the price back down to about $36 per barrel all within a year. In other words, as the oil price climbs, it and other goods and services from industry become less affordable. On the other hand, a sustainable economy can grow faster because energy cost no longer acts as a restrictive tax. With zero price per ton of sunlight or per barrel of wind, renewable energy costs become noncyclical and nonvolatile, and an even better basis for a good economy.

Fossil fuel companies are experiencing an existential blow from the public, markets and legislators. They are fighting back both by embracing the renewable energy sector, as Exxon did when it spent $600 million acquiring Craig Venter's algae activity, and by more negative means, where they disavow early climate research or pay institutes to misinform the public and politicians about the facts of the situation to serve their own short-term interests. It has become clear that in the balance between externalities and resilience, the latter is so far in favor of the best interests of the human race, with an overwhelming global vote, as evidenced by the extraordinary and unusual global accord in Paris at COP 21 in November of 2015. 194 countries voted in favor of the science and of doing something about the climate. The result of action as we are discovering is best demonstrated by resilience-friendly locations like Germany or California where there is now more employment and money in the system, the end of petrodollars and a decline in fuel logistics and a more equitable world energy geopolitics are a result.

The still-profitable fossil fuel lobby will continue to benefit from political access and economic advantage but let us be clear: The coal and oil companies are staring at potentially billions in stranded assets, worthless mining and drilling rights, and few new discoveries of scale for decades, and remaining oil reserves are in cold, wet places. At some future juncture, this will translate into declining revenues and earnings. This much is inevitable, even if it were simply from depletion of current reserves. The interests of the public are harmed by the lack of transparency, pollution, science censorship (recently in Canada and potentially elsewhere) and an embrace of dated technologies. This harm has been amplified by almost uninterrupted global growth, which brings with it huge volumes of consumed combusted oil and gas and coal. Add to their negatives the enhanced pressure caused by natural disasters, the ever more common wildfires, droughts and floods and the tendency to hang on to centralized power distribution structures that enslave subscribers to a helpless lifetime of expensive bills and there is only one way for progress to go.

Clean energy, by its nature, is distributed, democratic, transparent and without any of the externalities of CONG resources. The new world is embracing energy efficiency, distributed supply and lower energy consumption per unit of GDP. Appliances such as fridges, LEDs, air conditioners, motors, carbon-neutral fuels, electric and autonomous vehicles, to mention just a few, are now growing as a solid part of the "low-energy- consumption" side of the equation. Soon, the ability to store electricity, which is a technology on the verge of making an impact, will present an existential threat to the utility, as the electrical power to run a house and car suddenly become the domain of the roof and garden. CONG companies will be unable to survive the arrival of cheap electricity storage. They will also be up against a wall because of cheaper liquid transportation fuels and biodiesels from massively available cellulose, an industry not dissimilar to beer or wine manufacturing but with a ubiquitous and potentially enormous, nonfood feedstock that we have not been able to use until now.

There is a revolution happening in this industry and it's a good one. It creates jobs, sustainability, better health, geopolitical amicability, good targets for subsidies, much lower energy commodity price volatility, and much less waste and it will finally have shown humanity's best in its ability to impact the climate for the better, although we are aware that there is a need for some geoengineering to help make that into reality. As centralized systems yield to more self-control and distributed systems, aided and abetted by the massive impact from the communications sector in terms of individual control software and mobile applications, the future of the world appears to be one in which social justice, gender equality, individual fulfillment and higher quality of life are inexorable likelihoods. All the pain endured by humanity in the last 200,000 years can finally come to an end as we replace the darkness with the next phase: clean energy abundance through technology with massive beneficial social implications and a real resilience.

We are still in the age of transition and huge amounts of energy are still sourced from fossil fuels. Human beings have an emotional wistfulness for a cleaner, better state of being. When we understand that sustainability is not a choice but an essential quality that is also cheaper and achievable, we can better move forward toward the transition. The goal of this book is to stimulate a hunger for whatever improvements this vision of a better world might bring. A world based on real, disruptive, economic and practical technologies that carry few external costs would be a world that—like the two Nigerian states that chose a fishing festival instead of constant war— allows humanity to celebrate something on a global scale that enables the Anthropocene to finally yield to the Resiliocene epoch.

Chapter 2: The Externalities of Renewable Energy

The Largest Transition in Human History

What's the use of a fine house if you haven't got a tolerable planet to put it on? —Henry David Thoreau

Humanity is going through many big changes but arguably one of the biggest is an energy transition. Our use of energy has already transitioned through wood and dung to coal and oil and now we are fast transitioning again. This time it's global and not local. We are experiencing (there is resistance) a simply huge, multi-trillion-dollar, political and market-led transition from unsustainable energy to sustainable energy.

We commonly celebrate the long-life span of an elderly family member and reflect on the things that have changed dramatically in that time. This kind of reflection has really only taken off in the last two hundred years[2] or so because, prior to that, life did not change as fast, although, admittedly, there is a difference between advancement events, like inventing the wheel, and other natural historical events, like floods, volcanoes, earthquakes and meteors that have always affected us. The last century has given us a remarkable transition already, but arguably there is so much more to come before we obtain what might, appropriately, be called a "mature," sustainable quality of life, if that were ever to happen. That relative also is living to a ripe old age in today's conditions, so the average age of 40 years or fewer prior to 1800 may not have been long enough to span many historical changes anyway.

What this means, of course, is that even in the middle of the most dramatic century for technological change in almost all of human history, people alive today can still be forgiven for also seeing things as very stable. We are used to a certain conservatism in review of our times that can become anxiety or fear of change. Well, in this case, I am identifying a change in the way we "do" energy, which, remarkably, may remain almost invisible to the large majority of people who take it for granted and who, unaware, continue to flick a light switch. They little realize that historically the light source has changed from a flame to incandescent then to compact fluorescent and now to light-emitting diode (LED), which uses only 15% of the energy of the earlier equipment (and shortly, less again!). We are going through a revolution in the way energy is consumed. Germany and Europe, China and the United States deserve a lot of the credit for initiating this recent phase of history.

Energy demand and production are becoming more efficient. Significantly reduced electrical demand from modern technologies that provide lighting (LEDs), heating, ventilation, air-conditioning (HVAC) and refrigeration in properly insulated buildings and efficient vehicles means that incumbent power capacity can now service more users even though it is still generated by the CONG basket and, thankfully, an ever-increasing capacity of wind, solar, hydro and geothermal energy. A message I am sending is that our CONG habit is now destructively expensive, and the sustainable replacement is superbly and increasingly efficient and economic. The transition is well under way. In just 200 years, we have come from a much less developed stage of civilization,[50] and unfortunately this is a quality of life that almost 30% of the 7.4 billion[51] people alive today still live. We speak of the developed world, but it remains far from the vision of a sustainable world. If ever a trend was visible, the current transition of humanity to sustainability fits the bill. Thinking a bit about where we have come from is very instructive.

In the past, living was harder. Work was more manual and menial. We suffered dirtiness, lack of warmth or cooling and no light or information. Any extra hours of study or entertainment were provided courtesy of firelight and oil lamps. Clothes were washed manually. Historically, everybody had this existence. Prior to the industrial revolution (and after a nomadic and hunter-gatherer period), we lived where we were born and hardly traveled beyond a 50-mile radius, unless we were in a military or had a horse or a boat. Energy was almost only fire and motion. It meant cooking and winter warmth only and came almost exclusively from wood and dung. Wind and water were both harnessed for various purposes, such as milling flour from grain and working the bellows in a blacksmith and, of course, in sailing ships.

The Roman Empire is often seen to have collapsed because of internal corruption and the decadence of the ruling class. I quote this section of Jeremy Rifkin's book *The Hydrogen Economy*,[52] which illustrates that a civilization has a demand for energy just like an engine might and that entropy, the tendency of natural systems to become more random, rules here too. He shows that as the Romans shifted from being a conquering civilization to a colonial one, there was a huge reduction in wealth coming in from conquered Germanic tribes and an obligation to provide services. Rome had over one million inhabitants at its height, but as the colonies failed to provide any revenues, the burden of supporting the empire fell on agriculture. Migration to the cities, the increasing

size of government bureaucracies, welfare payments and public buildings sapped the ability of the empire to survive:

> *Rome was experiencing the harsh realities imposed*
> *by the laws of thermodynamics. Maintaining its*
> *infrastructure and population in a non-equilibrium*
> *state required large amounts of energy. Its energy*
> *regime, however, was becoming exhausted. With no*
> *other alternative sources of energy available, Rome*
> *put even more pressure on its dwindling energy leg-*
> *acy. By the fifth century the size of the government*
> *and military bureaucracy had doubled. To pay for*
> *it, taxes were increased, further impoverishing the*
> *population, especially the dwindling farm popu-*
> *lation. The empire, writes Joseph Tainter, began*
> *consuming its own capital in the form of producing*
> *lands and peasant populations.*
>
> *Weakened by a depleted energy regime, the*
> *empire began to crumble. Basic services dwindled.*
> *The immense Roman infrastructure fell into disre-*
> *pair. The military could no longer hold marauding*
> *invaders at bay. Barbarian hoards began to whittle*
> *away at the decaying Roman Empire, at first in dis-*
> *tant lands. By the sixth century, the invaders were*
> *at the gates of Rome. The great Roman Empire had*
> *collapsed. By the sixth century, the population of*
> *Rome, once numbering more than 1,000,000 had*
> *shrunk to less than 30,000 inhabitants. The city*
> *itself was reduced nearly to rubble, a stark reminder*
> *of how unforgiving the energy laws are.*
>
> *The entropy bill was enormous. The available*
> *free energy of the Mediterranean, North Africa, and*
> *large parts of Continental Europe, reaching as far*
> *north as Spain, and England, had been sucked into*
> *the Roman machine. Deforested land, eroded soil,*
> *and impoverished and diseased human populations*
> *lay scattered across the empire. Europe would not*
> *recover for another 600 years.*

We have already experienced a series of changes in our energy resource usage, moving from wood and dung to vegetable or whale oil and coal and then to mineral oil and gas.

Education is being transformed. Working from home instead of in office environments can help us be much more productive and comfortable. If done well, and there are significant doubts still, these changes can increase human happiness, a pivotal article of the American Declaration of Independence[53] and, generally speaking, all this could give rise to a "human constitution"—and I'm not talking just health. Can you say Utopia? It sounds too good to be true, but if it's possible, isn't it worth aiming high? With nine billion people, able to access communications, hot water, comfortable living conditions, efficient and rapid transportation as well as nutritious food with a sustainable, small footprint, we will have almost made it all the way to the highest quality of life[54] possible... but even the definition of that will evolve as our vision of it improves.

The huge number of scientific advances over the last 200 years includes many in sustainable technologies. Fuel cells, solar and wind power, farming, water production and the atmospheric impact of carbon dioxide (CO_2) were all well understood by the natural science cognoscenti in the late 1800s. We had learned a lot about how to live sustainably already by that time. But recognition of the need for sustainability and the vision of how to achieve it were far from developed. Since 1800 we have compounded our knowledge and owe our extraordinary increases in population, living standards, longevity and wealth over the last two centuries to this growing resource expertise in the supply of our main resources, food, water and energy.

This know-how has been largely taken for granted. Our generation has grown up in environments where cars, planes, fridges, washing machines and now computers are commonplace. Using these gadgets has rid us of many of the stresses our forebears had to endure (or were unaware they were enduring, just as we may be unaware of what we are still enduring!). The hard-won knowledge needed to understand and use this technology is essential, but we are also in an economy which effectively has no user-serviceable parts inside anymore. Our cars and appliances, which a handy person used to be able to fix at home with a screwdriver, have become so complex that owners can no longer hope to repair anything, and likewise we have all become dependent on community skillsets. All the knowledge we have learned has led to a more complex world. It's fair to say that some people rebel against this complexity and have a thirst for a simpler world.

This is especially so when the economy is a little weaker, many are struggling, and the unequal distribution of resources is extreme. There is no denying that this complexity also has the potential to afford us all a higher quality of life. Sustainable technologies are indeed even likely to bring back a certain simplicity as distributed energy and lower reliance on centralized power take hold. Technology is a double-edged sword, and managing it properly means determining what is the good side and then emphasizing it.

There is always resistance to change. A specific current challenge is that many expect renewable energy to bring higher costs during an economic downturn or a loss of profitability in businesses that are currently thriving while using cheap CONG energy and are therefore unsustainable. These forces put the brakes on progress and manifest as enemies of change. At its worst, we may only earn a sustainable environment to live in after the challenges of unsustainable living become so threatening that the political consensus finally tilts its way. This would be apocalyptic, like a climate crisis where a thousand Hurricane Sandys and Katrinas damage coastal cities of the world like a thousand nuclear attacks. Many business interests resist interrupting profitable activities to do something for the wider good. This is more circumstantial than malevolent, however, but it has now been seen repeatedly. The tobacco industry and increasingly the food industry are often used as case studies[55] to illustrate this.

In the 1960s, it was thought that fat alone was the main cause of obesity and heart disease, which, between them, kill so many people that they dwarf any other cause of death currently and are the number one cause of death globally. As of May 2017, the World Health Organization (WHO) estimated 17.7 million people died prematurely from cardiovascular disease (CVD), both strokes and heart attacks. Almost all of these deaths can be prevented by arresting smoking, drinking less, doing some exercise and eating better.

One negative diet factor is the consumption of sugar. In 1965, the Sugar Research Foundation (SRF) quietly carried out a review to assess the impact of sugar on health. Two rat studies done in the late 1960s, called Project 259, funded by sugar lobbyists both showed a link in rats between sugar consumption and bladder cancer and coronary heart disease (CHD). Although the authors wanted to continue the studies, the funders dropped both of them, and the industry successfully continued to blame fat for health problems. In a new study published in *PLOS*

Biology, it has been revealed that the sugar industry, like the tobacco and oil industries, discovered their product was bad for human health in a major way that they already knew about but downplayed so that their businesses were not impacted.

> *The sugar industry has maintained a very sophis-ticated program of manipulating scientific discus-sion around their product to steer discussion away from adverse health effects and to make it as easy as possible for them to continue their position that all calories are equal and there's nothing particu-larly bad about sugar.* —Stanton A. Glantz of the University of California at San Francisco, one of the *PLOS Biology* study's authors.

A study[56] completed in 2017 found that mice fed on a sugar-heavy diet were more likely to develop breast cancer. The Sugar Association, one of the largest industry lobbying groups in the United States called the study "sensationalized." This means that, like Exxon and the tobacco industry, internal industry science has known of the negative implications of their products for over 50 years and has misinformed and kept quiet about it. There is no question that this is a very difficult thing for an industry to cope with, but in the case of sugar and fuel, both industries had reason not to worry too much about decreased consumption, something that is not true today, with alternatives present for both. Tobacco is another situation.

It is now widely recognized that sugar is the main culprit in the obesity pandemic. Despite the criticism in New York about Mayor Bloomberg's "nanny capitalism," where schoolchildren were warned away from sodas rich in sugar, the outcome of any successful interven-tion is most likely to be more healthy kids. The CONG energy industry is likely to continue to resist sustainability in the same way.

This is why I describe the societal costs as externalities. They are not paid directly by the fossil fuel companies, but they accumulate and are paid by all of us in the end. It is important, also, not to demonize CONG companies, except where evidence exists that they have delib-erately tilted the playing field, like the tobacco companies did, against progress. Exxon's previous support of groups such as the Competitive Enterprise Institute, whose director Myron Ebell[57] was almost selected by President Trump to head the Environmental Protection Agency

(EPA). Ebell has promulgated denial propaganda and has been paid to do so by a group of companies,[58] with the strategic goals of confusing the issues, suggesting there is a debate over the science, keeping things stable for fossil fuel companies and suggesting that climate threats don't exist. To be fair, Exxon Mobil actually did at least verbally support the Paris COP 21 and its wider climate goals.

Those climate threats are now ever present, along with species extinctions, declining water tables, melting glaciers and polar ice caps and many other ominous manifestations. The high cost of fossil fuels and nuclear power, with their externalities, mean they are best replaced as soon as humanly possible. CONG is not as cheap as it is commonly thought to be. It is subsidized, and its effects are negative and expensive.

There is a simply huge and potentially cataclysmic transition going on, which is going to draw on all our accumulated experience and knowledge base. It is nothing less than the transition of our civilization from being unsustainable, with some convenient and extremely energy-dense fossil fuels that have been necessary to reach this point, to being an entirely sustainable global community that uses resources that don't damage the planet. This has become possible only because education and innovation created new technologies and materials. In a world where there is essentially freedom to act, private enterprise will embrace the benefits of sustainability until it takes root completely and becomes very profitable.

Capital, shelter, land, food, water and energy are all intertwined and fundamental for our continued civilized development. Each is represented by its own resources. The water resources we currently use—lakes, water tables, rivers, atmospheric humidity and rainfall—are threatened with overuse and wastage, suggesting limits. Australia, the Middle East, China and California,[59] among others, are experiencing severe droughts. Food depends on water, and we have seen it is possible to generate good food with drip irrigation and hydroponic techniques that use a fraction of the water as previously. If necessary, one can even obtain water without expensive reservoirs or piping infrastructure simply from the air by condensing it, even in the driest conditions, using energy from geothermal, solar and wind power to do so. The issue of intermittency (solar and wind power that works for an unpredictable period while other intermittent technologies such as tidal energy are more predictable) can be diminished if the role of any excess energy is to fill a reservoir with water or generate hydrogen.

Before the industrial revolution, we lived sustainably because we were in balance with or overwhelmed by the natural world. We burned wood that grew again quickly, and our communities were too small to have a deleterious effect on nature. Transitioning from nomad to farmer to industrialist brought longer life spans, wealth and huge population growth and destroyed that natural balance. In the process, we started to exploit new energy resources. Today we continue to move ahead and learn to identify and exploit better resources. The advantages of plentiful CONG energy resources were their economics and their convenient energy density. But they are unsustainable. Living unsustainably is where we are now.

I like the metaphor of the orchard and low-hanging fruit. We have literally consumed the major part of the low-hanging energy fruit at this point in history. Fracking and shale resources have reinvigorated our fossil resources,[60] but we are using energy on such a scale that, in reality, we are likely to face a transition to sustainable energy resources only because of CONG depletion. Sadly, we are navigating in the early 21st century where not everything is conducive to rapid transition. The financial collapse in the autumn of 2008 caused capital to fly to "safe" investments such as low-yield, low-credit-risk government bonds. After the dot com crash of 2001, the 2008 market crash caused investors, twice bitten, to be twice shy. Even ten years later, the investment community has lost much of its willingness to take risks.

The post-2008 deep aversion to investment risk resulted in the paradox that we may delay addressing the biggest risk of all—climate catastrophe—by continuing to emit greenhouse gases and manifest the other externalities for an extended period. We may end up compounding our situation. We may not recognize how precarious and risky our current resource conditions are and therefore how important it is to move beyond them. Historically, it was convenient for us to exploit any resources when we needed them. We didn't need to think about how nature would replenish them. Animals, people, plants, minerals all needed exploiting by using energy from CONG. Today, our emerging energy competence makes the vision of an age of sustainable and relatively unlimited energy solutions clear, economic, desirable and achievable for the first time. I'm just concentrating on the energy transition, but the wider community is also in the throes of making a transition in the domains of food and water. The technologies required for living sustainably exist today and are becoming cheap enough that they are becoming a major economic driver for change.

When we look closely at the consequences of CONG, it quickly becomes clear that they operate in different domains and that the costs in each are severe. These externalities have increased in intensity over time and have now collectively acquired critical status. Their effect is increasingly extreme, yet, because of the current structures, we still fail to see the impact of their full cost.

Clearly, nobody was jumping up and down at the dawn of the industrial revolution telling us that oil, coal, gas and nuclear energy sources were poor choices because of their costly externalities. We adopted fossil fuels because they were available at low cost at a time when few externalities had accumulated. The convenience of CONG was significant. Fossil fuel use accelerated because of demand from affordable steam and internal combustion engines and, later, electric motors as well as other technologies that could use electricity generated by CONG to leverage our ability to work and generate wealth. In chapters 3 through 10, I look at these eight externalities in more detail and examine externalities of renewable energy in this chapter. Each externality represents costs which are paid eventually—if not straight away—by the immediate consumer.

If we were more aware of this huge cost from our current energy resources, we would feel more alarm and a greater need to act. We are insulated from the full story by the complexity of the situation. A complex energy market generates confusion and offers the enemies of change weapons to misinform us. There is little dissemination of fact and little trust in facts disseminated.

Since we generally make decisions based on the cost of things, one way to understand the accelerated adoption of renewable energy by some countries over others such as the United States, is to see the cost of their CONG resources. Today (2018) U.S. consumers pay about half the cost per gallon of gasoline that Europeans pay. It is therefore less of a mystery why European voters and politicians have been moved to act on several fronts, including supporting a carbon tax and trade system as well as subsidies for renewable energy which have effectively kick-started the revolution. Figure 16 shows third quarter 2014 global pricing in USD for a gallon of gasoline worldwide.

The difference in these consumer costs is largely due to the UK gasoline tax, which accounts for over 10% of all UK government tax revenues. Bear this in mind as a potential reason for government itself to resist a new automobile energy regime. Thomas Friedman in his book

Hot, Flat and Crowded[61] suggests that the U.S. government could add a $1.00 "Patriot Tax" to a gallon of gasoline and use it to pay for a significant amount of clean energy benefits. If it was done at a time like the present, in pennies per gallon, imperceptibly while the cost of a gallon of gas is falling anyway, it might almost be invisibly done.

The price disparity extends to natural gas as well. There is a glut of natural gas in the United States currently keeping its price capped, whereas in Europe, where geopolitical concerns abound about self-sufficiency and buying Russian gas, gas retails for over $10.50 per million BTU (MMBtu) in Germany and the UK and only $3.00/MMBtu in the USA.[62]

It's also currently $16/MMBtu in Japan. Only a few years ago, the United States had plans to cover a shortage of natural gas with liquid natural gas ship transportation into the country. The implementation of fracking has reversed this calculus, resulting in U.S. gas exports to satisfy foreign demand.

The pricing also supports this configuration and underlines how much the United States had been willing to pay for gas only recently. It's quite likely that the same effect plays in respect to attitudes to solar power since different locations have different electricity prices. High electricity prices could easily be correlated to increased numbers of solar installations.

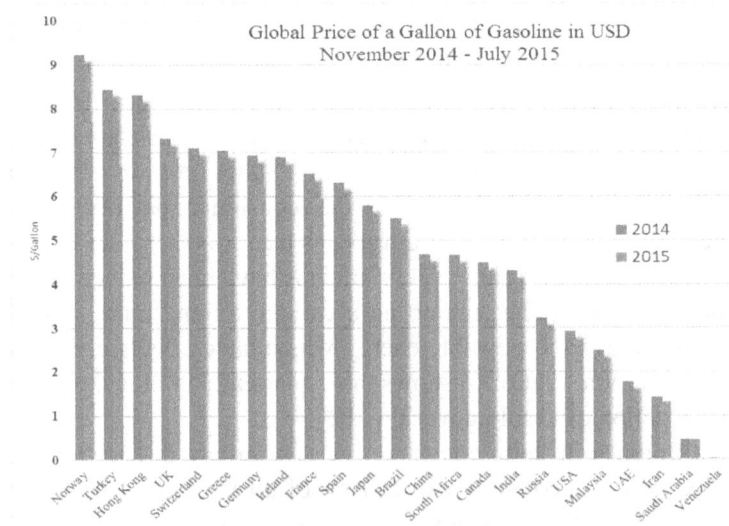

Figure 16: The varying cost of a 2013 gallon of gasoline in different countries around the world. Source: Data from Bloomberg, 3rd Quarter, 2014.

Despite the potential for price shock to local consumers wherever they are, both gasoline and natural gas prices are very low in terms of the real accumulated costs of CONG. Nobody pretends that a blockage of the Strait of Hormuz or some other geopolitical event in the Middle East will leave the oil price as low as $60 per barrel. Looking at today's real cost of these externalities, this book will not try to assess a specific internal cost for gas, coal, oil or nuclear power but will describe the impacts and compare estimates from other expert groups that do a good job to see whether an informed position or consensus clarifies the argument. Oil is acting as a governor on the growth of our economies. When the price climbs, it impacts economic growth.[63] Renewable energy will remove that blockage and release the opportunity for more economic growth because there are no logistics issues and no cost per ton or gallon. In fact, renewable energy technology is getting cheaper all the time as we go up a learning curve. It will find its level as production volumes match the market, so the getting cheaper part won't last forever. There is every possibility that renewable energy will totally replace problematic CONG in its own good time.

Environmentalism Is Important, but This Is About Economics

This is not a book about environmentalism. It is pro-sustainability and is about the economic opportunity offered by a transition to sustainability. This transition achieves all goals of preserving the environment and species while also taking humanity off its damaging global footprint. I have a vision of a better place, and it is based on an economic evolution using real technologies that exist now and that in many cases have existed for hundreds of years. It is based on the explored argument that the accumulated cost to society of CONG is greater than the cost of the sustainable technologies we are rapidly adopting.

Many sustainable technologies are already long in use. We are in a revolution of change toward sustainability. We have a huge capacity for innovation, which we have proven but which is largely forgotten or ignored. Things that didn't exist five years ago exist today because of our enthusiasm for improvement in our quality of life. Things are changing so rapidly that looking back a single generation is to find a completely different world. Adapting to these changes can generate economic growth, although the reverse is the common criticism.

This trend will persist, and the world will continue to change. Every step forward provides potential good and bad outcomes. The best tool for making easy decisions about what's good or bad is probably based on an excellent education and trained objective thinking.

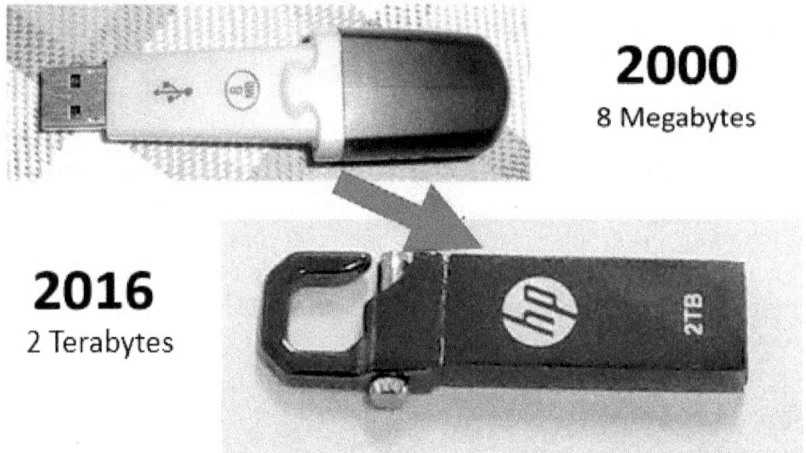

Figure 17: Sixteen years in this age is full of radical change. These memory chips reflect incremental innovations in technology, which, for example, has increased the capacity of thumb drive chips by 250 times. Source: NEF Advisors, LLC.

It's very clear that over time we are trending toward higher-quality living conditions. I really want to make the point that the vision of this book is not environmentalist. It's more economic humanist. Environmentalism is an appeal to the past, to a simpler time, and it is grim, rugged and cold. It's about wearing hair shirts and cold showers. It hates exploitation and profitability and blames the world's ills on big business (which certainly isn't perfect). It's a step backwards because its goals are retroactive and limited, and it gets our future very wrong. It works in guilt and fear. It believes that we've damaged everything permanently and need to repeal our progress. It thinks technology is just bad, and not a double-edged sword that needs maturity.

It doesn't trust humanity to make the right decisions (and there may be many reasons to show this is true, but it need not be the case). There is no room in the environmentalist approach to enjoy more energy use or more travel or to indulge ourselves more, all of which look to be eminently possible with clever use of sustainable technology. There is nothing wrong with an SUV—if it's got a powerful electric motor and a range of a thousand miles and can be charged from the garden solar array and wind turbine, or a thorium reactor.

The vision I speak of here is of an economic path to both a zero human footprint and a higher quality of life. For the puritanical environmentalist mind-set, this is synonymous with decadence or the impossible. This positive view more resembles the quest for fulfillment, for the happiness that the extraordinary authors of the U.S. Constitution integrated into that document. It's not a zero-sum game. Humanity is moving relentlessly toward a revolution in at least six domains: education, energy, communications, transportation, exploration and health. All these areas are benefiting from our ability to discover solutions to our challenges. It's something we will always do, and these are the most exciting times to be alive. This is a vision with detailed technical solutions, not a mindless rebellion.

Complaints Leveled at Renewable Energy

Now we look at general questions leveled at renewable energy that are useful to answer, and then we look into solar, wind, ocean power and hydroelectric in a bit more depth to sample the negatives of these renewable energy sources. The following represents some basic arguments against renewable energy that are still commonly used by the forces against change.

Fossil Fuel Energy Has Powered Civilization to Its Highest Levels

The work humankind needed to do was accelerated hugely by CONG. Traveling, cleaning, delivering, ploughing, sorting, harvesting, trading, building, constructing and manufacturing were all made much more efficient and speedy by engines working off CONG, and increasing efficiency over time had a huge impact. There are still spirited efforts to support CONG, as might be expected, such as Alex Epstein's misguided *The Moral Case for Fossil Fuels*, but the entire history of growth of the oil age has now stalled. There is no reason why renewable energy cannot take hold of the energy baton and continue the race to improve human quality of life. We have come to a point where we recognize CONG limitations as well as the advantages of renewable energy.

Negative Arguments Against CONG Are Limited and Evaporating

A recent *Wall Street Journal* article authored by Matt Ridley[64] stereotypically framed this point with a limited look at depletion, price and climate. Ridley argued that each of these CONG issues was diminishing in importance. To counter this view, firstly, there are not just three externalities visited on us by CONG but eight. CONG has passed its sell-by date and we need to move forward to a new world of emissions-free energy production. Ridley argued that alternatives have failed to price incumbent technologies out of the market, but I would counterargue that this is precisely what is happening. The convenience of solar and wind energy in not having a continual logistics overhead, being lower cost and having lower inflation exposure is rapidly becoming a compelling component of community resilience. I would also venture that other technologies have significant potential cost declines to come. Traditional nuclear has priced itself out of the market with the wrong designs. New build requires financial guarantees and subsidies, but if thorium molten salt reactors were to appear, then the combination of emissions-free, cheap fission energy would be available to fill demand and combat the CONG externalities in an optimal manner.

Despite a Reputation for Being Finite, Fossil Fuels Will Not Run Out

There are two issues here. Clearly, new technology comes to bear when oil prices are high because there is a greater incentive to find more CONG with new, ever-cheaper methods. This means that the CONG sponge is being squeezed ever tighter and the smaller fruits at the top of the orchard's trees are being found because the low-hanging fruit has already been consumed. The increased flow of oil is largely due to this sponge squeezing. The second issue is that the CO_2 (characterized by being purely carbon-12, the signature, nonradioactive isotope of carbon that marks fossil fuel emissions) released by burning the fossil fuels leads directly to climate change, so we cannot even burn the finite reserves if we want to avoid climate tipping points. If a way of absorbing the carbon dioxide can be found that would allow us to continue to burn CONG, then that would be better. Currently, there are some really good ideas on this front. Carbon sequestration by capturing the CO_2 and injecting it into wells is a nonstarter because there is too much, and the CO_2 is back in nature again, albeit in a very contained way, but it will still have an effect over time.

Renewable Energy Equipment Uses CONG Energy in Its Manufacture, So It Is Self-Defeating

Today we manufacture wind turbine towers out of steel and burn CONG to generate the heat required to smelt, melt, and bend the steel to our design. The energy required is enormous and part of the cost of production. Some say that wind turbines must spend years providing the energy that pays back what was put in to make them. By 2012, many European countries, including Germany, a manufacturing economy and global leader in the production of wind turbines and solar panels, had accumulated huge renewable energy capacities. Thirty-one percent of German electric-generating capacity in 2014 was renewable.[65] On some days, the renewable contribution to the electrical grid in Germany reaches 74%, from a mixture of biomass, hydro, solar and wind, which is really saying something for the fourth-largest global economy. This means that, reasonably, 31% (the actual German renewable capacity) of the energy needed to manufacture the steel for wind turbines now comes from renewable sources. Portugal enjoys days when, for an hour or so every year, and more over time, 100% of the nation's manufacturing is supplied with sustainable energy resources. An ongoing 75% of Portugal's generating capacity is now sustainable, a huge advantage for the country. This means that increasingly renewable energy equipment, not to mention all other products of industry, will be manufactured with clean energy as well, and that this criticism is only useful for a brief snapshot in time.

Researchers in the U.S. Pacific Northwest performed a detailed life-cycle study of 2-megawatt wind turbines and concluded that they pay back the energy required in their manufacture within a short 5 to 8 months.[66] This is nothing like the lengthy payback time that critics believe to be the case. It's also true to say that challenges experienced by the fast growth of renewable energy with, for example, integrating intermittency into a reliable grid with storage, as the German grid is doing, are extremely valuable ways of upgrading the infrastructure.

Energy Levels Generated Are Insufficient or Low Intensity

We've been spoiled by the energy density of CONG. One gallon of gasoline provides 115,000 British thermal units (BTUs) of energy.[67] Nuclear energy is even more energy dense, with a single kilogram of uranium fuel offering 45,000 kilowatt-hours[68] of electricity, or 153 million BTUs (1,200 gallons of gasoline). This convenience has allowed global economies to grow fast since the start of the industrial revolution. Yet the high energy density of CONG is part of the trap in which society finds itself currently ensnared. CONG is like a drug we are all hooked on whose effects escalate to compound our health problems. Solar and wind can "seem" anemic and insufficient in comparison. It's the subject of a tremendous machismo swagger that jokes about how renewable energy is the ugly duckling and will never be important. In the current context of the United States growing to become a larger producer of oil than even Saudi Arabia, revisiting a period in the 1930s when it previously held this crown, there is a tremendous pride in the energy community that is very misplaced. This only compounds the incredible mistake of finding more oil at this critical time. Oil has done wondrous things and it is certainly needed as a transitional resource, but we are facing limits in its use caused by its own externalities.

However, the high-voltage output from utility-scale solar farms is no less energy dense than nuclear power stations at the grid level. Biofuels present lower energy densities than CONG sources do. For example, ethanol contains approximately 70,000 BTUs per gallon (gasoline 115,000 BTUs/gal) and represents a lower-energy component of liquid fuels at the pump, so it takes a larger volume of ethanol to supply the same energy as a CONG source. If it's made from cellulose, ethanol is carbon neutral and a good source of BTUs, but it is not acceptable made as it is today from foods such as corn, beets and sugar cane (unless you are in Brazil). Another biofuel, biobutanol, is comparable to gasoline in its energy density and is fermented from less sugar. I am in touch with a company called Alliance BioEnergy Plus that has proven out its economic cellulose-to-sugar technology at scale and I am confident this solution will find its way into the mainstream within five years, although it has not yet gone commercial at the time of this writing.[69]

Renewable Energy Takes Up Too Much Land

The plain vanilla comparison is between a coal-fired power station and its solar or wind equivalent. The coal power station is often situated in an industrial area, conveniently close to its subscriber base. The railway trains full of specific types of coal that arrive to feed it are the essential support of today's 100% energy demand lifestyle. The base-load electrical power it produces defines the predictability and reliability of our modern lifestyle.

On average, an individual home is situated on a quarter acre, and only about one-tenth of the total area is necessary to generate enough solar power for that home and a 1-acre solar array can provide enough power for 32 homes.

Technology	Direct Area		Total Area	
	Capacity-weighted average land use (acres/MWac)	Generation-weighted average land use (acres/GWh/yr)	Capacity-weighted average land use (acres/MWac)	Generation-weighted average land use (acres/GWh/yr)
Small PV (>1 MW, <20 MW)	5.9	3.1	8.3	4.1
Fixed	5.5	3.2	7.6	4.4
1-axis	6.3	2.9	8.7	3.8
2-axis flat panel	9.4	4.1	13	5.5
2-axis CPV	6.9	2.3	9.1	3.1
Large PV (>20 MW)	7.2	3.1	7.9	3.4
Fixed	5.8	2.8	7.5	3.7
1-axis	9.0	3.5	8.3	3.3
2-axis CPV	6.1	2.0	8.1	2.8
CSP	7.7	2.7	10	3.5
Parabolic trough	6.2	2.5	9.5	3.9
Tower	8.9	2.8	10	3.2
Dish Stirling	2.8	1.5	10	5.3
Linear Fresnel	2.0	1.7	4.7	4.0

Figure 18: Summary of land use requirements of PV and concentrated solar power in the United States. Source: NREL.

It keeps our economies moving forward and we have come to take the 24/7 supply of electricity for granted. Renewable energy power for 1,000 homes takes up only 32 acres.[70] The report looked at data from 72% of U.S. solar power plants (2.1 gigawatts installed and 4.6 giga-watts under construction) as of the third quarter of 2012. A large fixed-tilt, 1-gigawatt-per-year PV plant requires 3.2 acres. Use of solar track-ers used more land (3.8 acres) because of spacing to avoid shadows.

NUMBER	UNITS	%	DEFINITION
3,796,742	Square miles		Total US land area square miles
196,961,745	Square miles		Total Earth (land and water) global area square miles
		1.93%	US land area as a $ of Earth's total area (land and water)
36,170	Square miles	0.9526%	Square miles of land needed for US primary energy demand if solar at 20% efficiency
14,142	Square miles	0.3725%	Square miles of land needed for annual electrical US Energy demand
779,640.00	PWH		Total annual global solar energy received Petawatt Hours
15,028.77	PWH		Petawatt hours and BTUs of Solar irradiation landing on the US annually
28.63	PWH	100.00%	Total US Energy Consumption 2015 in PWH (BP say 27.37 PWH)
11.20	PWH	39.40%	US Electrical generation in 2015 (EIA) Petawatts PWH
		0.1905%	US total primary energy consumption as a % of solar energy on US land area
		0.0745%	US electrical generation as a % of the solar energy on US land area

Figure 19: Calculations for the small amount of land in the United States needed if solar alone were the resource used for all the electrical energy consumed in 2015. Source: EIA, BP Statistical Review 2016, NEF Advisors, LLC.

Concentrated solar power works at a higher efficiency, always has double- axis tracking and needs less direct land (2.7 acres, but 3.5 acres if including unused portions inside the perimeter). A previous NREL report[71] suggested that if PV were the chosen generation method, all the electrical power consumed by the United State could be generated on as little as 0.6% of the U.S. land area.

We can easily check this NREL study by comparing the land area of the United States (3.796 million square miles) and the electrical energy (11.2 petawatt-hours [Pwh]) consumed in 2015. The issue, of course, with this solution is that solar energy, like wind, is intermittent, with capacity of approximately 20–30% depending on location. A solar system hybrid with storage that would permit power to be saved until it was needed would be a better solution, and there are many mechanical and chemical storage technologies that are all decreasingly expensive, but we have limited but growing experience with them still.

Comparing land use intensity doesn't make sense until you bring in a comparison with traditional forms of CONG energy. A study by Columbia University in 2009 showed, for example, that the land needed for a coal-fired power station was not just that occupied by the generating plant itself but also any mining land, especially surface-mined coal or subsurface coal, which was stored on extra land on the surface. Columbia University also prepared a study in 2009[72] that compared land use intensity by energy source. The conclusion of that study was that, including all direct and indirect land usage, South Western utility-scale PV requires less land than the average coal-fired power plant using surface-mined coal.

Research[73] by professor Alexander Mitsos of MIT and Corey J. Noone and Manuel Torrilhon of RWTH Aachen University in Germany discovered an interesting result regarding biomimetics, the mimicking of nature by engineers to obtain a significant improvement in output. Evolution is a powerful force for bringing to life configurations and patterns that are there because they do something optimally. Concentrated solar power towers have hundreds or even thousands of mirrors surrounding them. The pattern in which these mirrors, which are all mounted on sun trackers, are organized has a big influence on the amount of land taken up as well as the efficiency of the plant.

Figure 20: Sunflower seed-like distribution of mirrors around a concentrated solar tower. Source: Wikimedia Commons.

The images in figure 20 illustrate the similarity between the natural configuration on the left and the mimicked configuration on the right. It turns out that the curve of a Fermat's or Archimedean spiral, the same pattern described by a sunflower plant's seeds, increases the plant's efficiency a small amount and fits all the mirrors into 20% less land.

Entergy, the Arkansas utility and leading national nuclear operator, produced a study comparing the amount of land for a 1.8-gigawatt nuclear power station and wind and solar power facilities.[74] They did not look at the total fuel cycle for nuclear power and simply listed the site statistics of 1.7 square miles for the nuclear plant with two reactors, comparing it in favorable light with the 169 square miles for an equivalent wind facility (sourced from an American Wind Energy Association FAQ) and 21 square miles for the solar plant (using Sunpower panels on trackers). I suspect, though, that the mining land area, fuel processing area and transportation would not compensate for the wind space required and believe that a thorium reactor would be a safe way to beat the pressurized water reactors operated by Entergy.

The Equipment Is Too Expensive

Solar equipment has dropped in price in an incredible manner since the mid- 1970s. This chart from Bloomberg in figure 21 shows the extent of the decline to today's cost of less than $0.37 per watt.

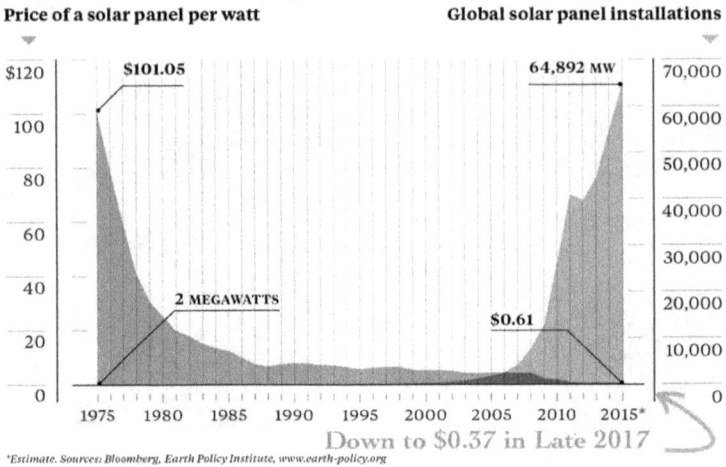

Figure 21: Decline in the cost of silicon solar panels per watt. Source: Bloomberg, Columbia University Earth Policy Institute.

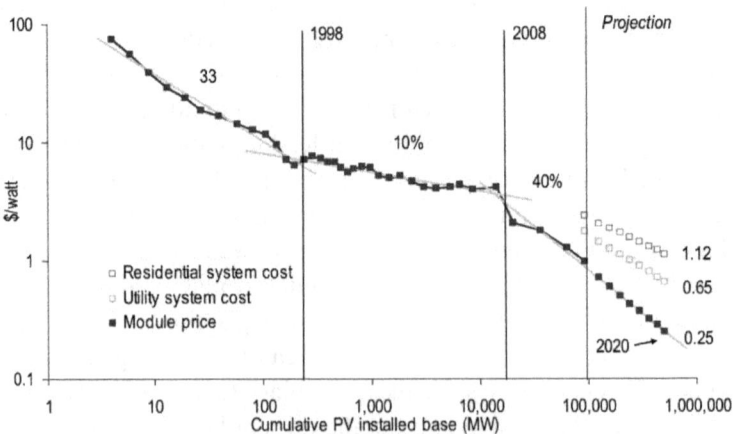

Figure 22: Decline in the cost of solar panels from 1972 to 2020. Source: Citi Research and Bloomberg New Energy Finance.

1,000,000 watt Solar Plant

Installation Costs

Capacity	999,000	205	Watts each	Sunrise 250						9/25/2007

Item	Description		Unit Cost	Cost	Unit Price	Markup	Price	Markup	Total Price
PV Modules	4,873		$4.10	$841	$4.51	1.10	$905		
Power (Wp, dc)	989,000	20,305	($/Watt)	$4,095,900	($/Watt)	1.100	$4,505,490	$409,590	$4,505,490
Shipping	LA to Site							$27,073	$27,073
BOS									
Hardware	225kW Inverter	5	$115,000	$575,000		1.10	$632,500	$57,500	$632,500
Array Structure	Inverter Room	5	$3,500	$17,500		1.10	$19,250	$1,750	$19,250
	Pipe & Material	1,218	$175	$213,201		1.10	$234,521	$21,320	$234,521
	Data Acq. System	5	$6,000	$30,000		1.10	$33,000	$3,000	$33,000
Switch Gear	SCE Requirements	1	$20,000	$20,000		1.10	$22,000	$2,000	$22,000
Tax									$394,890
Equipment Total			$/Watt (installed) =>	$4.94	$/Watt (installed) =>		$5.43	$520,233	$6,868,725
Installation					Supervision				
	40kW/wk x 28wks	Man Months	Cost/Month						
	Array assembly	15	$1,792	$27,290	$11,421	1.00	$38,711		$38,711
	Foundation/cement	15	$1,792	$27,290	$11,421	1.00	$38,711		$38,711
	Electrical	15	$2,304	$35,087	$11,421	1.00	$46,508		$46,508
TO Engineering	Site Engineering	1	$16,000	$16,000	$30,000	1.00	$46,000		$46,000
									$0
	Project Mgmt	7	$3,000	$21,000	$31,500	1.00	$52,500		$52,500
	Project Develop	7	$0	$0	$31,500	1.00	$31,500		$31,500
TO Engineering	Utility Interconnect	1	$12,500	$12,500	$3,000	1.00	$15,500		$15,500
	Monitoring System	2	$3,000	$6,000	$6,000	1.00	$12,000		$12,000
	Site Prep/Fencing	2	$50,000	$100,000	$6,000	1.00	$106,000		$106,000
	Security	14	$1,800	$25,200		1.00	$25,200		$25,200
	Contingency	7	$10,000	$70,000		1.00	$70,000		$70,000
Installation Total									$482,631
Total								TOTAL =>	$6,351,356
			$/Watt (Cost+Installation) =>	$5.44	$/Watt (Installation) =>	$0.44			
					$/Watt (Total) =>	$6.26			
1,000w module in 5 panels, 2 months each for assembly, foundation, electrical									

Figure 23: A 2007 solar model showing the cost per watt of solar panel installation at $6.36. Source: NEF Advisors, LLC.

From a recent solar research study by Citibank, figure 22 shows a continued decline in the cost of equipment to generate a watt of electricity from solar energy. Interestingly, the Citibank analysts separated out three distinct cost reduction phases. The first was the development phase, then there was an "industry development" phase, and the third phase is the mass production that we are currently experiencing. In this scenario, utility system costs come down to $0.65 per watt and modules to only $0.25.

In 1979, a joint venture between then Martin Marietta (Lockheed Martin) and Saudi Arabia chose two villages, al Jabaila and al-'Uyaina, close to Riyadh to install just 350 kilowatts each of concentrated solar photovoltaic power.[75] The project was the largest solar array in the world at the time and cost $46 million, or $65 per watt.

This spreadsheet in figure 24 from a 1-megawatt solar installation in late 2007 shows a figure of $6.36 per watt, more than three times today's total installed cost. As module costs have crashed, the balance-of-system (BOS) costs, which used to account for about 30% of the total installed cost, have come to be about 50% and are likely to climb a bit further. Inverters, labor, mounting systems and permitting are all dropping in cost less quickly than the solar panels themselves. Greentech Media explained in 2017 that volumes of solar equipment shipped since Sunpower, the California-based solar panel company, was founded had climbed by an astonishing 37,500% and it has only started to make an impact today.

By 2020 the panel cost is on target to drop to just $0.25 per watt.[76] This means that the panels in a 10-kilowatt array would cost only $2,500. In a sunny place, this is sufficient energy for all household and transportation requirements (30 kWh is needed for a 100-mile trip in a car). At least two solar companies are targeting costs below $0.05 per watt of capacity by the end of this decade. This is a cost of only $500 for a 10-kilowatt array and reaching levels that are well below the costs of fossil fuels.

Energy Type	2013/4 Cost Per Watt	2010 Cost Per Watt	% Difference
Coal			
Single Unit Advanced PC	$3.25	$3.29	-1.40%
Dual Unit Advanced PC	$2.93	$2.96	-0.74%
Natural Gas			
Conventional CC	$0.92	$1.02	-9.83%
Advanced CC	$1.02	$1.04	-1.92%
Fuel Cells	$7.11	$7.11	0.04%
Uranium			
Dual Unit Nuclear	$5.53	$5.55	-0.29%
Biomass			
Biomass CC	$8.18	$8.21	-0.30%
Wind			
Onshore Wind	$2.21	$2.53	-12.67%
Offshore Wind	$6.23	$6.21	0.31%
Solar			
Solar Thermal	$5.07	$4.88	3.90%
Solar Photovoltaic (Residential)	$4.56	$6.29	-27.49%
Solar Photovoltaic (Utility)	$1.85	$4.94	-62.57%
Geothermal			
Geothermal – Dual Flash	$6.24	$5.80	7.68%
Geothermal – Binary	$4.36	$4.30	1.35%
Municipal Solid Waste			
Municipal SolidWaste	$8.31	$8.56	-2.86%
Hydroelectric			
Conventional Hydroelectric	$2.94	$3.20	-8.16%
Pumped Storage	$5.29	$5.82	-9.08%

Figure 24: Cost per watt of different types of energy production equipment. Source: Energy Information Administration (EIA) and Cleantechnica.

The World Economic Forum has recently said that in more than 30 countries, solar energy is now the same price or even cheaper than fossil fuels without subsidies. From the perspective of a levelized cost of energy (LCOE), solar prices have collapsed since the year 2000. New contracts for solar electricity in India at $64 per megawatt-hour and in Chile for just $29.10 per megawatt-hour have broken through the competing coal and natural gas prices. Solar has also broken through the cost of wind.

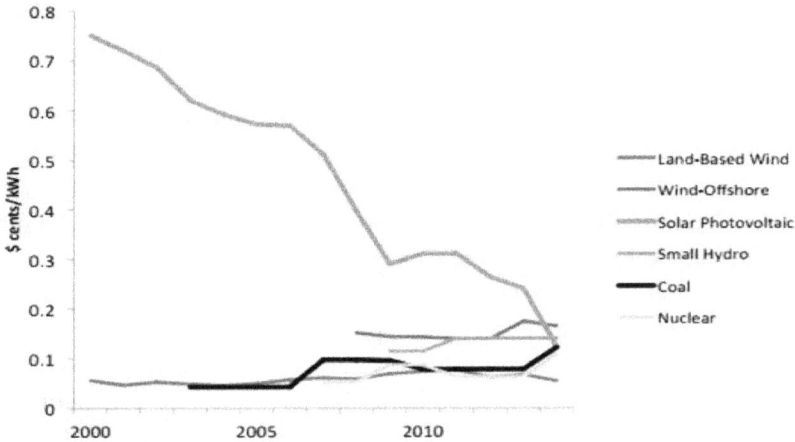

Figure 25. The levelized cost of energy (world averages). Source: OpenEI, Transparent Cost Database.

The increase in capacity installed worldwide is due to the economic as well as the clean aspects of the technology.

The World Economic Forum expects the 20% decline in costs means that two-thirds of the world will reach grid parity by 2020, and solar PV is expected to have a lower LCOE than coal or natural gas. GTM tells us that because of this, the world will enjoy about 400 gigawatts of solar capacity alone by the end 2017 and a terawatt by 2022, driven by China, the United States, Japan and India.

CONG costs are also liable to continue to increase as resources begin to dry up and legislation continues to mount. In the meantime, the cost of a coal-fired power station or a nuclear plant are not falling but staying stable. The data in figure 24 comes from the Energy Information Administration and you can see the decline in residential and utility solar and wind costs.[77] New technologies promise a continuation of lower costs. One example is the molded silicon cells of a company called 1366 Technologies from Massachusetts. It promises a huge reduction in solar costs. They have recently produced commercial poly-silicon with a higher efficiency, at 19.1% in November 2016. Higher efficiency is one aspect of reducing costs, but cutting out kerf loss, the silicon dust that sawing cells off an ingot produces, means that twice as much silicon gets used. Also, it means you can have a thinner cell.

Renewable Power Is Intermittent and Requires CONG as Backup
    The pillars of renewable energy have been installations of wind and solar power with geothermal a reliable third and ocean power a distant assist. Hydroelectric often doesn't count because of its terrible impact on the environment, although modern methods sidestep those impacts. If coal grew by just half a percent, it would still add more absolute capacity—coal is still very much king of the hill. Despite being more expensive initially, the use of both wind and solar power has grown at double-digit rates for well over a decade. However, when the wind does not blow, and the sun does not shine, there is no power. Worse, CONG generation methods are often used to provide spinning reserve or backup when the wind falls still or the sun sets. *Spinning reserve* means that a coal power station continues to burn coal to "spin" the generator just in case the wind falls away quickly. Fossil fuel resources, in this case, stay operational even when the wind is blowing so that they can ramp up and offset the lower electrical output of the wind or solar farm when they inevitably slow down. Geothermal energy, on the other hand, is up and running, with a capacity often higher than nuclear power in the high 90s at over 95% availability. Biofuels and biomass provide baseload electricity that is also carbon neutral; it's here today and there are many new configurations of renewable energy types that can be base load that are just around the corner.
    The secret ingredient that makes renewable energy sources independent of and superior to CONG is either a wind or solar resource that has no interruptions (space-based solar or high-altitude wind or solar chimneys, etc.) or a suitable storage method. With a storage mechanism, a solar farm or wind farm can produce electricity 100% of the time. In this case, the storage can act as the spinning reserve and the system would not require CONG. Both solutions exist to the problem of inter-mittency, and as time goes by more baseload/on-demand sustainable electricity options will be on offer. Efforts to reach 100% renewable capacity are deemed to be unrealistic by many operators in the energy industry until storage becomes a major part of the pie. The cost of cur-

rent storage systems is still very high and is why this industry is in its infancy currently. Growth is fast, however, and there are myriads of storage configurations coming online…including one we believe to be real in northern Europe, which will show its colors within two or three years.

When we were researching storage options, my fund, the New Energy Fund LP, discovered a company that had worked closely with the National Renewable Energy Laboratory (NREL) and had overcome the issues of providing solar electricity at nighttime and during overcast weather. They used large parabolic dishes to concentrate the sun's electromagnetic spectrum so that the temperature on the receiver, the focus, climbs higher than 1,832°F (1,000°C). This thermal energy is then transferred to an insulated ceramic storage vault, which glows white hot, like the inside of a kiln. You could call it a "reverse kiln."[78]

The power block (the engine that transforms the thermal energy into electricity) would continue to use the heat directly from the receiver until clouds or nighttime interrupted, and then it would switch automatically to the stored thermal energy without any output interruption. However, when the price of a barrel of oil declined below $10 in 1998, all interest in this project was lost and the engineers got jobs elsewhere. When oil prices made this technology feasible again, a new CEO reassembled the engineers and asked them to redesign their product as a utility-scale thermal storage solution. The New Energy Fund was an angel investor after the company visited us when oil prices shot up in 2006. They built a 146-kilowatt system in Tianjin, China, and had a demonstration day, with 70 attendees and speeches and lots of steam production. Our ceramic storage technology was proven to work and later won a $96 million contract for a prototype 16-megawatt installation in Gansu province in China. Disastrously, another investor thought this was an opportune time to execute what Wall Street calls a "cram down." Instead of offering cash at $3 per share, they offered the cash-starved company cash but at a very low price per share. This would have shifted share ownership into the hands of the investor. The existing management were forced to refuse the cash and instead closed the operation despite having a major contract. The new owner would be required to go all the way through developing the relationship again and failed to see that this was a reality.

Taking solar into space, something that we are on the verge of being able to do economically, is another way to solve the intermittency puzzle with solar. Parking a multi-gigawatt fleet of solar panels much closer to the sun would also be a great way to solve many other problems with solar. Similarly, using high-altitude wind is a way to overcome the intermittency of wind. Wind speeds in the upper atmosphere are much higher and hold enormous amounts of energy. Cities with appropriately designed helium- supported balloons might be able to generate all the power they require sustainably using this method. Areas encompassing these generating kites would simply become "no-fly" zones, which already proliferate on aviation maps.

Renewable Energy Externalities

Now follows a look at the externalities of renewable energy. While nothing is without risk, the move toward sustainability is itself a significant disruption. Even if these sustainable technologies were to remain more expensive (they won't), we deserve to spoil ourselves by installing them to obtain relief from the externalities of fossil fuels we already endure.

The message of this book is to identify the load of those externalities. We spend a lot of money on other things in our lives, battery electricity, a wedding, a new pair of shoes or a suit, a military, public healthcare, dog clothing, a space program. Why can't the community spend a bit more to achieve a historical and fundamentally huge step forward in sustainability? This section on renewable energy externalities is an honest listing of the bad things that we suffer because we embrace hydroelectricity, solar or wind. It's a bit tongue-in-cheek because even though there is a long collection of bad things, they all pale in comparison with the effects of the CONG externalities.

We are at a bottleneck in human history. What can we do to introduce a new period of growth? We can transition to sustainable forms of energy as quickly as possible. CONG externalities act as a governor and restrict the growth of our economies. When the price of oil goes up, it restricts economic growth. When it comes down, it's like a lower tax and more spending and growth can happen. Oil price rises correspond to U.S. economic recessions. That's not to say we owe nothing to CONG. Up to the present moment, CONG has powered us through enormous improvements in quality of life in all domains. Today however, the risks of CONG significantly outweigh any assessment of the downsides of using renewable energy. This section of the book is an attempt at an intellectually honest review of the afflictions perpetrated on us by renewable energy technologies.

The Evils of Solar Power

"The Sun will be the fuel of the future."—Popular Science, 1876.

We need to assess, just for formality's sake, the downsides of renewable energy. We have become so sensitive to the effects of new impacts of technology on our civilization that we have probably become far too touchy about some new things such as renewable energy technologies, which even if they are accepted to be less impactful, and even if they do require a lot of space, deserts have ample sun, are too nasty to be generally inhabited and have plenty of space. If rooftops or even road surfaces (25,000 square miles in the USA) that can continue to function as roof or road are used, then there is also plenty of space for energy generation as well as cost mitigation when hybridized as new.

There is no shortage of sufficient space in the world, but there are local shortages of space, for example, in cities or in densely populated areas or on roofs. It's very difficult for apartment dwellers, in the case where there are hundreds of units in a building, to address their own solar power. In New York City, apartment dwellings owned by rental companies experimented with some solar power on the roof. I contemplated taking an apartment at the Solaire, but the solar, visible and beautifully installed as building-integrated photovoltaic (BIPV) on the side and the roof, provides only about 5% of the energy the building needs. Of course, if it was totally refitted with the state- of-the-art equipment, this might become as much as 15%. As time goes by, solar technology is also increasingly efficient, reducing the space requirement and line losses, with no fuel supply logistics.

Solar installations need land-forming preparation, which often disturbs the surface of a property by removing topsoil and introducing semipermanent concrete anchor attachments or at least heavy weights to prevent wind damage. Some animals and plants can be affected if the area is a critical habitat. The panels also introduce shade, which is a change in the desert environment that can make life difficult for native species. The Ivanpah project on federal lands was aware that the desert tortoise[79] lived on the same location that they intended to cover with large, garage-sized mirrors. The LA-based Center for Biological Diversity testified against the project to protect the tortoises, which are protected by the Endangered Species Act. Construction started in 2010 in any case, and Brightsource, the developer of Ivanpah, asked the government for a permit to relocate any tortoises it found on its land. Initial surveys did not find many tortoises. On one piece of territory, they found 30 tortoises, but then, when it unusually rained, 173 tortoises suddenly appeared. Using pens, the tortoises were transferred to vacant land. In fact, 53 tortoises were born during the transition, and the company estimated that over $55,000 was spent per tortoise to account for the total relocation project.

Ninety percent of solar technology is benign silicon in discs like CDs laid out in large picture frame panels with wiring. Some solar technologies are made of dangerous materials. Cadmium and telluride (CdTe) combine in a thin-film solar technology and are normally poisonous commodities. First Solar, the principal manufacturer of these panels, has a policy for collecting, recycling and replacing cracked or worn-out panels that might be considered dangerous. Manufacturing processes for solar panels and the component value chain use power. As time passes, this power will be increasingly from renewable energy itself.

Figure 26: A Rough-winged swallow, killed after flying through the intense solar flux caused by the concentrated solar thermal power station in the Mojave Desert. Source: California Public Utilities Commission.

Some solar technologies use arsenic in a compound with the more efficient III-V multijunction solar chips. Compare this to the use of arsenic by mining companies and it becomes very clear that gallium arsenide is not a big problem.

Nobody has died from solar power as far as my research informs me, except by falling off rooftops during panel installations. In the complexity of connecting solar power to the grid, there are reports of engineers dying who have gone to repair grid equipment when the power was supposedly down only to fatally discover that it was still connected to a solar installation with enough power to kill them. Normally, the only source of power to a grid was the power station. Turning off the switch was simple and safe.

Today the power station may go down or the grid might have its own faults, producing a blackout, but just occasionally it would be good to remember that if there is a connected solar farm, there is also the possibility of a fatal electric shock from that source. This means that in the case of a power outage from one source on the grid, the safety of the repair crews needs to be assured by switching off any other local electricity sources as well. Roofs are often near high-voltage power lines and because they are high, there are deaths caused by simply falling every year, a common enough negative for any roofing activity.[80]

The photograph in figure 26 and others were supplied by Brightsource, the company that operates the Google/NRG Energy, 377-megawatt Ivanpah Solar Electric Generating System (ISEGS).[81] It is a concentrated solar power station sitting on 3,500 acres of public land near the Mojave National Preserve. The pictures were sent to the California Energy Commission (CEC). The bird, a rough-winged swallow, is an unfortunate victim of a new threat. The solar flux of the thousands of heliostat mirrors in the field around the Brightsource solar towers contains so much energy that it literally roasts any bird unfortunate enough to fly into the solar rays. Workers call them "streamers" because of the smoke trail they leave as they plunge to earth. This bird's wing and tail feathers have all been affected by severe heat and have been singed to the point that they are no longer useful for flying, if the bird survives.

This problem has been known since an original 1980s project at Dagget[82] in California. Normal desert sunlight has a flux of about 1 kilowatt per square meter, but when it's concentrated, it can have a flux of 50 kilowatts per square meter. In mid-2014, these systems were being tested, but when they are working full-time they can be expected to present much more dangerous flux to unsuspecting birds. You might ask why Brightsource would be willing to release such photographs given the concern of the CEC. It turns out that a second huge solar project to be built east of the Joshua Tree National Park is waiting for authorization and will perhaps have similar danger to birds, and to meet lenders deadlines, the company is doing all it can to cooperate. It ought to be mentioned that fossil fuels kill birds and we can't forget the ever-present images in our minds of birds covered in oil that die from spills. The BP Gulf of Mexico oil spill, one of the biggest in history, also killed over a million birds (Audubon Society), and oil facilities kill 500,000 to 1 million birds annually.[83]

In May 2015, word came from Brightsource that they had fixed the problem. The computer was programmed to "park" the 3,000 mirrors in a "standby" position so that they reflected light just above the target heating area on the tower. This formed a halo of light around the top of the tower that had an incredibly dense solar flux. Management had already spent millions looking after tortoises and were not in the mood to have the bird deaths overhang the project too. Sure enough, they simply programmed the mirrors to form a much wider "standby" with no more than 4 suns at any point, far less than a dangerous level for birds. This has been the case since January 2015 and now there are zero bird deaths. When the flux is focused on the tower while it is in operation the birds can see it and steer clear. Nonetheless the critics literally caused the cessation of building on the second project and thoroughly impacted operations on the first installation.[84] The solar manufacturing industry uses the same materials as the microelectronics industry. Silicon itself must be much purer (99.99999% to 99.9999999%) for microelectronics than it needs to be for solar applications. Silica is mined by finding it in sand or quartz and then melted at high temperatures to produce metallurgical-grade silicon that is 99.6% pure. Metallurgical silicon is then exposed to hydrochloric acid and copper to produce trichlorosilane gas, which is distilled to remove any remaining impurities such as chlorinated metals of aluminum, iron and carbon. Finally, it's heated with hydrogen to produce silane gas, which is in turn heated to make molten silicon, which is now used to grow monocrystalline silicon ingots. Silane gas can also be an input for amorphous silicon.

Polycrystalline silicon is made by pulling rods from molten silicon that are cooled and suspended in a reactor at high temperature and pressure. Silane gas is then introduced into the reactor and gets deposited on the growing crystal until it reaches the right dimensions. Silane gas is extremely explosive, and there are several incidents annually in the semiconductor industry. Production of trichlorosilane and silane gas produces toxic silicon tetrachloride, which reacts violently with water and causes burns, respiratory distress and eye irritation.

Silicon production is energy intensive and there is waste of about 80% of the metallurgical silicon. The ingots are sawed into wafers. A standard size used to be 120 microns thick. The saws are approximately that thick as well so 50% of the ingot silicon is reduced to kerf. Silicon particulate matter is produced during this process, posing inhalation challenges for production workers. The U.S. Occupational Safety and Health Administration (OSHA) has a regulatory guidance to keep levels of silicon dust low, but workers remain overexposed.

The reactors used to make silicon are cleaned with sulfur hexafluoride gas, considered by the Intergovernmental Panel on Climate Change (IPCC) as one of the most potent greenhouse gases known, 25,000 times more heat absorbing than CO_2. A string of toxic chemicals spring from these production methods. Sulfur hexafluoride reacts with silicon to make silicon tetrafluoride and sulfur difluoride, which can turn into tetraflourosilane and sulfur dioxide (acid rain). These are such poisons that specialized scrubbers are used in exhaust systems to limit release Accidental emissions of this gas can totally undermine any reductions in greenhouse gases caused by installation of renewable energy.

On November 17, 2006, a company called First Solar did an initial public offering on NASDAQ with a novel solar technology. It was a combination of cadmium and telluride, called CdTe. Efficiency for the technology in the 1970s was in the 8% to 9% range. By the time of the IPO, NREL had done much of the research and helped to boost those efficiencies to the 7% range at the cell level. In 2012 First Solar was manufacturing 2 gigawatts of solar cells in its U.S., German and Malaysian factories. By 2013 the cost of a watt of solar cell had fallen far enough ($0.59/W) to be competitive with silicon. In 2014, the company's record-setting cell had exceeded 20%, and commercial cells were at 17%, meaning that panels were operating at a competitive 15% and selling for $0.52/W, while similar efficient silicon panels were still selling for $1.85/W.

As a strategic play, First Solar engaged in large, utility-scale installations rather than rooftop installations. They were better able to compete in larger installations, but it raised a question in the minds of residents living, for example, close to the 230-megawatt Antelope Valley Solar Ranch One whether the 3.7 million solar panels were any threat to them.

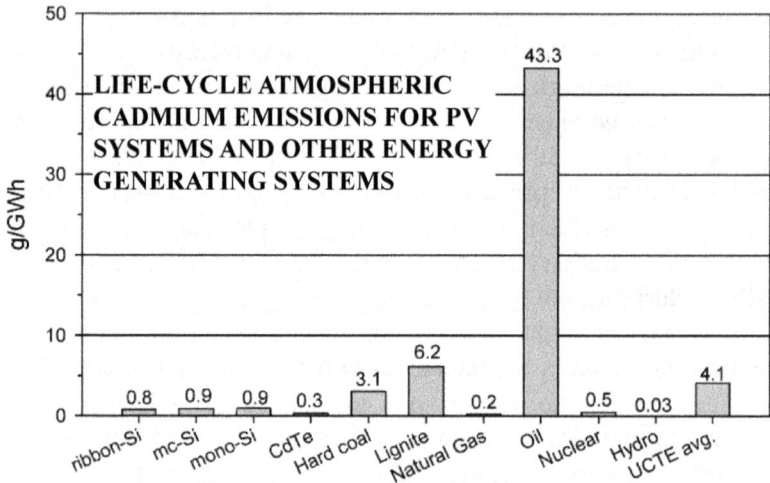

LIFE-CYCLE ATMOSPHERIC CADMIUM EMISSIONS FOR PV SYSTEMS AND OTHER ENERGY GENERATING SYSTEMS

(y-axis: g/GWh)

ribbon-Si	mc-Si	mono-Si	CdTe	Hard coal	Lignite	Natural Gas	Oil	Nuclear	Hydro	UCTE avg.
0.8	0.9	0.9	0.3	3.1	6.2	0.2	43.3	0.5	0.03	4.1

Figure 27: Life-cycle atmospheric cadmium (Cd) emissions for PV systems from electricity and fuel consumption, normalized for a southern European average insolation of 1,700 kWh/m²/yr. performance ratio of 0.8 and lifetime of 30 years. Source: NREL.

First Solar and the National Renewable Energy Laboratory, along with Brookhaven National Laboratory, supported a definitive toxicology study as early as 2003. Previous toxicology studies showed that cadmium was slightly more toxic than the combined CdTe. On its own, it is highly toxic and carcinogenic and can cause lung, kidney or liver failure. The European Chemicals Agency (ECHA) designated that combined with tellurium (CdTe) it was no longer classified as harmful if ingested or in contact with the skin, and the toxicity for aquatic life was reduced. CdTe is stable, benign and less soluble. It forms a crystalline lattice that is several orders of magnitude less toxic that cadmium alone. The encapsulation in glass further resists any leakage of toxic material from the panels.

The biggest threat to workers at any site is the risk of being cut by broken glass. The company expects breakage to be at 1% over 25 years, and thinking ahead, they have already put together a prefunded recycling program for any broken panels. In fact, there is 2,500 times more risk from nickel-cadmium batteries than from CdTe solar panels, according to NREL. Then, finally, comes the chart in figure 27, from a study at NREL by Vasilis Fthenakis[85] that neatly turns the question around, as so often is the case, and shows us that CONG turns out to be the villain in this picture. The chart shows that oil is the main culprit, with 43.3 grams of atmospheric cadmium per gigawatt-hour emitted compared to just 0.3 g/gWh for CdTe solar panels over the entire 30-year life span of the equipment.

Since so much land is now being shaded by solar panels, it's becoming clear that there are microclimate concerns. A team from Leicester University in the UK published a study on July 13, 2016, that examined a solar park in the UK on species-rich grassland. They investigated soil and air and vegetation and greenhouse gas emissions for 12 months, considering the underside and gaps between photovoltaic arrays. They showed changes in the microclimate and air quality. In the summer, there is a temperature drop of 9.36°F (5.2°C) and drying under the arrays, while in the winter, gap areas were 3.1°F (1.7°C) cooler. Variation in daily air temperature and humidity as well as photosynthesis was reduced under the panels. The implications for ecosystems, crops and soil carbon storage are increasingly important to understand.

Another angle of an increasing solar externality is the associated junk or solar scrap that is accumulating every year now that the scale of installations is so huge. India alone has a target of 100 gigawatts of new solar installations by 2022, which would be an average rate of 12 gigawatts per year, even though the rate is growing over time. Solar plants have a life cycle of 20 years. Good panels such as those by Sunpower degrade by as little as 0.1% per year and will last for a good 40 years. Cheaper, lesser-quality panels might degrade by as much as 1% each year, cutting their effective life to 20 years. The amount of material that will be scrapped is growing. Every megawatt of solar power involves 50,000 tons of steel, 100,000 tons of modules and 5,000 tons of cable. Every year, there is about 1 gigawatt of material waiting to be scrapped or recycled, or 2 million metric tons per year. The essential remoteness of solar parks makes this more daunting, and every year the problem is growing more severe.

The Kyoto Protocol states that use of hazardous substances in electronic devices must be reduced to zero, and the European Union classified old solar panels as electric and electronic equipment waste (WEEE) with specific measures for collecting, recovering and recycling them.

PV companies are expected to have a plan to ensure the collection and recycling of discarded panels. Frequent bankruptcy passes the buck to the community, but it's also clear that improving technology means the same solar installation will be capable of generating more power in its reincarnation with new equipment. Most of the steel and glass is recyclable, but there is a large amount of encapsulant and wiring plastics that need to be recycled. Plastics recycling is now capable of making fuel out of these parts. This externality might well be a mountain made out of a molehill. We are trying to find problems with renewable energy, but it's difficult to find something lasting and important. Perhaps wind power will lend us some really damaging social costs.

"An Yll Wynde That Blowth No Man to Good"

> *"America has the best wind resources in the world. Not harvesting America's wind would be like going to Saudi Arabia and not drilling for oil."*—Ditlev Engel, chief executive of Vestas Wind Systems.

Of all renewable energy, the most contentious is wind.[86] Wind stirs the most passion both ways, and documentaries are made of suffering communities at war with each other in the shadow of the big blades. "Big Wind" companies are made to sound as evil as Big Oil in their calculated pursuit of profits. I know people living north of London who are actively moving to cancel wind farm installations on the grounds of fears of wind turbine syndrome (WTS),[87] a serious health problem described by people who live close to the towering structures.

The Caithness Windfarm Information Forum (CWIF)[88] produces a list of the frequency of all wind turbine–related accidents globally confirmed by press reports. Renewable UK[89] also follow such data with reports on such topics as

- Radar and aviation security
- Scenery despoliation
- Property prices
- Health impacts from aerodynamic noise and shadow flicker

The CWIF reports find that blade failure is the most common prob-lem that causes accidents, with fire a close second and poor maintenance coming third. They found that, globally, total accidents since the 1970s numbered 1,549, a number that grows each year along with the number of installed wind turbines. Fatal accidents also are rising but at a much lower rate, with a total of 146 deaths in 108 accidents since 1970, and 14 in 2011, but more in 2012.

Blade Failure—Up to 2012, there were 289 incidents, with some cases of parts of blades being thrown up to a mile away from the turbine hub. In Germany, parts of blades have penetrated roofs and walls of nearby buildings. Renewable UK reported 1,500 accidents in the UK alone over the five years up to 2011, with some deaths and serious inju-ries. Unless there is an injury, there is no requirement for an incident to be reported. The wind industry plays down the incidents. In 2006, part of a wind turbine blade snapped off its hub and crashed into a field in high winds. The operator, Cumbria Wind Farms, said, "Nothing like this has happened there before," but they forgot to mention that in fact one month after the park opened in 1993, a similar accident occurred. A sim-ilar situation occurred with Scottish Power with a blade separation event in Whitelee. Three-bladed wind turbine blades are secured on only one end, unlike the arguably safer vertical access wind turbine designs. The Risø National Laboratory[90] in Denmark reported 15 turbine collapses in the three years from 2005 to 2008.

Fire—Fire can occur as a result of gearbox lubrication failure or friction within the nacelle or when bearings fail. Two hundred thirty-one incidents of fire have been reported. Most fire is restricted to the turbine nacelle but out of reach of fire fighters on the ground. In dry weather, there is a danger of wildfire. Wind turbines are also a magnet for light-ning strikes, which can ignite flammable blade resins. In October 2013, a crew of four mechanics was working for a service company charged with maintaining the 13 turbines at Deltawind's Piet de Wit wind farm in the Netherlands. They were in a gondola next to the nacelle of a Vestas V-66, 1.75-megawatt turbine when a fire, likely caused by a short cir-cuit, blocked the only escape to the stairs in the shaft. Two of the men jumped through flames to reach the stairs and saved themselves, while the other two men, only 19 and 21 years of age, were trapped and died. One jumped from the tower and the other was burned.[91]

Structural Failure—There are 148 instances mainly of collapsing turbines in storms but also component failure. This is a very expensive form of failure but mostly at arm's length from human beings.

Ice Throw—In icing conditions, wind turbines can fling a loose piece of ice a considerable distance, and as with aircraft wings, the performance of the turbine blade deteriorates as ice builds up. Turbines are equipped to detect imbalances caused by ice and normally shut down upon imbalance. The U.S. Occupational Safety and Health Administration (OSHA) details requirements for wind turbine workers to observe in icing conditions. Complaints about ice are common, and the fear is that rotating blades in melting conditions will fling heavy chunks of damaging and lethal ice long distances. This is alleged to have happened in Whittlesey in England, where lumps of ice two feet long were flung from a 410-foot wind turbine to finally collide with a carpet showroom and car park. Residents had the offending turbine shut down. A report by GE's wind turbine division[92] did alert users that ice chunks can indeed be flung several hundred yards. A Swiss study[93] made in a ski resort in 2007 showed that up to 5% of the ice on a turbine was able to travel 260 feet from the turbine. As experience grows with wind farms, the ability to protect the community that lives near them improves. Ice is thrown a maximum of 400 feet, and this is the tip of the iceberg, so to speak. A 2003 report cited 880 events between 1990 and 2003 alone, and another report published in 2005 described 94 incidents. Further reports in 2006 reported 27 incidents.

Transportation of Wind Turbine Components to Installation Site— Since 1970, 147 incidents, including a house being rammed through by a turbine tower section in Germany, a utility pole being knocked through a restaurant and a turbine section falling off in a tunnel. In one case, a $75 million barge with expensive turbine sections was lost at sea. Transportation is the largest cause of public fatalities, including the Brazilian bus disaster. In a single incident in Brazil in March 2012, a bus driver was behind a slow truck, hoping to overtake. He was indicating and thought the truck ahead of him was moving over to let him pass. He gunned the accelerator to overtake only to suddenly find himself faced with a 40-ton wind tower section being transported in the oncoming lane. It sheared off the left side of the bus, driver included. Fourteen passengers and the driver died on the spot and two more died later.[94]

Fossil Energy Payback—A common meme often circulated about wind turbines is that they never pay back the energy put into them during their manufacture. If this was true, of course, then there would be no renewable energy industry and all the growth we see would not be occurring. There are those willing to spread a false rumor to perpetrate misinformation.

ITEM	PAYBACK TIME (DA	% OF 20 YEAR
Energy	146.00	2.00%
"Carbon"	93.90	1.29%
Carcinogens	22.53	0.31%
Respiratory Organics	396.56	5.43%
Respiratory Inorganics	386.47	5.29%
Ozone Layer	393.42	5.39%
Ecotoxicity	209.62	2.87%
Acifidification/Eutrophication	205.61	2.82%
Fossil Fuels	168.32	2.31%
Land Use	Never	
Minerals	Never	

Figure 28: Payback times for a GE 2-MW onshore wind turbine, assuming capacity of 22% or 2,000 fully loaded hours of operation and an annual output of 4 GWh. Source: GE.

A formal life-cycle cost assessment of a typical 2-megawatt turbine conducted by GE reveals that, in fact, the energy is paid back inside 5 months, while the carbon is paid back within 3 months of a 20-year operational life. ISO standards require that this information be published for each model of turbine.

As time moves on, the quantity of renewable energy capacity in any grid increases, so it's quite likely, going forward, that renewable energy will be used more and more to manufacture wind turbines and solar panels and any other renewable energy machinery.

Of course, fossil fuel plants can never pay back on carbon emissions they emit, and unlike wind, the emissions happen over the operating life, while the wind turbines' much lower emissions are upstream of construction. Wind carbon emissions payback is very short and similar to that of many other types of renewable energy.[95]

Bird Deaths

"When you look at a wind turbine, you can find the bird carcasses and count them. With a coal-fired power plant, you can't count the carcasses, but it's going to kill a lot more birds." — John Flicker, National Audubon Society president.

BIRD DEATHS FROM DIFFERENT CAUSES

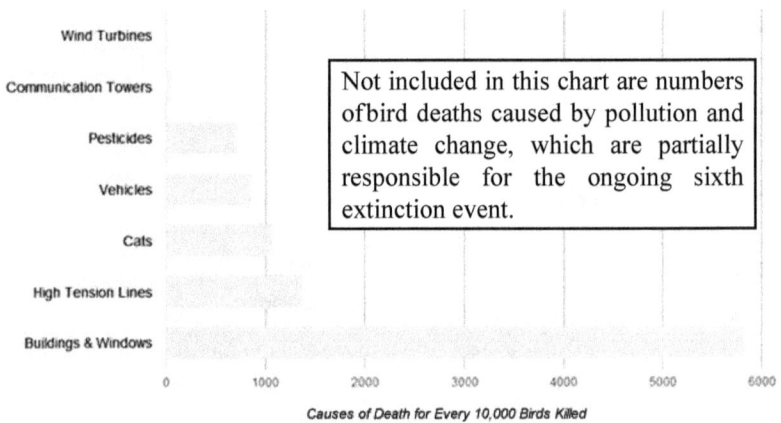

Not included in this chart are numbers of bird deaths caused by pollution and climate change, which are partially responsible for the ongoing sixth extinction event.

Causes of Death for Every 10,000 Birds Killed

Figure 29: Bird deaths from different causes, showing that wind turbines are the least of threats among many. Source: Bloomberg New Energy Finance, U.S. Forestry Service.

Sibley and Monroe estimated that there are about 9,703 species of birds.[96] They are found on all major land masses and over the oceans. Total populations are difficult to estimate due to seasonal fluctuations, but Sibley and Monroe accepted that there are between 100 and 200 billion adult birds in the world. Kevin Gaston and Tim Blackburn[97] doubled that estimate to 200 to 400 billion.

Birds are killed by wind turbines and solar installations, but it turns out that the numbers of birds already killed by pollution from oil and gas, buildings, high-tension lines, vehicles, cats, dogs and pesticides are so much greater that there is clearly a perception twist going on here, which is likely deliberate.

This is not to say that we should be complacent about bird deaths. It's a universally accepted fact that all parties are against any kind of animal mortality as a result of our energy activities. The presentation of it, though, ought to be based on the factual wider context of bird deaths from other causes.

The Altamont Pass was one of the first locations in the United States that was preserved for wind power due to the excellent winds funneled by the hills. At the time, bird deaths were not on the minds of the individuals who created this wind resource.

Even institutions who are protective of birds, the National Audubon Society, the U.S. Forest Service, the U.S. Fish and Wildlife Service and the Wildlife Society, all have commissioned studies that result in the same conclusions afforded by the above chart: Bird deaths by wind turbines do not remotely compare with the impact of cats, cars, power lines or buildings. As wind power increases its penetration, however, its currently small impact on birds will grow less than proportionately as operators learn how to avoid avian mortality by siting, colors on blades, kick-in speeds and other methods. Perception of bird deaths can halt wind turbine installations during the public planning phase, and then effective resistance can scuttle installation plans. It turns out, though, that wind turbines are responsible for only 1 in every 10,000 bird deaths.

Small birds are killed in the millions by housecats, while casualties of wind turbine tend to be relatively larger species of birds. Bigger birds, like the protected bald eagles and other birds of prey, are normally not the direct target of housecats and are more likely to be killed by a wind turbine than by a cat. Balanced against this must be the effect of coal and oil on birds of all kinds, as mentioned in the earlier solar report. Many energy technologies apparently are bad for birds, but wind and solar are far from being the worst culprits. In 2013 a study[98] by Smallwood indicated that the estimates of bird deaths by wind turbine may be under-stated for three reasons: Estimates depended on counting carcasses found under the turbines, and it was entirely possible that searches were done in less-than-efficient ways and in inadequate search radiuses. Additionally, carcasses could easily be removed by predators. This author's bird death estimate was 573,000, slightly higher than others.

A 2005 study by the USDA Forest Service, was an early indica-tion that wind turbines were a very small impact on overall bird popu-lations.[99] The National Audubon Society produced a study,[100] funded by the U.S. Fish and Wildlife Service, in September 2014 that took seven years to complete and that looked closely at 588 of the total 800 species of bird found in North America. Of these, 314 species are threatened in some way with a loss of environment by the end of the century. Climate change (therefore, CONG) is blamed for effectively destroying the ecosystem for 28 species. This data is not included in the chart above in figure 29. The bald eagle and state mascots are at serious risk because of climate change, which reduces birds' range and alters the life cycle of their food sources. Bird mortality from fossil fuel pollution and climate change represents a far higher risk than from wind turbines, as far as the Audubon Society is concerned.

A recent study[101] by the North American Bird Conservation Initiative (NABCI) highlighted climate and environmental impacts on 1,154 native bird species in Canada, the United States and Mexico. The study was compiled by experts from all three countries and accounted for population trends and breeding ranges as well as the severity of threats. Because of changes in the environment caused by humans, birds in every habitat, but especially ocean and tropical forest habitats, are of highest conservation concern: 432 species merit a level of "high concern" as a result of declining populations and habitat loss and climate change. Species with long migration paths have suffered 70% losses in the last 50 years. We are all familiar with some famous bird species that have gone extinct, such as the dodo, the great auk, the emu and of course the passenger pigeon. The oldest international nature conservation group, BirdLife International, says that since the year 1500, 140 bird species have gone extinct, and 22 of those in the last 50 years.[102] In terms of geological time, these species-level impacts are happening in a human instant. But the rates of extinction are accelerating.

A 2009 study[103] compared the bird deaths per kilowatt-hour of power generated by wind, fossil fuels and nuclear and concluded that there were 0.3 and 0.4 fatalities per gigawatt-hour for wind and nuclear and 5.2 for fossil fuel power stations. On this basis, the report says, in 2006 wind would have killed 7,000 birds in the United States, while fossil fuel power killed 14.5 million.

I want to use evocative language here. The legacy of the Earth's embrace of life and its eager occupation of different environments is something I believe we can so much better appreciate because we are intimately a part of that process. We are part of a huge evolutionary life miracle that we are only just now beginning to explore. Previous estimates for the number of species on Earth ranged from 3 million to 100 million. *PLOS Biology* published a report[104] in 2011 that was written by the Census of Marine Life scientists. It established a more accurate estimate of 8.74 million species on Earth of which 7.77 million are animals and only 953,4343 have been described. They used statistical methods to provide a more realistic estimate that nonetheless gave an error level of +/– 1.3 million. Bacteria and other small organisms were not counted. Eighty-six percent of all land creatures and 91% of ocean creatures have yet to be identified. Only 1.2 million species have been officially registered in the Catalogue of Life and the World Register of Marine Species. The detail of the success of the DNA molecule in evolving all these species in this life-encouraging earthly environment over billions of years will never be properly appreciated, but it is at risk from our misadventure with the chemical legacy of CONG and our despoliation of habitats, both marine and terrestrial. We know more about the 22 million books in the Library of Congress than we know about our fellow species on Earth. We are also putting many species in danger of extinction because of the use of fossil fuels in what's been termed the sixth great extinction event, currently under way.

Another great perspective on this is the work of a collector of natural sounds, Bernie Krause,[105] who has spent decades capturing the sounds of nature around the world in places as far afield as Alaska and the Amazon, the Arctic and Fiji, the Great Plains and Mexico's Chihuahuan grasslands. He also has an astonishing TED talk[106] in which he describes how he separates sound into geophony, or wind, water and Earth sounds; biophany, the sounds of natural organisms; and anthrophany, predictably, the sounds of human noise. What he has recently discovered is sobering. Recordings taken in the 1970s compared to recordings taken in the same location today show declines or disappearance of species. Nature is going silent over the Anthropocene. John Bakeless, in his book on discovering America,[107] talks about how early explorers were acutely interested in the sound of nature and developed a faculty of listening and observing to identify birds and insects. I remember our guide, on the last day of a 10-day Colorado river rafting expedition on the calmer 60 miles of the Colorado River just prior to Lake Mead, asking all 28 of the rafters to sit for 30 minutes and listen carefully to nature and then exchange what they had heard. Indeed, there was a sudden realization of insects buzzing, water chirping under the raft, wind in the leaves of trees, echoes of sounds around rock walls and birds, distant and close, calling for myriad purposes of alarm, food or connection. My point here is that although human impacts on the Earth's wildlife are currently severe because of our chemical CONG energy habit, moving to renewable energy reverses the situation over time, even if there are more humans around.

Birds are famously victims of the huge wind turbine blades. This is certainly true, and although bird fatalities from house cats, vehicles and building windows account for literally millions or billions more deaths, it doesn't excuse the wind turbine's impact. There is a very disturbing YouTube video of a large, elegant bird of prey being struck down by such a rotating blade.[108] In an awful European case, there was the death of a rare swift, the white-throated needletail, the world's fastest flying bird.[109] The poor, exhausted creature was spotted by a group of 30 bird-watchers, who had made a special trip to the isle of Harris in the Outer Hebrides of Scotland. The sighting was only the ninth time that the bird had been seen since 1846, in Essex, UK. The last time it had been seen at all was 1991. The assembled enthusiasts assembled in the appropriate location and waited for hours before being rewarded by sighting the bird. They were summarily horrified to see the rare bird, which had flown all the way from Australia, perhaps several times, knocked down and killed by the rotating blade of a wind turbine.[110] Efforts are made to relocate turbines out of birds' migration paths. Also, most songbirds migrate flying at a height of 2,000 to 4,000 feet, well above the tallest wind turbines, at least so far.

Between 2004 and 2009 in Colorado, Wyoming, Kansas, Oklahoma and Texas, just 85 unprotected migratory birds were deemed to have died due to exposure to oil and gas facilities owned by Exxon Mobil. The Justice Department fined the company $600,000, or about $7,000 for each bird killed.

Exxon pleaded guilty and cooperated with the department, spending a further $2.5 million to clean up the sites. It turns out that the fine was equal to twenty minutes of Exxon's profits, based on $8.6 billion earnings for the first half of 2009.[111] Other fossil fuel companies have been fined. BP paid $100 million for the impact of its 2010 Gulf of Mexico oil spill on migratory birds. Pacificorp, which operates coal-fired power stations, paid $500,000 in 2009 after 232 eagles along power distribution lines between its substations were found to have been electrocuted.[112]

Wind farms started to kill birds on a regular basis, prompting calls of hypocrisy against those claiming that wind was an environmental solution. Wind farms have been fined for killing birds, too, however. Duke Energy was fined $1 million for the deaths of 14 eagles and 149 other birds, including hawks, blackbirds, wrens and sparrows, between 2009 and 2013. Duke was also called upon to restore and do community service (how do you ask a large utility to do that!) and was placed on 5 years of probation while they put together an environmental compliance plan to prevent bird deaths. Interestingly, Duke then applied for a permit to kill eagles to help provide a context within which the system can absorb the inevitability of bird deaths. Another group, the Wind Capital Group, applied for such a license only to be embroiled in an argument over its granting by the Osage Nation in opposition. Many applications for this license have been filed. Environmentalists complained bitterly when President Obama's administration, eager for nonpolluting wind power, announced a new federal rule that allows wind farms to lawfully kill birds of prey.

There is some evidence that birds change their behavior when in the presence of wind farms. Lowther in 1998, studying a 22-turbine wind farm in Wales, UK, discovered that no birds were killed by the turbine and in fact they were seen to have shifted their activity to a different location. Some wind farms have no bird fatalities at all. A study[113] published in the *Journal of Applied Ecology* by Pawel Plonczkier and Ian Simms monitored migrating flocks of pink-footed geese using radar as they returned to the shores of Lincolnshire, UK. Monitoring the movement of the birds over four years, from 2007 to 2010, established that two new wind farms effectively caused the geese to change their flight paths. The proportion of goose flocks flying outside the wind farm locations climbed from 52% to 81% in this time, and even geese flying through the wind farm area had increased their altitude to climb above the turbines.

An Australian online group called RenewEconomy published an article that summarizes the whole bird situation quite nicely. Called "Want to Save 70 Million Birds a Year? Build More Wind Farms," it draws attention to the impact of CONG on birds. Replacing all fossil fuels worldwide, it says, would save about 70 million birds a year, establishing wind farms as a strong net benefit for birds. Author Mike Bernard[114] explains that wind farms kill less than 0.0001% of birds killed by human activities annually out of a total 1.5% of human-caused mortality.

Bats and Barotrauma — The other species that more recently became synonymous with death by wind turbine blade is bats. Most of the damage is done to migratory bat species in the autumn. Bats are famously known for their ability to echolocate hard objects in their environment, such as tree branches and cave walls, and, while they are feeding, even insects on the wing. They can detect moving objects better than they can stationary objects, so the high death rate from wind turbine blades was puzzling. Several explanations were proposed, but 90% of the bat fatalities involved internal hemorrhaging, just as might be expected with damage caused by sudden air pressure changes.

Birds have a more resistant respiratory anatomy and are killed by being hit by the blades, whereas bats do avoid the blades but come so close that pressure changes around the blades damage their lungs. The mammals have larger, more flexible lungs and hearts than birds. Birds have compact, rigid lungs with very strong pulmonary capillaries that can resist the high-pressure changes, even though the blood-gas barrier is thinner than those of the bats. An airfoil on a plane pushes against the wind, but a wind turbine blade is moved by the wind. In both cases, the airfoil cross section causes significant differences in air pressure. The greatest area of low pressure exists at the fast-moving (approximately 180 mph) tip of the blade and cascades downwind from the moving blade. A zone of low pressure can cause a bat's lungs to expand precipitously, causing tissue damage, or barotrauma.

A study[115] was paid for by fossil fuel companies like Suncor and Shell but also by wind turbine companies such as TransAlta Wind and Alberta Wind Energy Corporation in addition to academic institutions. The researchers found bat bodies from hoary and silver-haired bats killed at a wind farm in southwestern Alberta, Canada, and examined them for internal injuries. Of 188 bat bodies collected, 87 had no external physical injury. Very few bats had external injuries without internal bleeding.

In 2012, the National Renewable Energy Laboratory conducted pressure studies[116] on mice, which were used because their physiology closely approximates that of bats and discovered that pressures of only 1.4 kilopascals (kPa) were experienced by the bats at the blade tips in 11 mph winds but that it took 30 kPa to cause fatality in mice. There was no suggestion by NREL for an alternative cause of death, however. At low wind speeds, the pressures are even lower and yet it is at the low speeds that the bats fly, which further confuses the issue.

Intermittency — When the wind calms, electricity production needs to be backed up by a nonintermittent power source. On May 13, 2014, Germany experienced 74% of its electricity—an astonishing 43.5 gigawatts— successfully supplied by renewable capacity.[117] A year later, on May 15, 2016, when the demand at 2 p.m. that Sunday was 45.8 gigawatts, renewable energy supplied 45.5 gigawatts of power, or over 99.3%.[118] The world's fourth-largest economy not having to pay for fuel! Okay! It was on a Sunday and the wind, solar, hydro and biomass generation activities needed to be backed up by over 10 gigawatts of spinning reserve, and this was only made possible by the close international connections which allowed Germany to export its spinning reserve, which still represents about 20% of its total. In Denmark in May 2016 wind power alone provided 140% of the country's demand, and the excess was successfully exported.

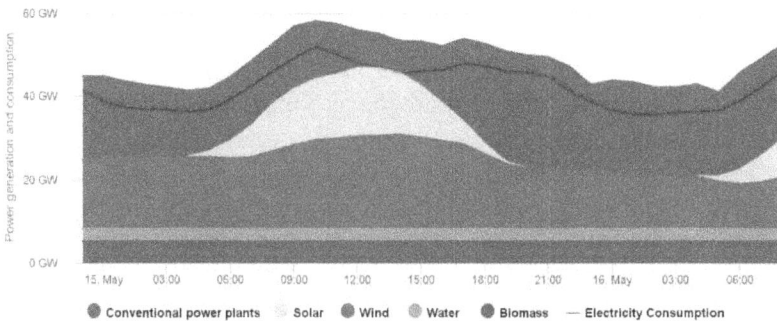

Figure 30: German renewable electricity production on May 15, 2016, was almost 100% of demand, not counting the approximately 10 gigawatts of spinning reserve, the power necessary to use if the wind drops or the sun is interrupted by cloud. This spinning resserve was exported to neighboring countries. Source: Agora Energiewende.

While there is no reason why a fossil fuel needs to be chosen to back up the wind, it just happens to be the current case that CONG fuels are the bulk methods most available to make wind and solar intermittency more palatable.

The short-sighted criticism is that wind doesn't cut pollution after all, but it all depends on which nonintermittent power source is used. Since wind intermittency is mostly offset by the use of fast-reacting gas turbines, instead of coal as backup power or spinning reserve, the impact on emissions can be minor. In the future, sustainable baseload renewable energy can act as spinning reserve. Almost every type of renewable energy can become base load with some tweaks. Solar can go into space. Wind can harvest the energy from almost permanent fast winds at high altitude. Almost all other renewable energy types, biomass, biofuels, geothermal, hydroelectric, and so on, are already base load anyway.

Noise—Like a propeller, wind turbine blades make a noise in contact with the air. Not surprisingly, this particular complaint turns out to be very much less annoying than it at first appears. It turns out that noise from other sources—traffic, aircraft, wind itself, household noises, industry, farming and so forth—is louder and more persistent.

When wind farms are going through the public planning stage, it's quite likely that the developer will ask local residents to sign a waiver for any noise irritation and give them an incentive to do so. Developers suggest local people accept this $5,000 check and if the turbines happen to be noisy, residents have no recourse. One of many states that has addressed this issue is Oregon, where a state noise ordinance reflects a specific regulation restricting noise from wind turbines. The law allows for noise to exceed what is considered an area's ambient noise level by a given amount, often the subject of controversy itself. Interestingly, in Oregon's case, the law that limits turbine noise is an evolution from one that once enforced industrial noise conditions and was part of the Department of Environmental Quality, which was closed down in 1991, before wind power became a state priority.

In 2009, an 85-page study was conducted on the subject for the Canadian Wind Energy Association and the American Wind Energy Association. The selected panel concluded that wind turbines do not make people ill because of noise. They did say the swooshing sound of blades could be irritating. Such a conclusion from such a source is hardly surprising, although the study panel members, a doctor, a vibration and acoustics expert from the UK, a professor of audiology and a biological engineer, all claimed to have been at arm's length and totally able to design the study themselves.

Eighteen studies were done between 2003 and 2014, not one of them saying there was any evidence that wind turbines did any harm at all. In 1918, there was a medical condition that at the time was not acknowledged to be real. It was a reaction to the hell of fighting on the First World War front often caused by the close impact of exploding shells. In fact, it was called "shellshock." The military term for it was "cowardice" or "desertion," and many otherwise perfectly good people were shot at dawn for supposedly letting the side down. In August of 2006, the UK Defense Secretary published posthumous pardons for 306 soldiers, 4 of whom were only 17 years old, who were executed this way.[119] I don't think that wind turbine syndrome will one day be recognized as a real complaint, but I wouldn't like to live close enough to a turbine to experience long-term noise effects either. In February 2015, the Australian National Health and Medical Research Council (NHMRC) completed a comprehensive study[120] on the effects of wind turbines and wind farms on people who live farther than 1,500 meters from the closest turbine. The study identified over 4,000 international papers on the subject of which only 13 suggested a possible relationship between the wind turbine and human health. They determined that the body of direct evidence was small and of poor quality but admitted it was a complex subject, as much of it is subjective opinion. NHMRC concluded there is no consistent evidence that wind farms cause adverse health effects in humans. The concern over the topic led them to recommend specific research to produce a body of high-quality observations of those who live within 1,500 meters of wind farms.

An interesting feedback loop that developed over millennia between rodents and owls has a future in reducing the noise of wind turbines. It would be useless if when an owl swooped in to grab its prey, the prey could escape by hearing the owl in the distance. Over time the owl's wings became covered in a fine velvety fur-like layer that absorbed the sound while having little impact on aerodynamics. Owls already had the advantage by using their highly developed eyes to spot movement at night, when their prey is less able to see. This has now been applied successfully to wind turbines by a team at Lehigh's P.C. Rossin College of Engineering and Applied Science, described in a paper called "Bio-Inspired Trailing Edge Noise Control."[121]

Resistance to Offshore Wind Farms

Cape Wind is the name of an offshore wind farm project that has been moving forward at a glacial pace, and though it appeared positive for the start of construction at the end of 2014, just a few months later, the cycle of delays began once again. There has been no offshore wind in the Americas, whereas many large installations have been completed in Europe. The UK has staked part of its energy future on very large offshore wind farms because of the huge reserve of energy there. It hopes to generate 18 gigawatts by 2020 and double that again by 2030.[122]

There is a paradox attached to the location of one of America's most affluent playgrounds, Cape Cod. White warriors of the U.S. clean energy army, people who in any other circumstance would do their best for renewable energy, were here arrayed against the installation of the first offshore wind turbine farm in America because of a fierce determination not to despoil their little plot of nature. This resistance to installation of something new is called "Not in My Back Yard" (NIMBY). Cape Wind Associates tenaciously hung on to the goal of installing 130 turbines, 400 feet tall each, which were originally supposed to be up and running by 2006. Opposition was fierce. An entity called The Alliance to Protect Nantucket Sound raised millions and paid staff members and a public relations firm in Washington. It purchased radio, newspaper and TV time and distributed flyers. It also engaged the support of wealthy landowners in the region such as Robert F. Kennedy Jr. and Walter Cronkite, who lent his distinctive, patriarchal, trusted voice to a radio advertisement for the group.

Many Cape Cod beach homes are as much as 7 miles from the proposed site, Horseshoe Shoal. At that distance, the relative size of the giant turbines is less than that of a dime held at arm's length. The indigenous holdouts were being marginalized and the project was closer than ever to going ahead at the end of 2014. Today, though, it has died after delaying lawsuits which seemed to never end. The site is in federal waters not subject to the same zoning laws as land-based projects. Private money was ready to be put up to pay the expected $750 million equity money. Complainants like these are well funded, whereas the Cape would benefit hugely over two decades at least of clean energy supply for a very reasonable cost. Any foundations placed offshore additionally act as a wildlife magnet, creating the equivalent of an artificial reef teeming with life. There are artificial reef projects achieving this in many locations along the world's coastlines using old ships, planes and other relics.[123] The project has now closed, but others have already taken its place.

However, the fact that the wind speeds and daily wind periods are higher at the coast means that even the U.S. coastline is about to become a major energy producer, and with the small wind farm inaugurated at Block Island in Long Island Sound, we can expect a long pipeline of new projects that answer the energy prayers of coastline communities across the United States, just as they have done so in Europe.

A Major Paradox

In a heat wave in the Southwest USA in June 2017, aircraft, notably the Bombardier CRJ airliner, hit their maximum operating temperatures of 118°F (47.78°C). In hot air, the density of the air decreases, and the lift generated by the wings is reduced. Airlines were forced to cancel over 40 flights. Larger planes like the Airbus and Boeing can handle 126°F (52.22°C). Ironically, even the solution to global warming, wind power, will produce less power if the air is warmer. A study[124] released in *Nature Geoscience* showed that it's not just changing climate patterns that affect global wind patterns. If air temperatures climb by more than 9°F (5°C), wind resources will decline in proportion by up to 40%. Hopefully, we can expect hotter days to be rewarded by more solar power and better wind designs to offset this apparent failing of future wind power availability.

Hydroelectric Externalities—Eternal Dam Nation

> *"We will let people in the future worry about it."*—Bureau of Reclamations Commissioner Floyd Dominy in 1936, when asked what the bureau would do if Lake Powell, behind the Hoover Dam, filled up with silt.

The original idea of hydroelectric was to find a convenient hydrological formation such as the Grand Canyon on the Colorado River and build dams that would block the entire river for electrical power. This steadily creates a huge artificial lake in the valley upstream of the dam. A river is an umbilical cord for nature and people. It transports huge quantities of water, life and silt. It is a carver of the geology, gouging valleys and forming oxbow lakes and creating beautiful settings. It connects fish species like salmon to locations where they spawn far from the ocean, and communities up- and downstream look to the river's abundance for water, food and transportation. A dam built across the river empirically interferes with the ecological value afforded by that waterway. Today we know that most of the world's 50,000 dams, and not just the 70-year-old Hoover and Glen Canyon Dams, are already too silted up and mired in a long-term drought for their intended purposes to be fully realized. Dam longevity was routinely exaggerated by proponents in the 1930s and 1940s.

Construction of the Hoover Dam was a huge infrastructure project that put a few workers back to work in the world's worst economic depression. The building of most dams results in the flooding of huge areas that often hold immovable cultural assets, carved walls, human history, archaeological sites, stone buildings, towns and villages, agricultural land, wonderful features, alcoves and monuments and the displacement of millions to locations that are challenging and far from the land of their forebears. This sacrifice is always made in the interest of development, but like CONG, few such externalities are initially considered. Enough silt arrives in Lake Powell, behind the Hoover Dam, the world's number 50 dam by electrical capacity, to fill 1,400 cargo ships a day.[125] This silt is good agricultural soil, and its accumulation behind the dam deprives downstream plants and animals that depend on its annual accumulation.

The melting winter snow swelled the old Colorado River to flow at more than 100,000 cubic feet per second (cfs), the equivalent of Niagara Falls. In the winter, that flow drops to just a few thousand cubic feet per second. This normal cycle was subsumed by the need for electricity from the hydroelectric generator that, as a consequence, now releases a nonseasonal, regular water flow. Sometimes, when the hydroelectric plant is used, water sweeps through the canyon below the dam like a tidal wave, washing away the silt and reliable flow rates the river knew for millennia.

The short-term benefits of dams are all in turn swept away by the detriment of accumulating silt. As the reservoirs fill with mud, the original river is likely to carve its place across the new plain and form a waterfall over the dam that will eventually erode the concrete until the river has regained its original course. The Matilija and San Clemente reservoirs in California are already full of mud. The Sanmenxia and Three Gorges Dams in China, among others, are doomed to this fate. The silty mud is much heavier than water as well, so there are mechanical considerations to consider too.

When I was living in Pakistan, we visited the Tarbela Dam,[126] built across the Indus River, 31 miles northeast of our home in Islamabad. One of the engineers had a daughter at the same school in England as my sister. Tarbela Dam is the 15th largest in the world by installed electric-generating capacity and is the largest of 40 dams in Pakistan. It was completed in 1974. It is the largest earth and loose rock dam in the world and the second largest by volume. It was designed principally to conserve water for farming purposes. It can generate 13 terawatt-hours annually, with an hourly capacity of 3,478 megawatts. This capacity is currently being increased to over 6 gigawatts by a Chinese construction group financed by the World Bank. This will make it the world's seventh-largest dam.

The flooded valley displaced 96,000 inhabitants of 135 villages above the dam. Resettlement claims have still not been finalized after 40 years. Knowing the silt load on the Indus River, rushing Himalayan snowmelt to the Indian Ocean, engineers calculated it had a life of 60 years before the 100-square-mile reservoir would be full of mud. This silt load is filling the reservoir faster than expected. The back of the reservoir is where the sand and gravel are deposited and forms a sort of backwater delta that advances toward the dam where the clay and finer silt is dropped. In 1983, Tarbela Dam's delta had advanced to 19 kilometers from the dam when it was expected to have come to only 48 kilometers by this time. By 1991, the delta was only 14 kilometers from the dam.[127]

Other dams are not so lucky. Sedimentation is the most serious technical problem faced by the dam industry. Fifty cubic kilometers of silt is trapped behind the world's dams annually. The Cerron Grande Dam[128] in El Salvador is experiencing a significant shortening of the life span originally discussed during its planning stages to just 30 years, or one-tenth of its original expectation. The World Bank completed a study on global dam silting in 1990 that found that 1,100 cubic kilometers of silt accumulated in all those dams. Most predictions for Lake Powell suggest that only two decades remain before the reservoir surface drops to its "dead pool" level, the lowest exit from the dam, which is 237 feet above the original riverbed. Between 1999 and 2005, the reservoir lost two-thirds of its volume. Even at dead pool, Lake Powell will still hold one-thirteenth of its capacity, or 2 million acre-feet of water. It will take about 55 years for this to be filled with sediment, but real dead pool is currently expected in the 2020s.

The Colorado River is one of a growing number of rivers in the world that no longer reaches the ocean. Demand for its waters continues to rise. Ever since the 1920s, its water has been divided up among the major southwestern cities of Phoenix, San Diego, Los Angeles and Las Vegas and it serves 30 million people in 7 states and Mexico as well.[129] Drought conditions in the American West have affected the rainfall and snowfall, reducing the volumes of water expected in the Colorado River further. As supply diminishes, demand increases. Studies show that as warming persists, evaporation increases.[130] If rainfall also declines, the residual water of the Colorado River drops disproportionately. As water levels decline, sediment that was deposited previously is washed down along with the new sediment, compounding the problem

Wet sediment weighs twice as much as water. In May of 2008, in Sichuan Province in China, an earthquake killed 70,000. Its epicenter was only 3.5 miles from the Zipingpu Dam,[131] which itself stood only 550 yards from a fault line. It had not had time to fill with silt but did hold 315 million tons of water. The dam itself cracked in the earthquake, forcing the authorities to empty the reservoir. Geologists confirmed that the dam contributed to the earthquake, advancing its timing and severity.

In the decade after the Hoover Dam was completed, there were 600 local earthquakes. The Colorado River system in the U.S. Southwest provides essential services, but technology will eventually fail. Hundreds of dams in the United States have destroyed rivers and ecosystems and represent an accumulated debt that will be handed down to future generations. Though hydroelectric is commonly referred to as green, the externalities it presents are severe, and there is much concern about methane released by the huge quantities of mud in the reservoirs, which offset the claims for carbon neutrality even as a 2013 study[132] by the World Bank estimated this to be about 1% of all new greenhouse gases.

No assessment would be complete without looking at these issues with regard to the largest dam in the world, the Three Gorges Dam, with the single largest powerplant at 22.5 gigawatt-hours of capacity, completed July 4, 2012, with the final turbine installation. Each of the 32 turbines has a capacity of 700 megawatts. In 2012, the dam generated 98.1 terawatt-hours of electricity, 14% of China's hydroelectric supply, and in 2013 it broke the world record for production with 98.8 terawatt-hours, more than the Brazilian-Paraguayan Itaipu Dam.[133] A ship elevator is built into the dam, like a large canal lock, to maintain shipping on the Yangtze River. The dam flooded a huge valley containing 1,300 archaeological sites, including the Hanging Coffins cliff site in the Shen Nong Gorge. The new reservoir covered villages and towns and farmland, displacing 1.24 million people. Funds intended for compensation payments to 13,000 farmers in the Gaoyang area mysteriously disappeared. Almost 40 cubic kilometers of water are held in the reservoir in an area of 390,000 square miles. Days after the first filling of the reservoir, hairline cracks appeared in the dam's structure but within design parameters.

The cost of the dam was 180 billion yuan ($22.5 billion) and the payback time is estimated at 10 years. At full power, the Three Gorges Dam avoids the use of 31 million tons of coal, or 100 million tons of CO_2, 370,000 tons of nitric oxide and 10,000 tons of carbon monoxide, along with a significant amount of mercury and all the logistics of mining. The dam increased the Yangtze's water traffic by six times, which also reduced greenhouse gases. The government intends to capture more water with about 16 more upstream and downstream dams to more than double the capacity of the Three Gorges Dam itself.

Silt is deposited behind the dam at a rate of 40 million tons per year, and this valuable resource no longer arrives downstream. Shanghai sits on a silt plain, and arriving silt strengthened the silt bed on which the city sits. The absence of this silt makes the city more vulnerable to flooding. Shanghai has sunk more than 6 feet already from a combination of groundwater pumping and the impact of its dense skyscrapers and high-rise buildings on the soft alluvial plane; subsidence continues today at a rate of half an inch per year.

The Three Gorges Dam sits on an earthquake fault line, and there have been frequent landslides. Wastewater above the dam now is deposited into the stagnant waters but has galvanized the production of wastewater treatment plants. Of the 6,388 species of plants that live in the region, over half are threatened. Forests have been depleted despite a program to reforest the area. Over 360 freshwater and terrestrial animal species were disrupted by the changes in the river's flow. The turbine blades also hurt the fish, including the Baiji or Chinese river dolphin, which is now deemed to be extinct. The Siberian crane, of which only a few thousand remain, wintered in the wetlands that no longer exist, and the Yangtze sturgeon fish is also endangered.

One of the main reasons for building the dam was to control frequent downstream flooding, responsible almost annually for death and destruction. In 1954, 33,169 people were killed and over 18 million were forced to flee floods that inundated 74,518 square miles, covering a major city of 8 million, Wuhan. Even as recently as 1998 a flood killed 1,526 people and affected more than 2.3 million, destroying 787 square miles of farmland. In August 2009, flooding that would have been disastrous was captured by the dam, raising the water levels in the reservoir from 145.13 meters to 152.88 meters within a week, capturing 4.27 cubic kilometers of floodwater. The reservoir can now be discharged during the dry season, helping agriculture, shipping, downstream dams and industry and dropping the reservoir from 175 meters to 145 meters in preparation for the rainy season.

If dolphins had Facebook, humans would be in a lot of trouble. What with the Japanese dolphin meat harvests that kill tens of thousands of individuals annually and nets worldwide that classify these friendly animals along with other creatures as "bycatch," the building of dams is another threat to their survival. On October 3, 2013, the Laotian government decided to move ahead with plans to build the 250-megawatt Don Sahong Dam on the Mekong River in southern Laos, for completion in 2018. This decision bypassed their obligation to go through the consultation process of the Mekong River Commission (MRC). Countries along the Mekong River are bound by the MRC agreement to hold inter-governmental talks before proceeding with any infrastructure projects on the river that will impact communities in Laos, Vietnam, Cambodia and Thailand. Laos is failing to observe this agreement, threatening the security of millions and flood control in the name of selling energy that actually can be generated by another, less- damaging project, the 30-megawatt Thakho project, which merely diverts the water flow for hydropower without blocking the river.

The Don Sahong Dam blocks the passage of fish migration, taking away the supply for the critically endangered Irrawaddy dolphin, a eury-haline (handles freshwater or saltwater) oceanic dolphin that lives in a 190- kilometer section of the river in the Veun Nyang/Anlong Cheuteal pool close to the Cambodian border. This area is also one of the world's largest inland fisheries, so communities along the Mekong River will be deprived of a source of food along with the dolphin. This animal has been a friendly participant in cooperative fishing with humans. There are records of court cases in the 19th century of fisherman complaining that the dolphins helped other fishermen more. The Irrawaddy dolphin still helps fishermen with nets and paradoxically it is these gill nets that are the main cause of dolphin drowning deaths. Mother dolphins gestate for 14 months and have only one calf at a time. Though the species numbers some 7,000 overall and is distributed in discontinuous subpop-ulations on the coasts, bays and river estuaries of Southeast Asia and northern Australia, local populations are small. The Mekong River population has only about 85 individuals. The dam will be sited just 1 kilo-meter above the animal's core habitat. Building it will involve breaking millions of tons of rock with explosives, which will change the water quality, create blast waves that will wreck animals' sensitive hearing structures, bring significant boat activity and habitat destruction typical of other dam construction cycles.

No mention of the ills of hydroelectric would be complete without the story of the Banqiao Dam.[134] No other dam failure in history caused so many casualties. There were several bigger dam failures but all due to wartime attacks rather than the force of nature. Russia helped the Chinese build the Banqiao Dam starting in April 1951 on the Ru River in Henan province. It was intended to control water supplies and flooding of the Huai River basin, which in 1949 had been severe. The dam was completed in 1952 but without using the normal hydrology information any feasibility study would exact in the early stage of such a project. The construction material was clay and was 24.5 meters in height. The upstream reservoir capacity was 492 million cubic meters, most of which was reserved for flood storage. After completion, cracks appeared in the dam but were repaired by the Russians who rechristened the dam "The Iron Dam," which is a bit like a reference to the *Titanic*.

One Chinese hydrologist was a consistent critic of the project and was twice fired from his role as advisor because he disagreed with the safety measures being taken. In 1975, a once-in-2,000-year rainstorm dropped a year's supply of rainwater in a day as a typhoon hit the region. This same typhoon broke 62 dams in the same few days. The Banqiao Dam broke and released 701 million cubic meters of water in 6 hours; 1.67 billion cubic meters of water were released at the upstream Shimantan Dam, and a total of 15.738 billion cubic meters of water were let go. This sent a 10-kilometer wave that was 3 meters high rushing downstream at 31 miles per hour that flooded seven counties, an area 15 kilometers wide and 55 kilometers long.

Thousands of square miles of countryside were flooded and telegraph communications were cut in the middle of a very severe storm system. Some 90,000 to 230,000 people died in the tragedy, many in the subsequent epidemics and famine. Some communities were completely destroyed, such as that of the 9,600 souls who lived at Daowencheng Commune, none of whom survived the wall of water. Eleven million residents were affected and almost 6 million buildings collapsed. Aerial bombing of dams to release water that was simply too much to hold was necessary as a safety control after the fact. Many downstream dams were badly impacted by the Banqiao failures. The Seventh Five-Year Plan for the Chinese economy prioritized rebuilding the dam system. The new Banqiao Dam was finished in 1993 and has a 30% higher capacity than the failed dam, and new standards of construction rule out any catastrophic future.

Today, new hydroelectric technology is designed to leave nature to itself and benefit from "run-of-the-river" techniques that don't disturb the major natural functions of the river and still yield a lot of energy.

Ocean Energy Externalities

The story of the ocean and renewable energy is a reversal, where renewable energy gadgets are repeatedly smashed on the rocks in storms or lost to the abyssal depths. Conditions in the ocean are extreme for equipment. Cyclical wave conditions wear and test equipment over time. Acids and saltiness effect a caustic environment that erodes and decays all except the hardiest, and expensive, of materials, as well as compromising electrical integrity. Hydrostatic forces which come about from currents, tides or during storms are extreme and put great stresses on blades and other equipment.

While much is made of the "damage" that offshore wind farms will wreak on coastlines around the world, in actual fact, much can also be said for the positive impacts of "artificial reefs," which is what the foundations of wind turbines actually represent. Such reefs are in demand on many coastlines as refuges for fish populations, which have been seen to expand when a ship is sunk, or hundreds of car wrecks or concrete blocks are deposited on a shallow ocean floor, forming other types of artificial reefs.

In the collection of hydro/ocean energy technologies, over 40 technologies generate power from thermal, waves, tidal currents and ocean currents. The ocean is so large that it more than digests the machinery impact of greasy mechanical systems, hydraulic fluids and lost metal parts, which litter the bottom. We are talking a tiny impact from renewable energy systems compared, of course, to that of shipping, coastline pollution, sewage disposal and garbage dumping. The ocean is suffering so much more from the CO_2 absorption from fossil fuel energy production that it's difficult to point to any negatives from machinery placed in it to generate energy.

The impacts of human activity are very visible, of course. The plastic gyres in the Pacific and Atlantic Oceans, reduced fish stocks from overfishing and the consequent impacts on the otherwise enormous and stable ecosystem— we damage the oceans at our peril. This is a big part of the Earth's footprint and nothing that we do within the ocean to obtain energy will challenge the damage that we have already done with fossil fuel energy, wastewater disposal, transportation and fishing.

Biofuels: Impact on Food Price Inflation

"Qu'ils mangent de la brioche." (Let them eat cake.)
—Jean- Jacques Rousseau, 1782, *Confessions* (incorrectly attributed to Marie Antoinette).

Renewable energy is mostly sun or wind based and makes electricity. If we are to transition away from releasing fossil carbon into the atmosphere, it's vital to replace these CONG liquid fuels with carbon-neutral alternatives. The first steps have been taken to replace liquid fuels like gasoline, diesel and aviation jet fuel, which is a kind of kerosene, of which 20 billion gallons a year are currently burned. These are all liquid fossil fuels. Essentially, in 2014 all of these fuels were derived from fossil fuels. To make a fuel out of something that doesn't contain carbon coming from fossils, you need to combine hydrogen gas from water with the oxygen from water and a carbon from the air. This sounds crazy, but we are becoming very clever at manipulating our environment's basic molecules. This is already happening but still at too low a volume and too high a cost.

A more reasonable approach is to pick a feedstock from something that's been growing recently, such as an agricultural commodity like molasses, coarse grains such as corn, sugar beet, wheat and sugar cane, or roots and tubers such as potatoes. The starch and sugars in these foods can be fermented and made into biofuels such as ethanol, biobutanol or even lipid oils that can then be turned into biodiesel variants. Grains, palm oil and oilseeds such as canola with its beautiful yellow rape flower, soy, nuts from the Jatropha tree and animal and sewer fats can all be processed into biodiesel or aviation fuel.

Global biofuel production tripled between 2000 and 2007

Billion gallons

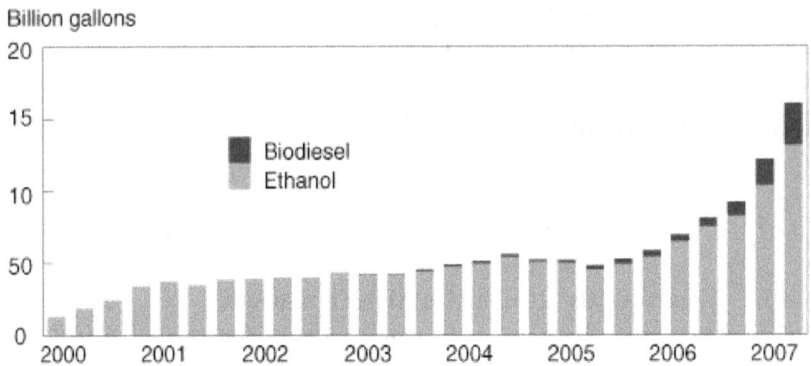

Source: International Energy Agency; FO Licht.

Figure 31: Progression of the volumes of biofuels globally since 2000. Source: International Energy Agency; FO Licht.

The problem with using food crops for fuel is that we live in a 7.4-billion- person world where 2 billion don't have enough to eat and even in the USA children go to bed hungry. Nothing sparks social unrest more than hunger. The quote misattributed to Marie Antoinette, "Let them eat cake," was very powerful precisely because it explained the unrest of the people in the context of French history in 1789.

Food today is a strategic commodity, subject to national strategic reserves that are designed to kick in if shortages occur to cut increasing prices, and therefore to allay perceived unrest. Marc Bellemare observed that high food prices cause social unrest but that volatile food prices are not associated with social unrest.[135]

Food is one of those core issues that will cause a break with law and order. Basic commodities are linked to social unrest in a visceral manner. The recent unrest in Ukraine is partly driven by high Russian gas prices.

Thomas Friedman's poignant comments about a WikiLeaks article[136] showing that Syrian agricultural officials were only too well aware of and also asking for help with the effect of the severe Middle Eastern drought on the Syrian civil war are just another illustration of this powerful relationship. In Mexico, corn flour is used to make tortillas, a flatbread that predates the arrival of Europeans in the Americas. Mexicans often eat 10 of them a day, and normally those 10 tortillas would cost about $0.30. In 2008, the United States was allocating more of the North American corn harvest to make ethanol than food. As it did so, the price of a bushel of corn rose and along

with it the price of tortillas to about $0.50 or more. This coincided with the rise in the price of oil to $147 per barrel that year.

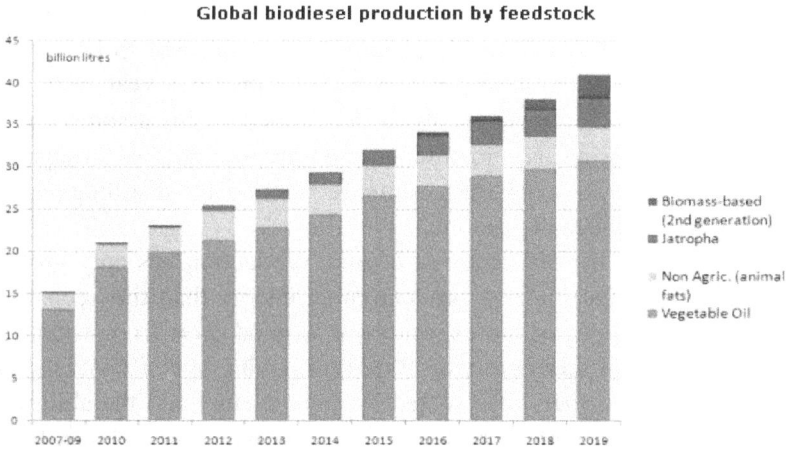

Global biodiesel production by feedstock

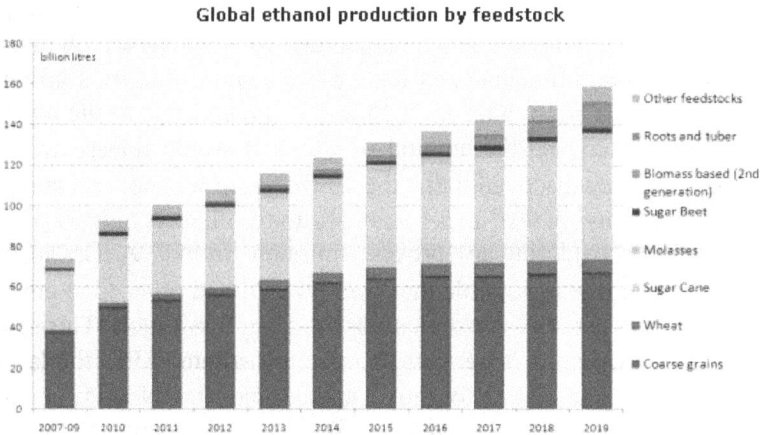

Global ethanol production by feedstock

Figure 32: Feedstocks used for production of ethanol and biodiesel from 2007 to 2019. Source OECD-FAO Agricultural Outlook.

Mexican corn was even being transported to the United States to take advantage of higher prices. It's not clear in the noise whether it was increasing demand for fuel or financial speculation that caused the rise in prices, but the social unrest was real.

To illustrate part of the problem with food for fuel, occasional studies show that the carbon emission advantages are not even that real. A study[137] released on August 25, 2015, investigated the total carbon emissions of corn ethanol in the United States. When fuels are burned, CO_2 is emitted. When corn ethanol is burned, says the conventional wisdom, the CO_2 emitted from the exhaust pipe is deemed to be carbon neutral because of the absorption of CO_2 the previous year when the corn was grown. Life-cycle analysis (LCA) makes this assumption. It focuses on the energy used to produce and transport biofuels throughout the entire sowing, growing, harvesting, fermenting and distilling phases as compared to those of gasoline.

Over time, the efficiency of ethanol production has only grown, and we accept that biofuels produce more energy than they consume in manufacture. Sugarcane especially is a good crop from which to make ethanol and sheds glory on Brazil for having over 70% of its liquid transportation fuels represented by cane ethanol and the ability to export millions of gallons as well. The study mentioned above, however, asks whether the assumption of corn ethanol carbon neutrality is justified. The LCA approach assumes that 100% of the carbon released when burning the corn ethanol is reabsorbed when the new crop of corn is grown. The authors analyzed the carbon capture by farms growing corn for biofuels from 2005 to 2013 and compared this to the amount of biofuel created and the amount of CO_2 that would release when it was burned. They discovered that the 100% figure was not real and that the offset of growing feedstocks like corn for biofuels is only 37%, and this is before considering any process emissions from CO_2. The authors concluded that it was possible that corn ethanol and other food ethanol production was actually worse than just burning fossil fuels. This study, though, came from the American Petroleum Institute (API), the largest American trade association for the oil and gas industry. Nobody is under more pressure from biofuels than the fossil fuel community.

Here are some details of a new approach to ethanol that could derail the use of gasoline entirely; it looks at a resource that scientists have had a tough time cracking—cellulose. Cellulosic materials include agricultural waste such as corn stover, stalks, sugarcane bagasse, straw, wood, grasses, garden cuttings, leaf litter, forest residue, wood mill waste, palm fronds, coconut shells, rice hulls, fruit peels, food waste, old newspapers and cellulose from municipal solid waste or even old furniture. Cellulose has been very problematic. The process of cutting the cellulose molecule into its sugar molecule constituents, also called hydrolysis, has been fraught with difficulty, so the easy money embraced ethanol and biodiesel from crops. Scientists even call the cellulose molecule and its evolved impenetrable nature "recalcitrant," highlighting exactly the difficulty posed by having all this potential chemical energy in direct sight but still unavailable. To alter a famous quote: "Energy, energy everywhere, but nor any drop to burn."

As shown in the global biofuel production chart in figure 32, the biofuel feedstock volumes have been growing strongly in recent years, which implies that the proportion of crops dedicated to fuel has been increasing. A report by the OECD-FAO Agricultural Outlook has shown expectations for the consumption of agricultural feedstocks up to the end of the decade in 2019.

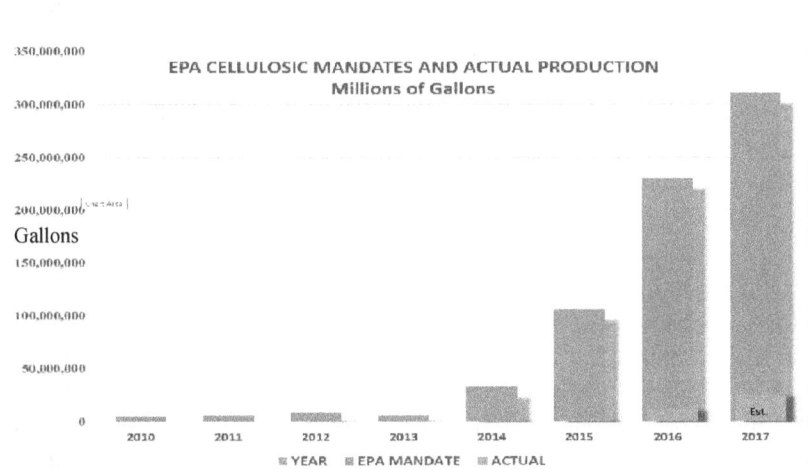

Figure 33: Environmental Protection Agency cellulosic ethanol mandates compared to annual U.S. production of the original definition of cellulosic ethanol. The author believes that within a few years, cellulosic ethanol production will increase because of ball mill technology. Source: EPA and NEF Advisors, LLC estimates.

It is clear that vegetable oils are expected to be the principal source of feedstock for biodiesel and that corn and sugarcane will be the major feedstocks for ethanol—at least in the minds of the OECD report authors. But I have new information. I think that cellulose will be the main source within a few years. In the United States, the Environmental Protection

Agency is tasked by the federal government under the Clean Air Act with setting the annual standards for the Renewable Fuel Standard (RFS) program. This means they mandate how much cellulosic, biomass-based diesel, advanced biofuels and total renewable fuels are to be produced or imported for the purpose of blending with the existing mineral gasoline and biodiesel supply. The EPA currently sets the total target for renewable biofuels at 15.21 billion gallons. Cellulosic biofuels come under the "advanced biofuels" category along with biomass-based diesel and other advanced biofuels, and the EPA is proposing a reduction in these volumes. The old volumes for cellulosic were 2.2 billion gallons, which was in 2013 reduced to 1.75 billion gallons. One of the reasons quoted for reducing volumes is the "limitations in the ability of the industry to produce sufficient volumes of qualifying renewable fuel."

This reflects the lack of production by all the existing cellulosic ethanol companies. The 2014 target is nonetheless 33 million gallons, or 0.01% of the total 2014 biofuels production. The RFS for corn-based ethanol is reduced as well, to 13 billion gallons from 14.4 billion gallons, much to the disappointment of farmers and corn ethanol manufacturers. Many cellulosic ethanol production plants have been partially funded by the Department of Energy and are being built by companies such as DSM in Holland, DuPont, American Process, POET, Abengoa, KiOR and INEOS.

Already the INEOS plant, the first to start operating in 2014, has suffered cost increases with contamination of its unique biomass gasification cellulosic process. The syngas produced by gasifying the cellulose can be used as food by a bacterium in a fermentation process that makes ethanol. The heat of gasification also provides electricity. However, the presence of hydrogen cyanide gas in the syngas kills all the ethanogen bacteria, so INEOS were obliged to spend money on an expensive scrubbing flue. It is observed that the other companies, which collectively want to produce about 90 million gallons of cellulosic ethanol, or three times the amount mandated by the EPA starting in 2015, are experiencing similar difficulties. So optimistic have these groups been that DuPont actually testified before the Senate Agriculture Committee in defense of the EPA RFS before it was cut, saying that the demanding mandate had actually served to focus minds and solve the production issues.[138] This, of course, is an example of the classic driver of technology: Progress springs from a driving need. The fossil fuel industry, however, leaps on the argument that because there is no working technology, the RFS should be scrapped. This is not how we want to go on.

Interestingly, despite all the EPA emphasis on cellulosic biofuels, 75% of each barrel of crude oil goes into the production of fuels. That market is worth about $400 billion annually in the United States. Less than 10% of the barrel of oil is used to make chemicals, but that 10% is worth as much as the fuels market on its own. This high potential for the chemicals market is under no EPA mandate to achieve higher penetration levels but is a wonderful indicator of the potential for the biomass-sourced chemical industry, which sells so many products, such as plastics, cleaners, degreasers, personal care products and polyurethanes. All of these products can instead be sourced from plants.[139]

In November 2014, another cellulosic fuel company, KiOR, went bankrupt. It had received an interest-free loan of $75 million from the state of Mississippi and within the political community was regarded as Mississippi's "Solyndra," after the DOE-funded solar company that went bankrupt following dumping of competing solar panels on the world market by the Chinese. Apart from Fisker Automotive and Solyndra, that government funding program was a great success. Although I agree that the government should not play the role of choosing technologies, providing capital to a range of corporations can be a boost for the industry. We also know that some very worthy ideas (solar chimney, solar power 24/7, no water required) are probably now politically frozen out of the loan guarantee programs because of these other failures, which represented a small minority of companies, and despite the significant overall success of the DOE funding project.

Because of the difficulty experienced by the biofuel industry to find a way to manufacture the mandated high volumes of cellulosic ethanol, on July 2014 the EPA broadened the "pathways" to cellulosic ethanol to include cellulosic diesel, cellulosic renewable gasoline blendstock, renewable compressed natural gas, renewable liquefied natural gas and cellulosic heating oil.

Needless to say, the much higher numbers of so-called cellulosic biofuels that were included in this new definition do not reflect the true production of cellulosic ethanol, such as the figures in the chart in figure 33.

Social Unrest

A study authored by Marc Bellemare[140] in July 2012 related that using food for fuel had increased planted agricultural acreages and also food prices. The food price chart that he uses is overlaid on a line representing social unrest that he found from LexisNexis. There is a very high correlation. It's also instructive to note that this chart shows the relationship between food and oil prices, where there is also a correlation. Transportation and energy used in farming are major contributors to the final price of food.

To break the food-to-fuel relationship, we need to stop using agricultural food crops as a source of fuel. Today we are aware of a major new method of turning all that cellulose into fuel. The biofuel industry and consumers will win, win, win. This new method is represented by a company called Alliance BioEnergy Plus, based in Florida, which has adopted a 146-year-old technology (it was used to grind flint for pottery makers in 1870) that makes familiar products such as talcum powder and other cosmetics as well as cement powder and that is widely used in the mining, cosmetics and chemical industries.

Maize Price Levels and Social Unrest 1990–2011.

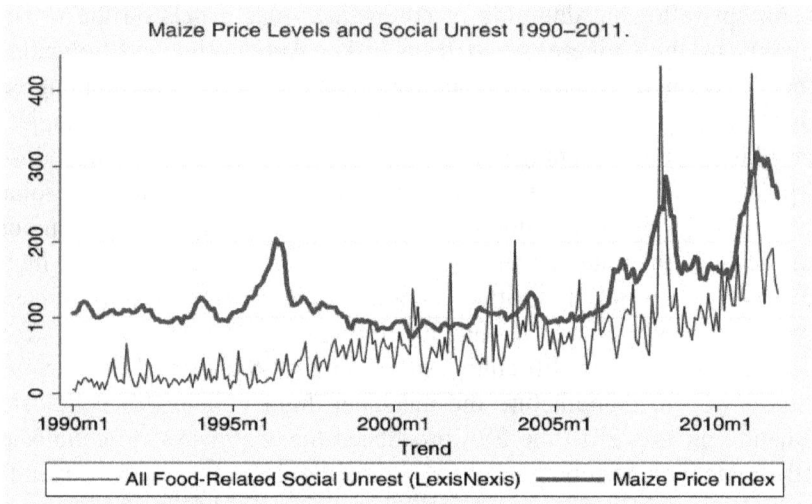

Figure 34: IMF Maize Price Index and Social Unrest, January 1990 to December 2011. Source: Marc Bellemare. "Food Prices and Social Unrest."

It can 100% convert any kind of cellulose into an extremely fine powder made up of approximately 70% sugars and 30% lignin. The process takes 12 minutes and needs no chemicals or enzymes or even liquids. It uses common clay as a catalyst and produces no waste. All the inputs are very cheap and so it produces very cheap sugar.

It would not be the first time that I have made a leap into the dark to predict the success of a methodology. I did it for thermal solar storage and copper on silicon solar cells and got very interested at the amount of extra sunlight that could be attracted to the surface of a solar cell with a basic wave guide. Such things excite me, and yet the marketplace is full of examples of where things that excited investors did not work. It's fair to say that Wall Street self-consciously avoids taking technology risk and doesn't claim to be a fortune-teller and reveal whether something will make it or not, and yet insights into success or failure must be informed by expertise.

Investors are responsible to themselves. *Caveat emptor*. At the higher-risk end of the market, money is poured into things that appear highly likely to occur.

This new cellulose method represents a logical next step for the inefficient acid and enzyme hydrolysis processes that have been growing in use up to 2016. Ball mill cellulose will be a success says this author and will finally be able to produce the billions of gallons of cellulosic biofuels mandated by the U.S. EPA. It will be able to reduce the pressure on food crops, and there is likely to be a decline in food prices that hope-fully will not lead to a deflationary spiral. Thirdly, it will make available a huge resource of cellulosic feedstock, until this point unusable except by combustion or gasification. A corn plant's cellulosic stalks, stover and leaves can make three times the ethanol of just the corn kernels. The cellulose in sugarcane bagasse can make twice as much ethanol as the sugar syrup pressed from the stalk. Such releases of biofuel production volumes will be remarkable and welcome. Few now expect cellulosic ethanol to gain any vantage hold of the market. Our insight gives us that comfort that it could not only be easily done but also be a pathway to replacing transportation fuels entirely and globally with carbon-neu-tral ag-crop waste. If 40% of the U.S. corn market is currently making ethanol for 10% of the U.S. gasoline market, then the plants from that 40% could arguably fill 50%' of car gas tanks, and the other 60% of corn plants could fill the rest. That's only scratching the surface of the cellulose that's available. Given the potential of this innovation to pour oil on globally troubled waters, it can't come too soon.

The Illusion of Carbon Neutrality in Biomass

In one of the classic misconceptions that plague human beings, it was always thought that biomass came from the ground. It made sense that the substance of the ground should enter into the plant, albeit organized, and then become the wood, food or leaves. Jan Baptist van Helmont, a Flemish alchemist and Doctor of Medicine performed an interesting experiment. Over a 5-year period in the early 1600s, he grew a willow tree in a pot. He measured the increase in the mass of the tree at 164 pounds while the soil had only dropped by 2 ounces, hardly enough to explain the mass of the tree. Though his understanding of chemistry at the time was too limited for him to understand that the tree's mass had come from photosynthesis of CO_2 and conversion of H_2O into the sugars and proteins that make up plant matter, that is, from gases in the air and water, he at least concluded that it must be from the water, delivered regularly over time, that the plant increased its mass.

Biomass is a hydrocarbon and holds energy. Plant matter from any source is a useful source of fuel, and since some of it grows within a year, it's also a great way to collect or sequester CO_2. Any energy from plants is, by this definition, deemed to be carbon neutral.

One method that addresses coal emissions is replacing the coal with biomass. Many coal-fired power stations and new biomass-burning power stations are burning biomass that has been made into briquettes or pellets. Oil burners are also being replaced for the same reason. The biomass is pulped by passing it through a hammer mill to create a dough-like mass. This is pressed into shape in a mold and can hold its shape because of the resin- like lignin that effectively melts then resolidifies due to the heat of the process. Led by the United States and Europe, this trend is certainly okay for providing more carbon-neutral electricity to subscribers because coal emissions are truly evil. Burning biomass also releases nitrogen oxides and sulfur oxides as well as volatile organic compounds, although in very much lower quantities than coal burning does. Transportation of biomass also results in emissions if the vehicles used are working on fossil fuels. In many cases, the particulates can be captured using electrostatic precipitators, cyclonic separators or filters, all of which add more cost.

A study[141] that looked closely at particulate emissions (soot, dust and smoke) was aimed at finding ways to reduce the negative effects of biomass combustion. The emissions are linked to breathing problems, irregular heartbeats and nonfatal heart attacks and 350,000 premature deaths annually in Europe alone. The U.S. Centers for Disease Control and Prevention (CDC) said that even a 10% reduction in emissions would keep 13,000 people alive. They found that emissions varied when the fuel was at different stages of combustion, meaning emissions can be minimized if burning is at high efficiency. This will likely lead to combustor design changes. Being carbon neutral makes pellets much in demand because the carbon they release is the right kind of carbon, recycled within the short period it takes to grow the biomass used. If you add water to many kinds of pellets, they expand back into their original sawdust and can be good bedding material for animals.

It is often the case that using biomass is deemed to be a carbon-neutral scenario. A new report[142] by the U.S. nongovernmental organization the Natural Resources Defense Council (NRDC) examines assumptions about carbon neutrality of biomass fuels. Just like burning coal or oil, carbon dioxide is emitted from the combustion of biomass. Much biomass is in the form of pellets or briquettes of biomass. If the biomass is an annual grass rather than a wood (usually perennial growth), then the carbon in it really does get reabsorbed the following year. If, however, the carbon comes from trees, which painstakingly lay down the carbon year after year until the tree is mature, and often with an accelerating absorption due to the geometry of tree rings in the later years of a tree's life, then it will not be until later in the century that new trees will mature and absorb that same amount of carbon.

At a time when avoiding adding new carbon to the atmosphere is even more critical, it is a major climate issue, but it is also an economic issue. The NRDC report suggests that burning biomass may be an even greater source per unit of energy of carbon because it has a lower energy density. It matters whether the wooden pellets are made from whole trees or waste wood trimmings and sawdust. Drax, the aptly named coal-fired power station in northern England, uses nearly a million tons of biomass a year now, since a program to become less carbon intensive was started a few years ago. The NRDC report, however, suggests that a quarter of this biomass comes from mature trees in which case the station's emissions will be more serious and carbon intense than from natural gas and only slightly less than coal until 2065. Only after 2080 would replacement trees start to compensate for the carbon grown over decades and burned today. If the biomass was purely carbon neutral, its emissions would be 34 grams per megajoule, only a fraction of the 280 for coal and 193 for gas. Unfortunately, Drax, which describes its biomass project as "the biggest carbon reduction project in Europe" is using biomass that is not carbon neutral. Fossil fuels used to make and transport the pellets make up the 34 grams of emissions per megajoule if it was carbon neutral. Its claims only make sense after 2080, long after the current spate of "business-as-usual" carbon emissions have caused the climate to add more than 3.6°F (2.0°C).

Liquid biofuels, blended or not with fossil gasoline or diesel, compete with food crops and there are already incipient public regulatory initiatives to ban the use of food to make biofuels. This will be quickly put right by our ability to use cellulosic sugars to make biofuels and leave the food on the plate. However, such regulatory change in the biomass-burning industry would cause stranded assets.

The European Union comes under the EU Emissions Trading Scheme (ETS) as a support subsidy mechanism for renewable energy. This piece of regulation assumes that biomass power is zero carbon or carbon neutral. Coal-fired power plants are required to purchase emissions permits for emissions above a certain threshold. If they use biomass under current conditions, then they have fewer emissions permits to buy, affording them real advantages. In the UK, the government also supports biomass in a carbon-neutral manner under laws that assume their use is green. Will legislation change in the near future and alter the mix of biomass burned in coal plants? This is a question that is likely to become increasingly critical as our efforts to mitigate the climate issue intensify.

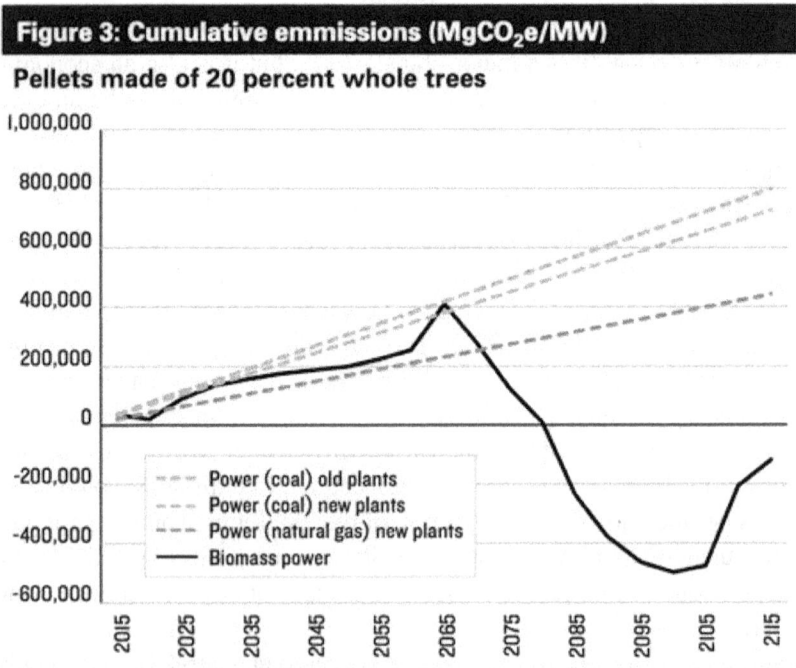

Figure 3: Cumulative emmissions (MgCO$_2$e/MW)

Pellets made of 20 percent whole trees

Legend:
- Power (coal) old plants
- Power (coal) new plants
- Power (natural gas) new plants
- Biomass power

Figure 35: Emissions of coal and natural gas and biomass from wood pellets made 20% of wood from whole trees. Source. NRDC report.

Chapter 3: Fossil Fuel Externalities: Climate

Figure 36: Earthrise photograph taken in December 1968 by Bill Anders, astronaut on
Apollo 8. Source: Image courtesy of NASA.

Scientific Debate Has Its Own Protocol

"Science is true, whether you believe it or not."—
Neil de Grasse Tyson. In 1968, *Apollo 8*, carrying
three astronauts, Frank Borman, Jim Lovell and

Bill Anders, was launched on a mission to circle the moon, break lunar orbit and return home. The astronauts had gone through selection and training for years, had seen colleagues die in accidents and had been at the pointy end of a hugely expensive mélange of all the physics and engineering expertise available to the world's biggest economy. Much of the drive to get to the moon had come from the space race, a super exertion for space dominance between Cold War rivals, the USSR and the USA. As part of the plan, they were going to collect experience about flight parameters to inform later missions. They needed to de-risk the lunar landings by making the journey there and back as easy as possible.

While orbiting the moon for the third time, Borman was carrying out a vehicle roll maneuver while fellow astronaut Bill Anders was taking pictures of the lunar surface. Suddenly, as the spacecraft's window rolled into the correct angle, Earth appeared above the lunar surface and caught their immediate attention.[143] The moment had huge significance. It was the first time that humankind had looked back on our own planet and had the chance to contemplate our place in the universe. I originally thought, like many, that this picture was taken at 90 degrees to the way I show it here, but then realized that it has much more impact in this, its original vertical orientation since that is the way the spacecraft was oriented, it is how the Smithsonian Air and Space Museum present the picture, and it also gives a lot more massive presence to the moon. The recording of the astronauts' conversation is priceless. They raced to load a color film into the Hasselblad camera to be in time for the next revolution of the spacecraft, and luckily the second revolution caught them ready to shoot. They took this picture of our beautiful, potentially perfect planet seated in the blackness of space.

Now starts an overview of the externalities of fossil fuels. I have settled on eight externalities of which arguably the most important is the damage that our coming century already bears from the change in climate precipitated by humans. This chapter outlines a necessarily incomplete list of events in climate science history that tells us when we became aware of the dangers of greenhouse gases (*Hint:* Long before it became politically contentious, so the good guys and the bad guys were equally concerned). Then I look at more recent history when we became aware of critical issues like the discovery that carbon dioxide captures heat (Fourier, 1824). Then I look at the issues that are commonly raised by climate deniers and respond to each of them using the huge array of scientific literature available. Today, it's all at our fingertips if you know where to get it on the internet.

Much of the media commentary on climate displays a sometimes-deliberate lack of understanding of what science is. It's vital to this discussion to accept that science has made some big mistakes of its own in the past and probably will continue to do so. Scientists are human beings, too, and have problems conceiving tests or experiments, coping with ethics and methodologies, collecting, processing and interpreting data, arriving at conclusions that can be politically inflammatory and simply failing to communicate. Science aspires to be a perfect discipline, but it is often a very imperfect art. This part is important: When it's done right, however, science speaks for everybody. This bit is critical: If science says something is true, it's true for everybody. The skeptical nature of science is to doubt any explanations, theories or hypotheses until good evidence shows that it is true, testable, repeatable and understood. Part of the scientific method is that tested conclusions and hypotheses are made available in the scientific literature so that other scientists in the same field will repeat observations and analyses in a practice called "peer review" that confirms or questions existing published positions. This is normal protocol, and to the extent that scientific conclusions are "debated," it is within these confines, where peer review can criticize or applaud a conclusion constructively and occasionally turn what was settled convention on its head because the new explanation is simply more rigorous.

The word *skeptic* is often used to describe those who would deny the science. It's more accurate to apply that term to scientists because they ask questions and start from scratch to wonder how something operates. Climate is a chaotic system. There are many variables, and small changes in one part of the system can result in big changes elsewhere. The requirement to understand chaotic systems eventually results in a growing understanding of how different factors work together, but only after tireless and assiduous research over many years, with mistakes being corrected along the way. This way, science is actually a self-correcting methodology.

Science makes mistakes as a matter of course. An incorrect hypothesis can be later corrected. A critic that denies what science is showing must go through the scientific protocol to have any scientific impact. Often the denialists use the media as a fast method to air their positions, which is not a scientific thing to do, but which may not be the goal: The deniers often have no need for rational explanations anyway because their main goal is simply to confuse those non-experts who lack the analytical tools to be objective.

Science never claims to understand everything but aspires to explain more. It is not dogmatic but ready to be replaced by a better interpretation of the evidence, which has happened frequently. In fact, we understand all too little, especially considering how much time has passed. Looking at the universe, we are full of questions but have few answers. Science at least provides a growing window on truth by dint of the nature of the evidence gathered and the process of analyzing it. Scientists are not afraid to say they don't know the answer because, with study of the issue, they may be able to discern an answer that fits the observed facts and slowly improve on this answer. Saying we don't know is the first and best step. Some scientific endeavors have been the subject of research for a long time, and not all but essentially most everything about those subjects is settled. It doesn't mean it can't be changed if better explanations arise; those are just the reliable, working explanations for now. Explanations get better and better over time, and this polish makes them increasingly powerful as explanations for nature.

This means that if somebody disagrees or believes there is a "debate" on a matter that is already "settled" by science—like the issues of gravity, evolution or anthropogenic warming of the planet—they are simply wrong. This doesn't mean that you can't question the scientific conclusions. Questions are welcome, and if the science is settled, those questions have answers backed by high-quality evidence or may lead to new hypotheses and an improved understanding of nature. The media can be guilty of perpetrating an incorrect judicial metaphor by offering a prosecution and defense or of offering different "opinions" or "sides" in a "debate" under the guise of fairness and balance. An example of how scientific claims and unscientific claims relate to each other is debating whether humans have influenced climate or whether zebras exist. There is a false equivalence. A similar approach denying established climate change is semantically the same as saying that 5 is a bigger number than 20. Opinions have no place in observations of fact, especially when the evidence is good, and the interests of humanity are at stake. We need to know what's happening and what we can do about it. There is no debate.

It should also be realized here that it is only since science became a discipline in the modern world that humanity has been able to empirically determine more about the universe. As mentioned in earlier chapters, the benefits science has brought are huge. For a species that has been around for a quarter of a million years, human beings have been in almost total ignorance of the basic facts of their environment for almost all of that time. Most humans who have lived have died at about age 27 not even knowing the nature of what stars, the sun, planets, galaxies, let alone planet Earth were about. Later I look at the way over the last 200 years in which humanity has only just begun to emerge from this state of ignorance and has started to understand more about our circumstances.

Being able to tell the difference between bad and good science requires a level of objective thinking and an understanding of the context. This capacity for objective thinking is the real value of an education.

A Terrible Choice

> *"You have made me see this matter so plainly and palpably that if Aristotle's text were not contrary to it, stating clearly that the nerves originate in the heart, I should be forced to admit it to be true."* "Dialogue Concerning the Two Chief World Systems," by Galileo Galilei, 1632.[144]

There's a curiously simple, commonsense way to examine the climate challenge. In 2007, Greg Craven, a high school science teacher, placed a video on YouTube called "The Most Terrifying Video You'll Ever See"[145] to illustrate. If you accept the premise of this argument, you can skip this whole chapter without needing to become an expert on the details of whether everything we have discovered about weather and climate is true or not.

There are four possible approaches and outcomes in Craven's analysis, shown in figure 37, and even if you can't understand the complexities of climate science, this method is simple and self-evident. If it's true that climate change exists, and you do nothing, shown in the red square, you deserve everything you get! Deniers are situated squarely in the red box with their heads in the sand, ignoring evidence. In the green box, we actually believe we are in a real climate predicament and consequently do something about it. If it turns out that the climate alarm was false, the effort wasn't needed, and an economic dip is a small price to pay, and in fact there is probably a boost from the economic activity. If we do nothing and there was no climate issue, then the blue box rules, and we can relax because we didn't spend all that money, but we would be very lucky, and we wouldn't have the benefits of all the R&D of taking action. There's a funny joke in which a climate denier says, "What if it's a big hoax and we create a better world all for nothing!" The yellow box is the responsible solution.

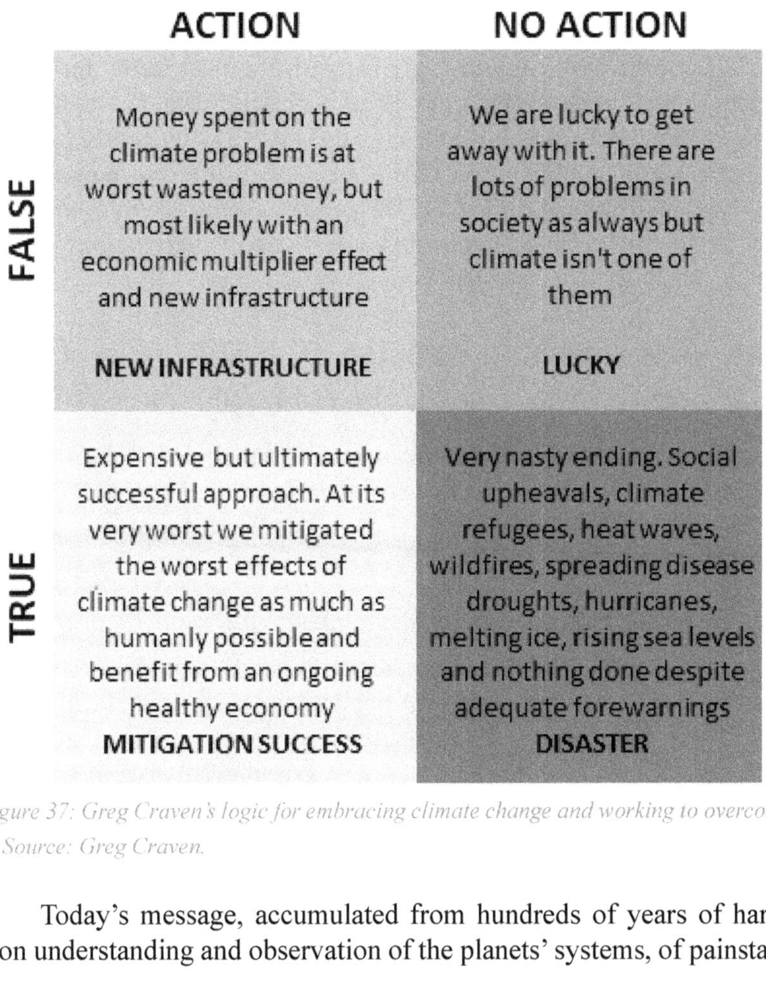

	ACTION	NO ACTION
FALSE	Money spent on the climate problem is at worst wasted money, but most likely with an economic multiplier effect and new infrastructure **NEW INFRASTRUCTURE**	We are lucky to get away with it. There are lots of problems in society as always but climate isn't one of them **LUCKY**
TRUE	Expensive but ultimately successful approach. At its very worst we mitigated the worst effects of climate change as much as humanly possible and benefit from an ongoing healthy economy **MITIGATION SUCCESS**	Very nasty ending. Social upheavals, climate refugees, heat waves, wildfires, spreading disease droughts, hurricanes, melting ice, rising sea levels and nothing done despite adequate forewarnings **DISASTER**

Figure 37: Greg Craven's logic for embracing climate change and working to overcome it. Source: Greg Craven.

Today's message, accumulated from hundreds of years of hard-won understanding and observation of the planets' systems, of painstaking hypotheses and careful logic, points toward a significant ecosystem collapse in a hot, foodless, disease-ridden, waterless, ice-free world, where sea level rise accelerates and blots out much of the coastal and low-lying assets humanity has labored over centuries to develop and where climate migrations threaten to overwhelm those places where living is still possible.

I don't fear that we are in the red box because technology and our intellect can help us, but I see a lot of observers who are very anxious about this, not realizing that we are moving toward a sustainable Earth at quite a clip already. The top two boxes consider a situation that doesn't exist, but of course if there is no climate threat, then there is no issue. This means it is black and white: We either address it and stand a chance or must prepare ourselves for the worst kind of economic dislocation.

This is vitally important. As early as the year 2000, the World Health Organization (WHO) published a report[146] that investigated the impacts of global warming and announced that 150,000 people had died of global warming effects due to additional mortality from malaria, malnutrition, diarrhea and drowning. The people who made this report were not idle job preservation bums but earnest scientists trying to establish the facts and scope of the challenge, so we can fix it.

How We Learned to Understand the Weather

> *"We're the first generation to feel the impact of climate change, and the last generation that can do something about it."* —Jay Inslee,[147] governor of Washington State.

The roots of many renewable energy principles often go back many hundreds of years. We had better keep our heads about us and remind ourselves what we have learned. A hundred years ago, we already knew about greenhouse gases, those constituents of the atmosphere that absorb long-wave reflected solar heat. If those greenhouse gases increase in proportion, as they are, they will cause our climate to change. Climate repercussions are just one of many consequences of our using fossil energy and are a major driver for the adoption of renewable energy that does not release greenhouse gases. Even if "climate" were taken out of the picture, I am looking at seven other expensive externalities that show the downside of CONG. Each is specified in coming chapters, but for now, let's concentrate on the climate.

A Very Brief History of Climate Science

"I'd rather have questions that can't be answered than answers that can't be questioned." —Richard Feynman, U.S. physicist.

Climate science methodologies, models and technologies have improved hugely over time. They have made it possible to test a greater number of difficult hypotheses with increasing accuracy and to cause models of the Earth's climate to become ever more accurate. The importance of the information revealed has further motivated research. Now we are in possession of damning, high-quality evidence about human complicity in the deterioration of the Earth's climate.

1500 to 1800: Gases and Orbital Mechanics

Philippus Theophrastus Aureolus Bombastus von Hohenheim, aka Paracelsus, in the 1500s, discovered that something in air was flammable.[148] Then, in 1671 Robert Boyle noticed a flammable gas was emitted when acids and iron filings were mixed together. In 1766, English scientist Henry Cavendish recognized that hydrogen was a separate gas and named it "phlogiston."[149]

In 1609, Johannes Kepler, an astronomer, published a book on planetary motion, which he called *Astronomia Nova*,[150] that concentrated on the planet Mars. It described laws that governed how planets can have elliptical orbits and not simply circular ones. This was the first sign that increasing distance of an orbiting planet from its sun might alter the amount of energy landing on the planet at various parts of the orbit, resulting in cold and hot weather cycles.

Kepler was initially an assistant to Tycho Brahe, a Danish astronomer who was famous for accurate measurements of stars and planets and who proved that comets were not in the Earth's atmosphere. Brahe was a colorful character. He had a brass nose after losing the tip of his own in a duel over a mathematical formula when 20 years old. He possibly inspired Shakespeare to write *Hamlet*, held about 1% of the entire wealth of Denmark after his father died, and had saved the Danish king from drowning. Suspected of sleeping with the queen, Brahe was forced out of service to the Danish king and went to Prague where he built up the world's largest compendium of astronomical measurements. On the death of his boss, a measure of confusion about inheritance allowed Kepler to assume control of the man's data and continue the research. Brahe himself is supposed to have died from politeness: It was customary etiquette at a banquet not to visit the bathroom until the king or host had done so, and so he remained at his place until his bladder literally burst; he died later from the ensuing complications. Both Brahe and Kepler had matched astronomical data with theory about orbits. This material was to be important later for Sir Isaac Newton in his work on gravity.

In 1761, Mikhail Lomonosov trained the telescope in his small observatory in Saint Petersburg, Russia, onto the planet Venus as it transited the solar disk and noticed the arc of its atmosphere, later to become part of the evidence for the importance of CO_2.

In 1783, Antoine Laurent de Lavoisier became famous for realizing that all chemical reactions happened with a "conservation of mass." He performed many of his experiments inside sealed glass containers to reveal this truth of nature. For example, burning wood resulted in ash, which was lighter, but the released gases, including CO2, still inside the glass container made up the difference. As a French court tax collector, Lavoisier and 27 of his co- defendants were sentenced to death by guillotine during the French Revolution. An attempted appeal against his sentence resulted in a statement by the judge, who said, "The Republic needs neither scientists nor chemists; the course of justice cannot be delayed." Lavoisier was guillotined at the age of 50, and Joseph Louis Lagrange,[151] a fellow scientist, lamented the beheading by saying, "It took them only an instant to cut off this head, and one hundred years might not suffice to reproduce its like." Shortly after killing him, the revolutionaries admitted their mistake in a note to Lavoisier's widow.

Lavoisier had renamed phlogiston hydrogen[152] because it generates water when it burns. Hydrogen is a very basic element, with a single proton and electron; it represents 90% of all matter in the universe. Just over a decade after its discovery, it was used as a means of lifting an observation balloon, "L'Entreprenant,"[153] during some of the many battles of the French Revolution.

Nitrogen was discovered[154] independently by Scottish chemist Daniel Rutherford, Swedish chemist Carl Wilhelm Scheele, and British scientist Henry Cavendish (again) all in the same year 1772. It makes up 78% of the atmosphere and got its name 18 years later in 1790 when French chemist Jean-Antoine Chaptal[155] found it in nitric acid. It should be said that nitrogen is only its English name and the word preferred by Lavoisier, *azote*, is still the name used in many other countries and languages.

Joseph Priestley and Carl Scheele (each working independently) are also credited with the isolation and discovery in 1774 of the most abundant elements on Earth and third most abundant in the universe.[156] A few years later, Antoine Lavoisier named it "oxygen" and showed that it amounts to 20.9% of the atmosphere. Its presence in the water molecule H_2O means it makes up 89% of the ocean. The U.S. Geological Survey tells us that water makes up 60% of our body weight,[157] and so for those of us who are weight conscious, much of it is actually oxygen.

CO_2 was first identified in 1750 by Joseph Black, a Scottish chemist. Then physician Joseph Priestly[158] called CO_2 "fixed" air and supplied the first soda water to navy ships as a potential cure for scurvy. He was even considered as the science officer for Captain Cook's second exploration expedition.

1800 to 1900: Capacity of CO2 to Absorb Heat

In 1824, Frenchman Jean-Baptiste Joseph Fourier[159] asked: If the sun warmed the planet up, why didn't the planet freeze again when all the heat bounced into space? Fourier realized that atmospheric gases intercepted part of the reflected heat, preventing it from escaping to space. He used a model consisting of a pane of glass set on a box, thus giving birth to the term "greenhouse effect." While the sun's energy reaches Earth in a broad spectrum of wavelengths and is absorbed by the material it irradiates, most of the heat reflected is light in the longer wavelengths or in the infrared spectrum.

There was an American version of the great British Cambridge scientist Sir Isaac Newton. Isaac Newton Jr. lived in Connecticut and later in New York State. He had 13 children, one of whom was Eunice, an unusually beautiful young woman who married Mr. Elisha Foote, a judge, inventor and mathematician, in August 1841. They lived in a house that's still there as of 2018 in Seneca Falls, New York. She managed to get trained, even this early, as a scientist, inventor and a women's rights campaigner who carried out experiments on how the warming effect of the sun was enhanced by the presence of CO_2 in the atmosphere. Eunice Newton Foote wrote a paper[160] that was read aloud to an audience in Albany, NY at the 10th meeting of the Proceedings of the American Association for the Advancement of Science in 1856 by Professor Joseph Henry of the Smithsonian Institution, who said of her, "Science is of no country and no sex. The sphere of women embraces not only the beautiful and useful, but also the true". She obtained patents and was published in other work about the electrical excitation of gases.

Just a few years later, John Tyndall[161] thought that transparent gases would let infrared light pass right through them. In 1859, he confirmed that nitrogen and oxygen do indeed act like this, but on testing methane, it turned out to be very opaque to the heat. He discovered CO_2 was similarly opaque. Even though the amount of CO_2 was just a tiny part of total atmospheric gas, he realized that, just as a thin sheet of paper can block more light than a hundred feet of water, CO_2 was absorbing large amounts of heat. He is now memorialized by the Tyndall Centre for Climate Change Research[162], which has offices in Newcastle, Oxford, at the University of East Sussex in Brighton, Southampton, Manchester and in Norwich at the University of East Anglia. Tyndall also published on other science in the same journal as Eunice Foote in 1856 but failed to reference her work in his own 1859 paper.

Thomas Chrowder Chamberlin was a geologist, headed up the glacier division of the U.S. Geological Survey and was also president of the Chicago Academy of Sciences and head of the Geology Department at the University of Chicago. In 1899, he proposed that an increase in the concentration of atmospheric CO_2 could change the climate.[163] He also concluded that the Earth was much older than the current standard then, which came from Lord Kelvin's assessment of 100 million years. He also speculated, with genius, that the sun was extracting energy from the inner workings of the atom well before this discussion was mature. A Mars and a lunar crater both carry his name as well as several University of Chicago buildings.

1900 to 1950: CO2 Levels Interrupt the Next Ice Age

"Once a photograph of the Earth, taken from out-side is available, a new idea, as powerful as any in history will be let loose." —Fred Hoyle, astrono-mer, 1948.

In the early 1900s, Svante Arrhenius,[164] a Danish Nobel Prize–winning physicist, shined infrared radiation through a tube filled with CO2, decreasing the amount of CO2 in the tube. He found that very little of the infrared reached the other side. Only a tiny trace of CO2 made the tube opaque to infrared radiation. Arrhenius reasoned that a doubling of CO2 in the atmosphere would raise temperatures by 9°F (5°C). He was the first to suggest that human-produced CO2 might offset the oncoming cold of a new ice age when it finally arrived. The human population of Earth in 1800 was about 900 million versus 7 billion today.[165] Nobody thought humans could change anything, let alone the climate.

Another common scientific thought of the past was that the balance of nature would always be restored. Dr. James Lovelock proposed quite recently something that echoes that idea. He named his idea the Gaia hypothesis,[166] and it concerns the way in which the Earth's conditions could be maintained within positive feedback loops producing a stable normality of atmospheric composition and temperature. Earth's systems are linked together and maintain an equilibrium or homeostasis like a living organism. The Gaia hypothesis has lost ground today, but it was a strong hypothesis more matched to thinking at the end of the nineteenth century than to the last quarter of the twentieth century. A simple example is where the planet warms up, water vapor increases, causing more clouds to form, which in turn reflect more of the sun's heat back into space, resulting in more cooling. The thought was that nature was stable, benev-olent and harmonious. The calm and stable order of the universe was also part of the religious thinking of many communities. Natural disasters, earthquakes, floods and so forth were transient and always reverted to calm. Today we see an acceleration in the quantity and strength of natural disasters, suggesting that reversion to the mean is no longer the case.

Serbian scientist Milutin Milankovitch[167] had a twin sister. They were the eldest of a family of seven children. Three of his brothers died young of tuberculosis and his father died when he was eight. With deli-cate health, the children were home schooled, and he excelled in mathematics. He graduated from the Vienna University of Technology as an engineer in 1902 with top marks. After completing his doctoral thesis, he started working for an engineering firm in Vienna. He obtained six patents and did extremely well in his profession. When the idea of Kepler's elliptical orbits was considered in terms of solar insolation, it gave color to his thoughts about how an ice age might come about. He published his first paper, called "Contribution to the Mathematical Theory of Climate,"[168] on April 5, 1912, which was quickly followed by others in 1913 and 1914. He accurately established the mathematics of solar insolation and described how ice age frequency matched the natural orbital mechanics discernible at the time. His work made it possible for the first time to analyze the Earth's climate in more detail over long periods as well as the climates observable on other planets. He showed impacts from three physical cycles to do with orbital mechanics: The Earth tilts on its axis and every 41,000 years, nods a little, and then goes back more vertical. It also has a precession cycle that lasts for 22,000 years that affects the exposure of the surface of the Earth to the sun and can be compared to the gyrations of a spinning top.

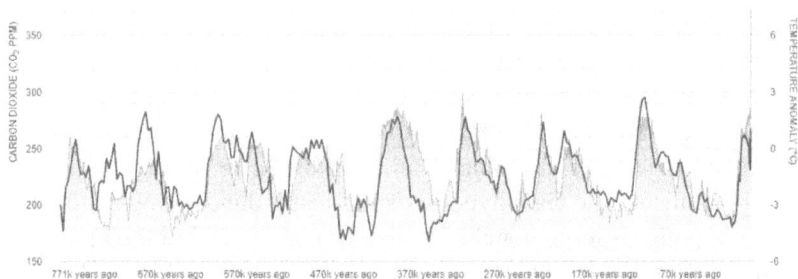

Figure 38: This research was made using ice cores from an ice formation in East Antarctica called the Law Dome. You can see the 100,000-year cycles It was selected for ideal conditions for reproducing atmospheric CO2 data going back 800,000 years, due to its lack of melting and very deep stratification. The gray line represents the temperature record and the blue shows the CO2 levels. Source: The Carbon Dioxide Information Analysis Center (CDIAC).

Finally, it has an orbital cycle lasting 100,000 years that changes the shape of the orbit from circular to elliptical, which influences the amount of heat that reaches Earth. At the extremes of the elliptical phase, the planet is farthest from the sun and therefore cooler. Sometimes the cycles cancel each other out; other times they accentuate the effect. The combination of effects means that the coldest periods are expected to occur every 25,000 years. Figure 38[169] shows the levels of CO2 in the atmosphere from bubbles in ice cores analyzed back almost 800,000 years. It depicts the regularity of the influence of the Milankovitch cycle on Earth's orbital mechanics. An excellent video made by Peter Sinclair covers this point.[170] Something called "global wobbling" happens. This was partly confirmed by research conducted on very thin layers in clay formed by annual water runoff in lakes.[171] Currently, we are about 16,000 years away from the next ice age and would normally be in a gently cooling phase. The current warming acts in contradiction to the orbital cycle, which is further evidence of its anthropogenic forcing.

Milankovitch got married, and on his honeymoon, the First World War broke out. He was imprisoned by the Austro-Hungarian authorities because he was Serbian but was released after only a few days because of his university and engineering connections. The passion and excitement of his research comes out in this quote from his diary about his first day in prison:

> The heavy iron door closed behind me I sat on my bed, looked around the room and started to take in my new social circumstances. ...In my hand luggage which I brought with me were my already printed or only started works on my cosmic problem; there was even some blank paper. I looked over my works, took my faithful ink pen and started to write and calculate.... When after midnight I looked around in the room, I needed some time to realize where I was. The small room seemed to me like an accommodation for one night during my voyage in the Universe.

Eventually he was released on a promise that he would bury his head in books and nothing else. In 1920 in Paris, he finally published a well-received book[172] on the subject of Earth's climate.

An example of the prescient science that observed how the CO_2 resulting from use of fossil fuels could change the Earth is this brief article from the *Rodney and Otamatea Times*, a New Zealand newspaper founded by Australian Charles de La Roche in 1901. Figure 39 is an excerpt from the August 14, 1912, edition, which discusses the potential impact of coal burning on the climate. The 2 billion tons of coal burned annually compares with the over 9 billion tons a year of anthracite and lignite blend that is burned annually today.

After the First World War, Russian, Vladimir Ivanovitch Vernadsky recognized that total human industry, the cumulative effect of all the car, ship, plane, architecture, factory, railway and manufacturing activity in the world, was approaching the scale of geological processes.[173] Geology is constantly making new sedimentary and volcanic rock, jostling the tectonic plates and containing the huge energy coming from within the planet.

On a global scale, geological events release large amounts of CO_2 (far less than humans today). Significant amounts of anthropogenic greenhouse gases were capable, said Vernadsky, of affecting the climate.

The Rodney & Otamatea Times

WAITEMATA & KAIPARA GAZETTE

PRICE—10s per annum in advance
WARKWORTH, WEDNESDAY, AUGUST 14, 1912.
3d per Copy.

Science Notes and News.

COAL CONSUMPTION AFFECTING CLIMATE.

The furnaces of the world are now burning about 2,000,000,000 tons of coal a year. When this is burned, uniting with oxygen, it adds about 7,000,000,000 tons of carbon dioxide to the atmosphere yearly. This tends to make the air a more effective blanket for the earth and to raise its temperature. The effect may be considerable in a few centuries.

Figure 39: The Rodney & Otamatea Times of August 14, 1912, a New Zealand newspaper with a prescient science article. Source: David Brand., New Forests Pty Ltd.

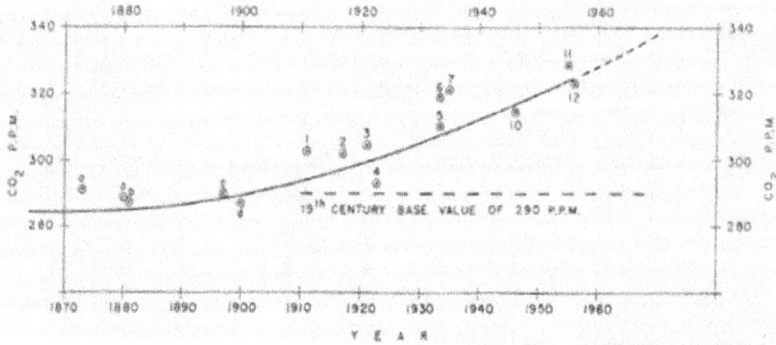

Figure 40: Guy S. Callender's 1958 chart showing increase in CO2 since 1870, describing a base just over 280 parts per million, the level at the time was 320 ppm (today over 400 ppm) and hinting at the increase to come. Source: Thayer Watkins. San Jose State University.

In the 1930s, English engineer Guy Stewart Callender argued that the proportion of CO_2 in the atmosphere was already increasing and that its effect was beginning to be seen. He produced charts showing this disturbing increase, including the one in figure 40.

In 1940, German-American astronomer Rupert Wildt, who was born in Munich and who grew up there during the First World War, became an expert in planetary atmospheres using spectroscopy to show absorption bands that identified atmospheric gases. As a Yale professor, he theorized that the Venusian atmosphere contained large amounts of CO_2 and that this would cause surface temperatures on the planet to be hotter than the boiling point of water. He was later shown to be correct. He has an asteroid and a moon crater named after him.

1950 to 1970: CO2 Identified in Detail as the Climate Culprit

Gilbert Plass, whose career in academia during the Second World War was to attend Harvard, get a physics PhD from Princeton, become a physics instructor at Johns Hopkins University, work for Lockheed Aircraft Corporation and Ford and then accept a professorship at the University of Texas and finally Texas A&M University as a head of the Department of Physics, hosted a classical music radio program called "Collector's Choice" on KAMU-FM for 20 years and was the leading stamp collector in the United States. In 1956, he discovered that water vapor and CO_2 are sensitive to different absorption bands of the electromagnetic spectrum and that heat was captured by CO_2. He was interviewed in May of 1953 by *Time* magazine, and this is a quote from the article:

> *In the hungry fires of industry, modern man burns nearly 2 billion tons of coal and oil each year. Along with the smoke and soot of commerce, his furnaces belch some 6 billion tons of unseen carbon dioxide into the already tainted air. By conservative estimate, the Earth's atmosphere, in the next 127 years, will contain 50% more CO_2. This spreading envelope of gas around the earth, says Johns Hopkins Physicist Gilbert N. Plass, serves as a great greenhouse. Transparent to the radiant heat from the Sun, it blocks the longer wave lengths of heat that bounce back from the earth.*

In later years, Plass[174] made three predictions related to CO2: that a doubling of CO2 would result in warming of 6.5°F (3.6°C), that CO2 levels in the year 2000 would be 30% higher than in 1900 and that the planet would be about 1.8°F (1°C) warmer in that century. In 2007, the IPCC estimated a climate temperature impact of 3.6°F–6.3°F (2.0°C–4.5°C) for a doubling of CO2. The volume of CO2 was up 37% since preindustrial times, and the planet had warmed by 1.8°F (0.7°C) between 1900 and 2000. He concluded that human activity would raise the average global temperature by 1.98°F (1.1°C) per century.[175]

After the Second World War, a by-product of nuclear weapons research was the process of carbon dating.[176] Atmospheric CO2 is usually made up of stable carbon isotopes carbon-12 (^{12}C) and carbon-13 (^{13}C). However, there is a small amount of radioisotope in the form of ^{14}C that decays over time.

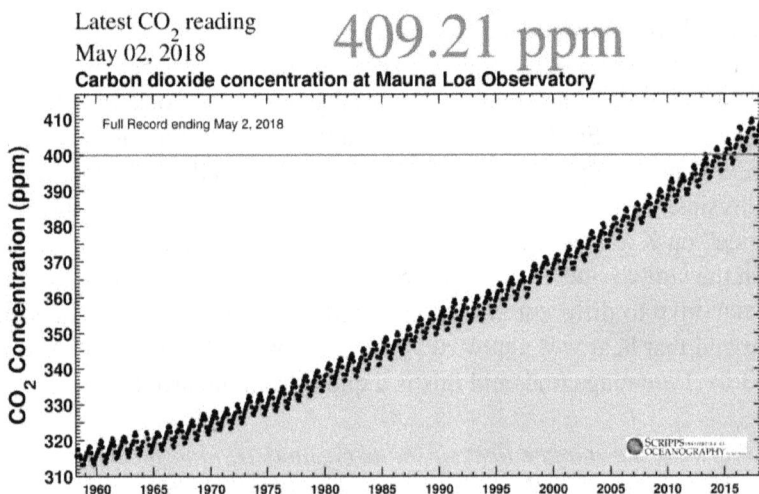

Latest CO_2 reading
May 02, 2018 409.21 ppm
Carbon dioxide concentration at Mauna Loa Observatory

Figure 41: The Keeling Curve, a plot of CO2 parts per million (ppm) in the atmosphere. The increase of CO2 is directly correlated with higher temperatures, higher population, greater GDP, species extinction, and the list goes on. In May 2013, it broached the 400 ppm level for the first time. It peaks in May and bottoms in September of every year. It has gone above 410 in April 2018. Source: Scripps Institute of Oceanography.

Carbon-14 is made when cosmic rays collide with nitrogen atoms in the upper atmosphere, transmuting the nitrogen to ^{14}C. It combines with oxygen to make $^{14}CO_2$ at a constant rate. There's always the same proportion of both carbon isotopes in anything organic. It is ingested by plants and incorporated into organic material. Plants are eaten by animals that incorporate the ^{14}C in exactly the same ratio as the plants. When a plant or animal dies, the ^{14}C continues to decay at a steady rate over millennia. The half-life of ^{14}C is 5,730 years. After this time has passed, half of the ^{14}C has reverted to stable carbon, ^{12}C. The remaining half of the ^{14}C takes another 5,730 years to decay and so on. Looking at how much ^{14}C is in the material effectively tells us when it stopped adding ^{14}C or died. By assessing the rate of decay, we can date organic materials or ocean water (which also incorporates CO_2).

Fossil fuels have long ago lost all the remaining ^{14}C in their molecules, and so when they are burned, the CO_2 released has no trace of this radioactivity. Hans Seuss, a chemist in a research post at the U.S. Geological Survey, discovered that there was extra, nonradioactive carbon, ^{12}C, in the atmosphere that could only have originated from burning fossil fuels. This detail about the relationships of the carbon isotopes was called the Seuss effect. Roger Revelle, a scientist and administrator of the Scripps Institute in California, hired Seuss to continue these studies at the Scripps Institute, where, in a 1958 shared paper, they both deduced that the ocean absorbed most of the extra, stable, anthropogenic carbon.

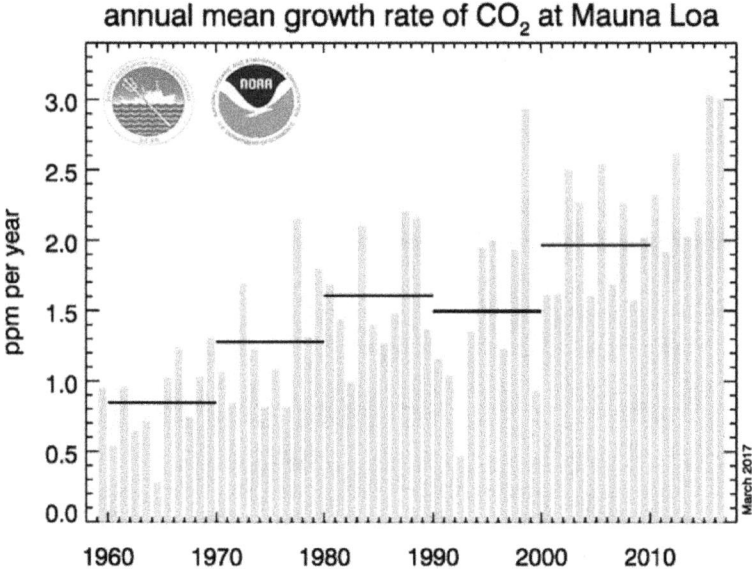

annual mean growth rate of CO_2 at Mauna Loa

Figure 42: Annual record of CO2 increases, showing the gradual acceleration as years go by, recorded in the pristine altitudes of Mauna Loa, Hawaii. This information explains the accelerating curve in the Keeling Curve in Figure 41. Source: NOAA.

Their teamwork led eventually to the formation of the World Meteorological Organization (WMO), an agency of the United Nations eager to improve agricultural, scientific and defense activities. The International Geophysical Year (IGY), formed in 1957, was another example of the worldwide cooperation of thousands of scientists from 67 different countries. Such bodies were natural precursors to the United Nation's Intergovernmental Panel on Climate Change (IPCC).

In July 1956, Revelle also hired David Keeling to work at Scripps to record the amount of CO_2 in the atmosphere. Keeling set up measuring instruments in Hawaii and at the poles and was relentless in his pursuit of accuracy and in isolating every source of error. His results paid off with a carefully created record of annual CO_2 levels that is still used today. His son, Ralph Keeling, is now the director of the CO_2 program at the Scripps Institute of Oceanography in La Jolla, California. The end result of all this research was the conclusion that human beings were responsible for large and significant amounts of CO_2 in the atmosphere, that anthropogenic effects are real.

While CO2 is naturally part of the atmosphere, humanity has contributed a quantity of atmospheric CO2 in a matter of decades that would have taken hundreds of millions of years to be laid down by geological mechanisms. Furthermore, this CO2 is accumulating and causing more of the sun's heat to be retained in the atmosphere. A May 1956 *Time* magazine article titled "One Big Greenhouse"[177] echoed this early awareness of the accumulation of CO2.

Figure 42 shows another hidden characteristic of the data, the inexorable, alarming acceleration of the increase of CO_2 as each year passes. This suggests that the changes resulting from more CO_2 will accelerate as well. Average surface temperature has increased already by 0.75°F (0.42°C) and is accelerating and risks adding another 5°F (2.22°C) by the end of the 21st century.

On January 1, 2018, a *Guardian* newspaper article[178] written by Benjamin Franta, a PhD, physicist and history of science student at Stanford University, examined the activity of Edward Teller, the much-revered father of the H-bomb. In December 1957, Professor Teller agreed to address a meeting of the American Chemical Society. He was a real celebrity at the time, and the right-wing flavor of the industry meeting would have been confused when Teller started to speak about carbon dioxide and the global climate. He even mentioned that business as usual would result in rising ocean levels and melting ice caps.

On its 100th birthday, in November of 1959, the American Petroleum Institute and Columbia University's Graduate School of Business welcomed 300 attendees to the Lowe Library for a symposium they named Energy and Man. Robert Dunlop, the institute's president, addressed the government and industry officials with an optimistic paean to oil, the prime energy focus of the century. Edward Teller was one of the day's five speakers, by now ostracized by the scientific community for his treatment of J. Robert Oppenheimer, the leader of the Manhattan Project, who did not want to develop the hydrogen bomb. Teller had been enthusiastic about building the bomb because he thought the Russians would obtain it anyway and so the United States ought to have it first and also because he recognized that fascism only listened to one argument: force. Teller's talk was about the energy patterns of the future, and his words warned heavily of the growing threat he saw:

> *Ladies and gentlemen, I am to talk to you about energy in the future. I will start by telling you why I believe that the energy resources of the past must be supplemented. First of all, these energy resources will run short as we use more and more of the fossil fuels. But I would [...] like to mention another reason why we probably have to look for additional fuel supplies. And this, strangely, is the question of contaminating the atmosphere. [...] Whenever you burn conventional fuel, you create carbon dioxide. [...] The carbon dioxide is invisible, it is transparent, you can't smell it, it is not dangerous to health, so why should one worry about it? Carbon dioxide has a strange property. It transmits visible light but it absorbs the infrared radiation which is emitted from the earth. Its presence in the atmosphere causes a greenhouse effect. [...] It has been calculated that a temperature rise corresponding to a 10 per cent increase in carbon dioxide will be sufficient to melt the icecap and submerge New York. All the coastal cities would be covered, and since a considerable percentage of the human race lives in coastal regions, I think that this chemical con-*

tamination is more serious than most people tend
to believe.

Thus, was the oil industry warned, on its hundredth anniversary, of its civilization-destroying potential. One can only guess at the response of the audience to this out-of-color warning.

In 1964, the U.S. government asked the National Academy of Sciences if it would be possible to strategically affect the weather and affect agriculture or defense. One conclusion was that it was possible to change the climate without intending to, or "inadvertent modifications of weather and climate,"[179] which cited CO2 as a factor.

Gordon McDonald, a meteorological professor at USCD, realized that there were inadvertent weather changes caused by increasing CO2. He was aware of Keeling's work and the geoengineering of a sort that was already under way.

In 1965, when the CO2 level was 319 parts per million, Revelle and Keeling wrote a paper for the President's Science Advisory Committee titled "Restoring the Quality of Our Environment"[180] that had a 23-page appendix on CO2. It predicted a 25% increase in CO2 by the year 2000. When the year 2000 arrived, the result was 369 ppm, a rise of only 15%. The 25% increase was finally realized in May 2013, when the number passed 400 ppm in the Arctic and then in 2014 in Hawaii, and there was a rise of 27.12% since the start of the Keeling series. On April 12, 2016, it was 408.5 ppm.

U.S. President Harry Truman established the Science Advisory Committee in 1951. After the USSR launched *Sputnik* in 1957, President Dwight Eisenhower moved the committee to the White House and rechristened it the President's Science Advisory Committee. Revelle and Keeling's report was the first ever made to any government about the climate, and Roger Revelle made the presentation to Democratic President Lyndon Baines Johnson in the Oval Office in November 1965. On February 8th, just three weeks after his inauguration, the president, in a special message to Congress, highlighted the dangers of the increase of CO_2 and global warming:

> *Air pollution is no longer confined to isolated*
> *places. This generation has altered the composition*
> *of the atmosphere on a global scale through radio-*

active materials and a steady increase in carbon
dioxide from the burning of fossil fuels.

The level of CO2 in the atmosphere was now known to be a critical factor. In a documentary called *Do the Math, the Movie*, by Bill McKibben, who runs a not-for-profit company called 350.org, McKibben mentions a James Hanson paper[181] that concludes that any excess of CO2 over 350 parts per million is "not compatible with the planet on which civilization developed and to which life on Earth is adapted."

1970 to 1990: Birth of the IPCC

"And it is astonishing but true that our civilization whose imagination has reached the boundaries of the universe, doesn't know to within a factor of 10 how many species the Earth supports. What we do know is that we are losing them on a reckless rate. Between 3 and 50 each day on some estimates. Species which could perhaps be helping us to advance the frontiers of medical science. We should act together to conserve this precious heritage." — Margaret Thatcher in her prescient 1988 speech to the UN.

In 1970, after the 1967 first mission to Venus, an Earth-sized planet, using the Venera 4 space probe, which managed to send back the first 15 seconds of signals from another world, the Russians built a much more rugged Venus probe. It was called Venera 7 and actually landed on the surface of Venus on December 15th of that year. It wasn't an easy mission, however, because the parachute failed to slow the craft sufficiently, resulting in it tipping on its side on landing, weakening the signals it returned to Earth. However, the craft sent back 53 minutes of telemetry, including 20 minutes from the surface, the very first signals received from the surface of another world. The spacecraft detected a hot, high-pressure, noxious atmosphere, where clouds drizzled sulfuric acid rain and lightning strikes were 100 times more frequent than on Earth.

One of the more significant results was the realization that the atmospheric gas was 97% CO_2 and that temperatures there were 887°F (475°C) in a pressure of 90 Earth atmospheres, similar to being 3,000 feet under the ocean. The data confirmed that humans could not live there and that there was no liquid water on the surface; more importantly, it offered an example of atmospheric warming when a greenhouse gas is in high concentration. Although Venus is much closer to the sun—only 67.24 million miles away— the extra heat from this proximity does not explain the total heat at the surface of Venus. CO_2 in a runaway greenhouse effect results in the planet's surface being hotter than that of Mercury, which is half the distance to the sun (35.98 million miles).

On April 28, 1975, *Newsweek* published an article[182] written by Peter Gwynne, a freelance science writer from Massachusetts, that collected evidence that there was a reduction in global temperatures since the Second World War of about half a degree Fahrenheit. The National Oceanic and Atmospheric Administration wrote a report in 1974 that catalyzed attention on this subject. Columbia University scientists were seeing more snow cover in 1971–1972, and NOAA again said they'd seen a diminution in the amount of sunlight reaching the ground over a similar period. Since average temperatures in ice ages were only 7°F lower than in 1975, the idea that Earth had already gone some 16% toward an ice age was meaningful. The link between population, food production and a cooling climate was "chilling." It was known that aerosols and other pollution could augment the cooling effect; however, there was also a consensus among scientists at the time that not much data was available about so many variables—the sun, CO_2, the oceans, aerosols, pollutants, the ice in the Arctic and Antarctic and orbital mechanics. The National Academy of Sciences study admitted as much, saying that "the knowledge of the mechanisms of climatic change was at least as fragmentary as our data."

In today's climate denial fest, this report has been regurgitated and used as evidence to convince the less-informed observer that scientists can't be trusted because one decade they're saying one thing and just a few short decades later....The truth is that just like so many areas of life, climate models, data and its availability from satellites, ice cores and other proxies have improved so that the understanding of the issue is more sophisticated. This one-page report became the most cited article in *Newsweek*'s history! Comedian Dennis Miller even brought the

Newsweek article onto Jay Leno's *The Tonight Show* in 2006. Today's deniers have turned up the volume on this subject so much that on Wednesday, May 21, 2014, Peter Gwynne published another article[183] titled "My 1975 'Cooling World' Story Doesn't Make Today's Climate Scientists Wrong" on *Inside Science*, where he elucidated his frustrations about how the original article had become a focus for denial of warming. Scientists at the time had woefully incomplete data and had got it wrong. Today, the consensus benefits from a hugely improved data set derived from the upgraded technologies and thinking that have gone into the subject in the intervening 40 years.

Doubting politicians, under pressure from interest groups, were inclined to initiate independent scientific due diligence on the science established up to this point in order to determine what was real and what was not. The first World Climate Conference[184] was held by the World Meteorological Organization (WMO) and expressed concern that "continued expansion of man's activities on Earth may cause significant extended regional and even global changes of climate." The WMO called for global cooperation to address the possibilities.

On June 20, 1979, there was an event on the roof of the White House: President Carter celebrated the installation of 32 solar water heaters that would take away much of the fossil fuel energy required to heat water in the building. He said at the opening:

> *In the year 2000 this solar water heater behind me, which is being dedicated today, will still be here supplying cheap, efficient energy.... A generation from now, this solar heater can either be a curiosity, a museum piece, an example of a road not taken or it can be just a small part of one of the greatest and most exciting adventures ever undertaken by the American people.*

The call to the adventure is visionary. That vision would have made America into the foremost sustainable nation on the planet and a big exporter of refined sustainable technology designs. The economy would have found resilience and would not have been choked by rising energy commodity prices. None of the externalities that are up for payment would be weighing over our future. As it turned out, one of the first things President Reagan did when he succeeded Carter was to take down the solar water heaters. Three panels did indeed end up as museum pieces. Two found their way to the Smithsonian Institution and the Carter Library. A third was donated to a museum in China, at the Himin Solar Energy Group Co., in Dezhou, which today builds over 80% of the solar hot water heaters in the world using a pretty much unchanged basic design that originated in the United States.

In 1979, an independent report[185] was filed by the Jason Committee, a distinguished scientific group founded in 1960 with younger scientists not part of the national laboratory establishment. This report had been prepared for the DOE and considered issues as broad as the Arab oil embargo and coal mining. It concluded that at current rates, atmospheric $CO2$ would double by 2035. They built models describing the changing chemistry of the atmosphere and predicted temperature rises of 4.5°F to 5.5°F (2.5°C to 3.0°C) overall, with polar amplification leading to changes of 18°F to 21°F (10°C to 12°C) at the poles, or about four to five times the average increase. Up to this point, data alone had provided damning evidence of *Homo sapiens sapiens*'s complicity in the dramatic and unique changes in greenhouse gases. Now scientists would start building models with computers and running scenarios. Many U.S. administrations and governments elsewhere in the world have requested objective verification about climate change. Every study confirms greenhouse gases are both anthropogenic and lead to dangerous weather changes that are amplified for the world's poor as well as for animal and plant species that cannot adapt quickly enough to suddenly rising temperatures or ocean levels. This science is all peer-reviewed. Skepticism and criticism are desirable and normal. Science is defined by skepticism. By observing phenomena and collecting observations, an explanatory theory is put forward. This theory or hypothesis can be tested by collecting data. If the data supports the theory, then that presents support, not proof or complete understanding, simply support for an explanation about nature. This simple idea about science so often appears to be misunderstood.

The Jason Committee's science was checked and verified by the global science community who were following and observing the phenomena and forming theories. This contrasts with the technique employed by climate deniers, who search for phenomena to fit an existing worldview. Climate issues were understood, and predictions made on the basis of this tested understanding more than 50 years ago, by disciplined, peer-reviewed scientists. Their predictions have been accurate so far in startling ways. Global temperature averages are already +0.9°F (+0.5°C) warmer, and Alaskan temperatures are up 3.78°F (2.1°C), or about four times the average. This is right in line with the 1979 scientific Jason Committee prediction.

The Jason report reached the Carter administration in the White House, where Frank Press, the senior scientific advisor, asked for a second opinion. This opinion was called the Charney report and was written by the National Academy of Scientists (NAS), where MIT atmospheric scientist Jule Charney was its president. It supported the conclusions of the Jason report. The Charney report summarized reports by all the major players in the space: The World Meteorological Organization (WMO), Geophysical Fluid Dynamics Laboratory (GFDL), NASA and the National Oceanic and Atmospheric Administration (NOAA). A press release was issued that underlined that climate change will result from increased use of fossil fuels. It demonstrated a strong consensus about the issues. This was 1979.

In 1985, there was a scientific conference in Austria held by the United Nations Environmental Program (UNEP), the World Meteorological Organization (WMO) and the International Council for Science (ICSU) called the Villach Conference. It concluded that the greenhouse gases already emitted meant that the first half of the 21st century would witness a rise in the global mean temperature greater than any increase in human history. In 1987, the WMO held their 10th conference. They now recognized a need for an objective, international, coordinated scientific assessment of the effects of increasing concentrations of greenhouse gases on the Earth's climate. The WMO Executive Council asked the secretary general of the WMO and the executive director of the UNEP to establish a group that could summarize the annual climate change information.

On June 10–11, 1986, the U.S. Senate Committee on the Environment and Public Works began two days of hearings convened by Senator John H. Chafee (R-RI). The subject was "Ozone Depletion, the Greenhouse Effect, and Climate Change." Chafee himself said that the scientific evidence was telling them that we have a serious problem. Al Gore, at this time a newly minted senator, said, "There is no longer any significant difference of opinion within the scientific community about the fact that the greenhouse effect is real and already occurring." We first hear the names of scientists who are now familiar to us, such as NASA's James Hansen and the future chair of the UN IPCC, Robert

Watson. Their subjects are global warming and sea level rise. The *New York Times* and *Washington Post* ran cover-page articles giving voice to these scientists. Hansen raised the volume two years later by calling for action, and he testified in more Senate hearings that were even more in the public eye, saying, "It is time to stop waffling so much and say that the evidence is pretty strong that the greenhouse effect is here."

The discussion about the ozone hole highlighted a topic that the public could easily digest and showed that there was an impact caused by humans on the atmosphere that did show planetary consequences. The 1987 Montreal Protocol was adopted by the nations of the world. The treaty called for flexible and adaptable approaches to ridding the atmosphere of certain gases that affected the ozone layer. By 2007, UNEP declared that the results were nothing less than spectacular. Unfortunately, the climate discussion was not going to be as amenable as the ozone layer, but there was a remarkable speech by a British lady.

Margaret Thatcher made a speech to the UN in 1988 which contained most of the talking points that we use today when discussing climate and a responsible approach to addressing the challenges using IPCC and British science support. She highlighted that the model of the 1985 Vienna and 1987 Montreal Protocols against the use of ozone-destroying chlorofluorocarbons would be a very useful way to address the environmental challenges that science was just identifying. Her undergraduate degree in chemistry focused on the X-ray crystallography method of examining the structures of important molecules. Her professor, Dorothy Hodgkin, won a Nobel Prize in Chemistry and was friends with Rosalind Franklin, the X-ray crystallographer whose work revealed the double-helix structure of DNA. This was no unimportant science environment. One fine day, Hodgkin and friends piled into two cars and drove to Cambridge to see the research firsthand. Franklin did not receive the Nobel Prize for this, one of the most notable exceptions in Nobel Prize history.

Thatcher was far prouder of being the first qualified scientist to become prime minister of Great Britain than she was to be the first female prime minister. Her science background led her to have such a preternaturally early vision of the importance of the ozone model on the climate and, more importantly, of the impact of man-made climate change. She foresaw the institution of the UN's Intergovernmental Panel on Climate Change (IPCC) and probably caused it to be a permanently established office, able to update and improve our scientific understanding of the climate. She felt that the climate threat was a far greater issue than the arms race.

Solar radiation, orbital mechanics, ocean plankton, volcanic processes, forestry, greenhouse gases, climate victims, pollution, burned fossil fuels, Arctic and Antarctic ice melting, bare mountainsides, the albedo effect, tipping points, energy efficiency, lean-burning engines, use of market-based incentives (reduced tax on lead-free gasoline), water vapor, eutrophication, the use of the space program to help collect data on the planet with a European Space Agency climate program and industrial despoliation were all part of her analysis. She even came up with the Easter Island[186] example that serves even today as a metaphor to the wider world's issues. She knew that there were an extra 3 billion tons of CO2 added to the atmosphere annually, over and above the natural quantities.

She was also aware that the CO_2 lingers in the atmosphere for hundreds of years. At the time, there was concern about the impact of tropical forest destruction. She was very aware that at her birth there were only 2 billion humans on the planet and that when her grandson was born, there were 6 billion. She also saw that nuclear power had significant environmental advantages and that everybody in the world is subject to the effects of these changes. She understood that science could predict and that this was an essential function of knowing our world better and being able to adapt. She recognized that reason was humanity's special gift and that it was only reason that allows us to live with nature and not ruin it or be wiped out by it.

1990–Present: The Fossil Fuel Industry Fights Back

> *"We don't want to believe what we know."*—Yann Arthus- Bertrand, in his TED talk, "Fragile Earth and Man-Made Climate Change," 2015.[187]

While CO_2 is naturally part of the atmosphere, humanity has also contributed a quantity of atmospheric CO_2 in a matter of decades that would have taken hundreds of millions of years to be liberated by geological mechanisms. Furthermore, this CO_2 is accumulating and causing more of the sun's heat to be retained in the atmosphere.

In 1992, the United States was a signatory under George Bush Senior of the Rio de Janeiro UN Framework Convention on Climate Change. This was the first Conference of the Parties, or COP (the Copenhagen convention in 2009 was COP 15). The IPCC has since 1990 produced several reports. In its first report, it pulled together the initial batch of data that described global temperatures over the last thousand years. This first go-around described a significant warming during the medieval period, and the data used originated from tree ring (dendrochronology) information alone.

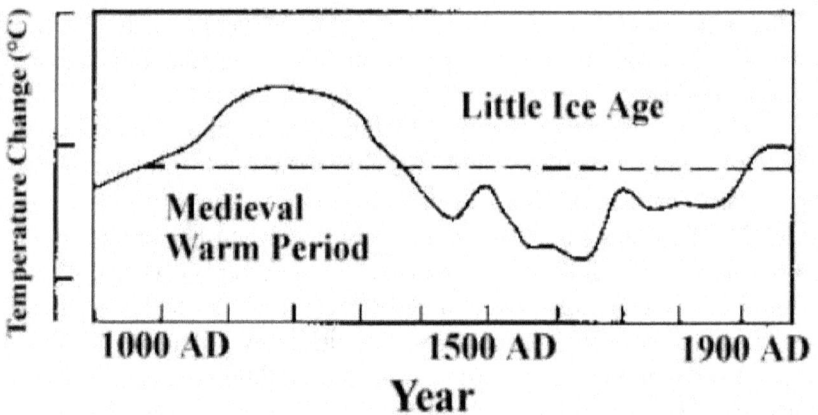

Figure 43: The initial IPCC global temperature chart showing the medieval warm period derived from dendrochronology studies (tree rings) alone. Source: IPCC First report, 1990.

Interest groups for whom global warming is an inconvenient reality, namely, the oil and gas industry, frequently refer to this supposedly warm period as evidence that the Earth has recently been this warm before and that, therefore, there is nothing to worry about. At this time, different opinions about climate began to populate the media and polarize industry.

A research fellow at the Center for Science and International Affairs at the Harvard Kennedy School of Government, Benjamin Franta, observed in 2017 that the oil and gas industry in the shape of the American Petroleum Institute (API), the largest fossil fuel lobbying group in the United States, relied on models generated by economists Paul Bernstein and W. David Montgomery,[188] who argued that expensive measures to address climate would result in job losses and economic damage that would outweigh any environmental benefits. Their research failed to mention the impact of fossil fuel externalities or the fact that the cost of renewable energy was beginning to plummet as the industry organized.

Unfortunately, Bernstein and Montgomery's biased interpretation was massively successful, effectively squashing many early measures—carbon dioxide controls in 1991, Clinton's proposed BTU tax in 1993, the goals of the COP 2 in Geneva in 1996 and of COP 3 in Kyoto in 1997—and impeded the deployment of the Kyoto Protocol in 1998. API lobbying worked repeatedly to overcome U.S. and global climate initiatives and eventually resulted in the decision by George W. Bush to extract the United States from the Kyoto Protocol. The trust of the public had been abused by this easily established bias. Independent scientific analysis was sorely needed.

As far back as the 1860s, President Lincoln needed a source of scientific verification as industrialization and scientific knowledge began its historical upswing. He inaugurated the National Academy of Sciences (NAS) in March 1863 as a sort of Supreme Court of scientific questions that were critical in government decision making. The academy is recognized for performing its duties with the highest level of scientific integrity.

In 2005, early in George Bush's second term, when both houses had a Republican majority, Congress commissioned the NAS to perform an analysis of the Medieval Warm Period and the hockey stick chart. In 2006, the report[189] was released and said the following about the chart controversy:

> *The basic conclusion of Mann et al. was that the late 20th century warmth in the Northern Hemisphere was unprecedented during at least the last 1,000 years. This conclusion has subsequently been supported by an array of evidence that includes both additional large-scale surface temperature reconstructions and pronounced changes in a variety of local proxy indicators, such as melting on ice caps and the retreat of glaciers around the world, which in many cases appear to be unprecedented during at least the last 2,000 years.*

This exonerated Michael Mann and led to improved charts, which were prepared with new proxies and with the help of many new scientific teams from around the world. Increasingly, the more accurate temperature chart became the familiar "hockey stick" chart, a shape, not surprisingly like other charts, presenting rising CO_2, human population and industrial activity. Subsequent investigations, using more evidence acquired under strict scientific conditions, to describe the temperature increase more accurately had the effect of flattening the medieval warming because these results incorporated data from ice cores and as many other indicators as possible.

In 1997, in Kyoto, Japan, the United Nations Framework Convention on Climate Change (UNFCCC) meeting resulted in an international agreement committing 192 signatories to internationally binding emission reduction targets. This accord recognized that developed countries had been generating greenhouse gases (GHGs) for 150 years and so it was weighted toward their preferential participation. The commitment period started in February 2008 and ran until the end of 2012. Rules for implementation were adopted at a Conference of the Parties (COP 7), the meeting of the UNFCCC in Marrakech, Morocco, in 2001.

Recent COP meetings have aimed at agreeing to measures to keep global temperatures from rising by more than 2°C by obtaining a new international climate treaty to follow Kyoto. Hardly a single government on Earth exhibits denial behavior about climate, but they find it difficult to come to any agreement. The buildup to COP 15 became acute, with over 100 heads of state attending. In the 17-year history of the framework talks up to that point, there had never been one attended by so many important players as COP 15. Even George W. Bush was there and was not in denial about the science. Interestingly, George W. Bush was largely responsible for the level of support for renewable energy subsidies in the United States when he signed the Energy Policy Act of 2005. He created the subsidy program that made Texas, not California, the largest wind power state in the United States.

Predictions for a worst-case 10.8°F (6°C) rise in temperature this century have been supported by another report from the Global Carbon Project,[190] a group of 31 leading scientists from seven countries who say warming will make the Earth effectively uninhabitable by 2100. COP 15 attempted to regulate global warming with various tools, including carbon markets, reforestation and renewable energy. The IPCC says that developed countries by 2020 will need to reduce their CO2 output by at least 30% over 1990 levels. Emerging markets can grow their economies but must pledge to move away from a greenhouse gas economy.

As an example of another government wanting to better understand climate change facts, the UK also wanted to have an objective eye on the subject. In July of 2005, Gordon Brown, then the Chancellor of the Exchequer, had the UK Treasury lead a major review on the economics of climate change to obtain a comprehensive understanding of the challenges for the UK and globally. The report was prepared by Treasury economists with the help of independent academics as consultants. It was not peer reviewed, but the subject matter and presentations were, being outlined in the literature in the months prior to the release. It was finally released as a 700-page document on October 30, 2006, by economist Nicholas Stern, who was at the time the chairman of the Grantham Research Institute on Climate Change and the Environment at the London School of Economics (LSE). The institute was named after Jeremy Grantham, who authored the "Unburnable Carbon" report[191] that emphasized that the action needed was not to burn any of the carbon if possible and to leave it in the ground. Stern was also the chairman of the Center for Climate Change Economics and Policy (CCCEP) at Leeds University and LSE.

The report was widely read and discussed. It stated unambiguously that climate change was the largest market failure ever seen. It proposed a series of fixes such as environmental taxes for the economic and social disruptions and concluded that there were significant benefits from strong early action that far outweighed the cost of doing nothing. Without action, Stern said, the costs of climate change will be equivalent to 5% of GDP annually forever, with serious deleterious effects on water, food, health and the environment. Every year that passes this cost rises to 20% of GDP. Stern explained that an 8°F–10°F (5°C–6°C) temperature rise is a real possibility and that 1% of GDP would be needed to invest annually to offset climate change. By June of 2008, Stern had increased his target cost for stabilization of the climate to 2% of GDP to keep it stable at 500 parts per million of CO_2. Strong early action on climate change would reduce the cost of dealing with it.

The climate has been impacted by us in myriad ways. We often think that replacing burned or cut-down forests will help mitigate the effects. In fact, replacing broad-leaved deciduous trees with evergreens since 1750 has only helped the climate warm up. It's to do with the Earth's albedo. Just as white ice melts to dark-blue ocean and sunlight that was previously reflected into space by the ice is now absorbed by the dark water, conifer tree forests are a darker green compared to the lighter leaves of forests of birch or oak trees, and the effect on solar irradiation is similar. A study[192] written by Dr. Kim Naudts and her team, of the Laboratory of Climate Science and Environment in Gif-sur-Yvette, France, has discovered that mass deciduous tree planting can alleviate this situation.

Europe's light-green forests declined by 190,000 square kilometers between 1750 and 1850 alone. Then, ironically, greater use of coal halted the deforestation, and since 1850 conifers replaced indigenous deciduous forests by adding 386,000 square kilometers and covering 10% more land than before the industrial revolution. Conifer trees helped to cause a warming of 0.22°F (0.12°C) over Europe, which is equivalent to 6% of the global warming now blamed on burning fossil fuels. Planting trees to combat climate change is a common response of governments that have committed to making a difference. China has a project to build a "Great Wall" of trees[193] covering over 400 million hectares of land.

Forestry managers wanted fast-growing, more commercially valuable trees such as the Scots pine and the Norwegian spruce. This has been seen as a good thing until a paper reconstructed 250 years of forest management history and found that less carbon was stored by humans and more by nature when it ran wild. Forest litter, deadwood and living soil contain a lot of the carbon that modern forestry techniques remove. The result is that planting forests does indeed cope with emissions, but the darker forest impact on the Earth's albedo also retains the heat that the CO_2 recovered by the forest would have produced.

Not long after his election and 16 years after George W. Bush pulled the United States out of Kyoto, Donald Trump pulled the United States out of the Paris Climate Agreement, becoming one of only 3 countries, along with Syria, obviously in chaos with its civil war ironically caused partially by climate change, and Nicaragua, which thought the measures did not go far enough, out of 196 not to sign. The points made by Trump—that efforts to address global warming would be harmful to the United States and would not reduce the risks or prevent U.S. poverty—echo the same points brought up by Bernstein and Montgomery in the 1990s. They are incorrect, biased industry reports designed to camouflage the multi-trillion-dollar economic opportunity of remaking economies to face the climate threat, with ignorant, nationalist protectionism. Luckily, science and the rest of the world are able to work together.

The famous picture "Earthrise," which later appeared on the cover of *Time* magazine, was able to placidly articulate something wonderful, beautiful, timeless and awe-inspiring about being human and the universe. Nothing so powerfully speaks to our priorities as this realization, only possible in the last few decades of history. We became sensitive to an idea that promised we could all be one human family and live sustainably. Today, when we look closely at our resources, there is enough evidence for us to realize that everybody and every living thing on the planet is in fact able to thrive and that this is both desirable and possible. Technology means that this vision is not just an environmentalist's wishful thinking but a pragmatic and necessary possibility that is arguably less challenging than much else that we have set ourselves to achieve.

Halfway through 2015, CO_2 emissions peaked. CO_2 is still being added to the atmosphere, but global coal consumption is going down. Europe and the United States are leaders in obtaining more GDP per unit of burned energy. The Paris Climate accords are riveting attention onto country emissions limits, and global capacity of renewable energy is climbing steadily.

The Carbon Cycle

> *"I wish it weren't true, but it looks like the world is going to blow through the 400-ppm level without losing a beat."* —Ralph Keeling, Scripps Institute (the world breached the 400-ppm level in May 2014).

Carbon and its molecule carbon dioxide (CO_2) are the mainstays of the greenhouse gases and a cause of much hurt otherwise. The name *carbo* is Latin for "coal," and we know that ancient civilizations used it as a fuel. It has three isotopes, ^{12}C, ^{13}C and ^{14}C (radioactive, with a half-life of 5,730 years), and several allotropes exist, including graphite, diamond and graphene, whose properties are extreme, from hard and transparent to soft black and opaque. It is tetravalent, which means it has four electron bonds available, making it a very convenient structural component with other elements and leading to its being one of the most dramatically varied organic molecules in the universe. All living tissue is sculpted with carbon as a scaffolding, and evolution has elicited its features to generate complex DNA code. This amazing element has resulted in the most complex feedback loops and long-term relationships. One such case is a plant with pollen almost out of reach at the bottom of a long tube in the center of its flower. The plant needs to be fertilized, however, and so there had to be a corollary organism. It turns out that organism is a moth with a tongue just long enough to reach that pollen. Protein, fat, bone, muscle, brain, wood, petals and bacteria—all have carbon as the basis of their structures.

Initially, coal was a wonderful substance. It is energy dense and convenient to transport. It provides millions of winter fires in houses all over the world and powers ships, trains and cities. I remember as a child traveling on steam railway engines and being in King's Cross Station's Potteresque Victorianism, with the heavy chugging and whistles in the great glass-roofed hall. It provided a sense of security, that everything had a good purpose and was leading to something good, like the destination platform at the other end of our journey.

Almost as if she knew it was going to be necessary, nature started to sequester carbon hundreds of millions of years ago. Dead plants and animals in the trillions fell into layers that developed and thickened at the bottom of oceans and lakes and that were then pressurized by the growing overburden. Organic rocks like limestone, coal, slate and chalk are made up entirely of the carbon from these times. The Earth's climate has seen times when CO_2 was a much greater part of the atmosphere and other times when it was hardly there at all with huge consequences. On the one hand, we have evolved to use carbon and it has boosted our civilization, our quality of life and our ability to ponder the universe. On the other hand, we are currently failing to transition away from carbon fast enough, given the vision of the consequences for the planet.

Carbon is exchanged between plants, the atmosphere, the ocean and land. Each acts as a sink for carbon, absorbing and releasing it, depending on conditions. The Earth's climate has been through some rough times in the past. We know that there have been times when the Earth was covered in ice, the "snowball Earth." Geologists have considerable evidence and we also know it's been much hotter. We have been working on how the planet's carbon cycle works for hundreds of years, and the science has become much clearer. Now we know that CO_2 has been increasing and is a collector of heat. Since this story now has the fate of humanity linked to it, its importance has become accentuated.

There is a greater need to discover more about the carbon cycle in detail, and we have learned more. Less romantically, it is especially true in a world where carbon emissions by different countries are closely scrutinized and carbon taxes or credit systems are being implemented. Human impact has been cumulative.

Initially, deforestation was a major contributor; then fossil fuel energy became the main culprit. The CO_2 in the atmosphere represents only half the CO_2 we know to have been emitted. The other half was absorbed by two huge carbon reservoirs, the terrestrial biosphere, or mainly plants, and the oceans, which absorb the carbon in approximately equal proportions. These carbon sinks behave differently depending on the quantity of CO_2 emitted.

Economic data provides estimates for CO_2 emissions from fossil fuel combustion and land use changes. Atmospheric CO_2 is measured directly already. Models are prepared to help quantify the impact of the ocean and atmosphere. Plant respiration rates and productivity, chemical reactions and biological activity in the ocean are also factors. Large natural systems also play a role in influencing these systems.

The Carbon Cycle

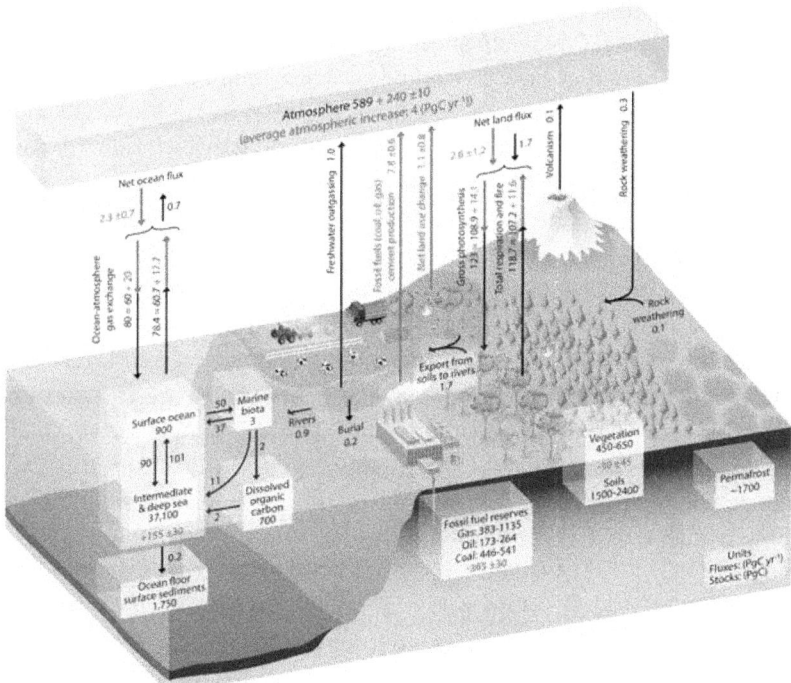

Figure 44: This diagram illustrates the carbon fluxes in red with carbon sinks, fossil fuels, vegetation, ocean and air. It does not show the 60 million petagrams of carbon (PgC) in rocks or methane clathrates. Source: IPCC Fifth Assessment Report, Chapter six.

In the 1970s, there was more CO_2 in the system than indicated by emissions. This was due to unusual conditions arising from the wet, cool La Niña system prevailing in the tropics. This caused more carbon to move to the land sink. Equally, the Pinatubo volcano eruption in 1991 affected global climate as large volumes of volcanic particles were emitted into the upper atmosphere, partially blocking the sun. This cooled the surface of the ocean, allowing it to absorb more CO_2. Also, continued Brazilian and Indonesian forest clearing, the result of political incentives, released carbon into the atmosphere in huge quantities.

We know how much carbon we burn, so we know how much is released into the atmosphere. How much stays in the atmosphere is a good indication of the activity of the ocean and land carbon sinks. Plants, the atmosphere, the ocean and land each act as a sink for carbon. This global carbon system works with water and nitrogen cycles, too, and is made up of several subloops, such as plants and the atmosphere, in which plants will take in CO_2 from the atmosphere, along with water, and then emit oxygen and store some carbon in their tissues. Joseph Priestley, Antoine Lavoisier and Humphry Davy all revealed that the carbon was there and how it worked. Carbon is in all biological compounds and exists in many minerals too. It also exists in the atmosphere as carbon dioxide (CO_2) and methane, or natural gas (CH_4), and is commonly emitted in combustion as the gas carbon monoxide (CO).

Carbon monoxide is flammable and breaks down fast in nature and is generally man-made. Methane has a 25 times greater effect on keeping heat in the atmosphere than does CO_2 but luckily only exists for a short time and in small volumes. The fifth IPCC report goes into the carbon, methane, nitrous oxides and oxygen cycles, and I have borrowed some charts and images from the publication. The carbon cycle existed long before the mass of humanity was large enough to influence it. Dying plants and animals placed large amounts of carbon in the soil and oceans over stupendous periods of time. Plants use sunlight and photosynthesis to add the carbon from CO_2 in the atmosphere to hydrogen and oxygen in H_2O, water, resulting in an extra oxygen atom, which is released into the air. If ever an example was needed to illustrate how living things can change the environment, plants (cyanobacteria) are credited with the original and gradual change of the entire atmosphere to one in which oxygen is now 21% of the gases present. We tend to think of the big carbon sink being fossil fuels formed over millions of years hundreds of millions of years ago, but the even more enormous carbon sink is the oceanic deposition of sedimentary chalk and limestone rocks with the carbonaceous remains (calcium carbonate) of living creatures over huge time.

Movement of the carbon between the different reservoirs, soils, ocean and rock continues all the time, but very slowly. Between 15% and 40% of the CO_2 will stay in the atmosphere for up to 2,000 years. This means that higher levels of CO_2 emissions are going to persist for thousands of years into the future unless we actively remove it. Different parts of the carbon cycle take different lengths of time to cycle. When it gets into the atmosphere, CO_2 gets mixed rapidly within a year. The uptake by plants of carbon dioxide happens relatively quickly, the time of the longest-lived plants, trees, between 1 and 300 years. Soil cycles take from 10 to 500 years. Ocean cycles can persist from 100 to 2,000 years, and sedimentary cycles are more than 10,000 years to millions. Rock formation and erosion take so long they gave birth to the phrase "geological time."

What is more important than this structure, however, is the quantities of carbon in each of the sinks and the changes between them. Figure 44, from the IPCC Fifth Assessment Report, shows both quantities. Diagrams of the carbon cycle are one of the commonest pictures in all the sustainable media and, after going through about 20 of them, produced by schools and universities, the main figures start to stand out. The atmosphere is the critical sink that we are considering currently, although the ocean sink is intimately geared to it. It is that light-blue bar at the top of the diagram. We can see 589 plus 240 petagrams of carbon (PgC), often also referred to as gigatons, describing the carbon in the atmosphere.

The other figure we are familiar with is currently 400 parts per million of CO_2 in the volume of the atmosphere. Over the years since the start of the industrial revolution in about 1860, this ppm number has risen from 280 to 400 parts per million, an increase of 120 PgC, or 42%, which is approximately equal to half of the extra 240 PgC having come from human activity in that time. Now, every single year, another 4 PgC of carbon from human activity are added to this number. The average atmospheric increase of 4 PgC is a net number resulting from the various inputs and outputs of carbon from the various sinks, also visible in figure 45.

If ever an example of the quantity of cellulose available was needed, the 450 PgC of vegetation represents the biomass that is mostly the cellulose molecule. Equally, the fossil carbon ranges from 1,940 to 4,000 PgC and indicates the scale of fossil fuel reserves. Adding this much carbon to the atmosphere is what will happen when we burn all this fuel, which clearly will increase the carbon dioxide parts per million by many hundreds of percent if just 120 PgC represents an increase of over 40%. Temperature, seal level and the concentration of CO_2 have been closely associated for up to 20 million years. Scientists have managed to reach back for data as far as 800,000 years using the trapped bubbles in Antarctic ice,[194] which represent the summer and winter snowfall all that time ago. They pretty much understand how CO_2 has varied over that timescale.

A technique that has extended our knowledge of CO2 levels back even further was developed by Dr. Aradhna Tripati, originally a member of a Cambridge University research team and then at UCLA with her team, Christopher D. Roberts and Robert A. Eagle.[195] They realized that the ratio of the element boron to calcium in the shells of ancient marine protists called foraminifera reflected how much CO_2 was in the atmosphere as far back as 20 million years ago. They double-checked their method by comparing it to the known CO_2 numbers gleaned from the ice cores going back 800,000 years and found that the boron-calcium ratio tells us about atmospheric CO_2 with a good accuracy, with an error of only 14 parts per million. This gives us the confidence to make statements about atmospheric CO_2 going back 20 million years.

CARBON FLUX	PgC
Net Ocean	-2.3
Fossil fuels plus Cement	7.8
Net Land use	1.1
Net Land Photosynthesis/Respiration	-2.6

Figure 45: The net flux of CO2 showing the anthropomorphic contribution, which amounts to a complete destabilization over time. Source: NEF Advisors, LLC.

Strikingly, the results show that the only time in the last 20 million years that there were CO_2 levels at 400 parts per million was between 15 and 20 million years ago, in a period called the Middle Miocene. I have to say it again! The atmospheric concentration of CO_2 has not been this high for 15 million years!

The planet was dramatically different then. Temperatures were 5°F to 10°F (2.8°C–5.6°C) hotter and the sea level was 75 to 120 feet higher.

Tripati demonstrated that the rise in sea levels was due to a rise of an extra 100 parts per million of CO_2 in the atmosphere.

There had been no permanent ice cap in the Arctic before this and it has been there subsequently, for 14 million years before the present, and the ice is now melting again at such a rate that the North Pole is expected to be free of ice on a regular basis in the coming 20 years. Decreases in atmospheric CO_2 coincide with times when the ice sheets expanded, and sea levels dropped, which happened between 14 and 10 million years ago, and again 3.3 to 2.4 million years ago, in the Pliocene. Tripati is pushing the limits of her research to extend back 50 million years, where there were times when no ice sheets existed at either pole and large deserts extended into what are temperate parts of the globe today.

If we go back the full 56 million years to the Paleocene-Eocene Thermal Maximum (PETM), we arrive at a time when global warming was at its greatest since the dinosaurs got wiped out 65 million years ago.

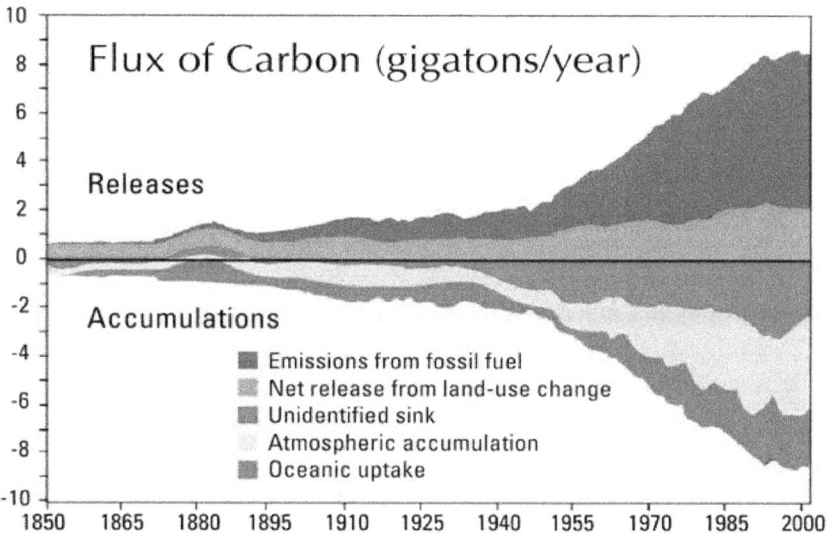

Figure 46: Diagram showing the alarming increase in carbon flux since the mid-1900s and how the ocean and atmospheric sinks have absorbed the excess. Land use changes reflect the slashing and burning of primary forests in South America and Asia. This is very similar to a figure in the IPCC's Fifth Assessment Report and is a stark reminder of the levels of CO2 that are in flux. Source: Woods Hole Oceanic Institute.

If the remaining fossil fuel carbon is released, we will reach 600 to 900 parts per million, which is a level that the Earth may have endured 50 million years ago. Unaided by humans, nature normally takes 5,000 to 20,000 years to add just 100 parts per million of CO_2 to the atmosphere when it is warming. Humans took only 120 years to change the level of CO_2 by 100 parts per million. Tripati's research team used a technique similar to that which Michael Mann used to calibrate his temperature proxies.

Michael Mann had the 150-year temperature record and calibrated the tree rings to this known period prior to extrapolating it to determine previous temperatures. His team then repeated this with the many different proxies they had at their disposal to obtain an ever more accurate picture.

Tripati used the 800,000-year ice core records of CO_2 parts per million and mapped her foraminifera records to this period, then extrapolated them back as far as they could go. The Tripati report explains:

> *The last time carbon dioxide levels were apparently as high as they are today—and were sustained at those levels—global temperatures were 5–10 degrees Fahrenheit [2.6 °C–5.0] higher than they are today, the sea level was approximately 75 to 120 feet higher than today.*

In just the last two centuries, human activities have added to the already dynamic natural carbon cycle by burning and emitting carbon from huge quantities of biomass and fossil fuels such as coal, oil and gas. Burning the fuel bonds the carbon with two oxygen atoms to make CO_2 gas. Adding new CO_2 to the atmosphere has increased the capture of the long-wave radiation normally reflecting off the surface of the Earth. Since ice bubbles taken from different levels of deep Antarctic ice formations contain a tiny sample of an ancient atmosphere, we can see how the atmospheric content of CO_2 has changed over almost a million years. Changes in the CO_2 levels are making big impacts on weather and ocean chemistry. The exchanges of carbon between the different reservoirs happens due to various chemical, physical, geological and biological processes. Humans burning fossil fuels added another process to a system that was otherwise in good equilibrium. As the Google scientists realized, if we transitioned quickly to a renewable energy economy the carbon flows would stop, but we'd still have to deal with the accumulated emissions of the last 150 years.

Photosynthesis is the process by which plants split the water molecule using sunlight and add carbon to make sugars like glucose and thereby also make cellulose, nature's structural polymer. Carbon also dissolves directly into water, whether in rain, waves or lake surfaces. Carbon dioxide in water forms carbonic acid, making the water more acidic. Once the carbon in the ocean has become carbonic acid, it then becomes a carbonate, increasing the capacity of the ocean to take in more CO_2 once more. It is from this that we know that all the carbon fossil fuels are really the product of solar activity in plants over hundreds of millions of years slowly sequestering carbon year after year safely out of the Earth's atmosphere and replacing it with oxygen. Human activities have contributed a lot more atmospheric carbon both by tearing down huge swathes of forest in various parts of the world where farmers prioritize food growing capability and by emitting it from burning fossil fuels and making concrete. These changed uses of land contribute to the large quantities of carbon going back to the air.

Carbon is found in all organisms, alive or dead (500 PgC) as well as in soils (1,500 PgC). Almost all this carbon is organic carbon coming from living things. One-third of soil carbon is calcium carbonate, the product of shells. Living things that absorb carbon from the environment, mostly plants, are autotrophic, whereas animals are heterotrophic because they need to eat something to obtain the carbon. Seasonal biological cycles control the supply of carbon in the terrestrial biosphere. The little teeth-like stitches in the Keeling Curve in figure 41 show these seasonal cycles.

Burning biomass and respiration take CO_2 from the ground back to the atmosphere, and rivers help take it to the ocean. Since 1989 we've known that soil respiration has accelerated and contributed more than 10 times more carbon to the atmosphere than humans have. As the Earth gets warmer, the previously frozen northern tundra melts, triggering a potential tipping point, releasing both methane, with its 25 times greater greenhouse effect than CO_2, and more CO_2. It is likely that this can no longer be reversed.

Nearly 1.6 million square miles, or 3% of the Earth's surface, mostly in northern Canada, Alaska, Europe and Russia, are peat bogs or peatlands.[196] Peat is made of sphagnum and other mosses that grow in wet conditions. The peat acts like a sponge, holding large amounts of water, and often creates an anoxic environment that prevents decomposition. This way, the peat slowly builds up over centuries, as annual growth exceeds decomposition.

Peat is mined, burned as a fuel and even serves as a flavor in some Scotch whiskies, but because it's so permanent, it ends up containing more carbon than all the world's forests and almost as much as is held by the atmosphere. There is a tipping point here too. Global warming creates conditions that can dry out a bog in a drought. When dry, former bogs are prone to fire, and like some coal mines that ignite, peat bogs can remain alight, smoldering for months. In Indonesia, bogs drained for agriculture burned, leaving a haze in the atmosphere that caused local health problems and that could be seen from space. These peat fires were releasing more carbon in a day of burning than all the European Union. The global atmospheric carbon content would be significantly higher if the carbon in all peat bogs were released.

Core samples of peat show black layers where ancient wildfires affected forests and released large amounts of the sequestered carbon back into the air. Typical of the way carbon works, thousands of years of being laid down year by year can be burned off in minutes. The large wildfires in Canada's Fort McMurray in the spring of 2016 were partly burning bogs as well. Sometimes these fires continue to burn for months at a time before finally being extinguished by snow or rain.

Oceans store the second-largest amount of CO_2 after the geological sinks. Organic carbon is dissolved in the ocean's surface, which is the most rapid carbon exchange with the atmosphere via evaporation, wind, wave and rain action. Previously, acid rain has been caused by sulfuric acid, formed by coal emissions, combining with rain and it turns stonework black. This acid contributes to weathering of exposed hard surfaces such as rocks and buildings but generally ends up in the ocean. Subsequent to controls on sulfur in coal and fuels, these effects have ceased.

The deeper part of the ocean holds much more carbon, again, and holds it for long time periods. The ocean currents take thousands of years to complete their circuits around the planet, and in that time, there is mixing from dead animals and other matter sinking from the surface to the depths as well as mixing of the different layers. In the ocean, CO_2 dissolves in the H_2O and becomes H_2CO_3, or carbonic acid. It's an acid because it releases H+ ions that then cause the H_2CO_3 to become CCO_3-, or bicarbonate, which in turn releases another H+ to make carbonate, CO_3--. A similar chemical operation actually echoes this within our own blood systems as a long-ago legacy of the fact that we came once from the ocean. As our cells respire, they give up CO_2 to the bloodstream, where it becomes a bicarbonate until it gets to the lung alveoli, where it is triggered to turn back into CO_2 once again. As the CO_2 in the ocean turns to bicarbonate, it allows the ocean to take up more CO_2, increasing the capacity of the ocean sink but also increasing the ocean's acidity. The more the ocean acidifies, the more that calcium carbonate dissolves.

This, unfortunately, is where our oceans are right now. Calcium carbonate is the substance used by all manner of ocean creatures for shells, and as the ocean becomes more acidic, it becomes more difficult for them to build the strong shell protection they evolved over millions of years, opening potential avenues for predators that did not before exist. Matter falling to the bottom of the ocean becomes carbon-rich sedimentary rock eventually. The ocean's capacity to take up carbon is limited and may be much reduced in future.

A 2007 study[197] from the University of York in the UK, by Dr. Peter Mayhew, looked closely at 520 million years of the fossil record—almost all of it—and specifically extinction events. He noticed that when temperatures climbed by 5°F to 6°F (1.6°C) there were extinction events. His team matched diversity with temperature estimates and found solid data to relate high diversity with low temperatures and vice versa. They also suggested that today's steady temperature climb could trigger a mass extinction where 50% of animal and plant species disappear. The unabated addition of anthropogenic CO_2 results in such temperature rises, and this association with extinction events has caused considerable concern. In the past, it was only volcanic activity that could have released enough CO_2 to make this difference, and this would have occurred over thousands of years. Today we are managing it within a few human generations; the U.S. Geological Survey explains that humans are responsible for over one hundred times greater rate of CO_2 release than volcanoes. Critics jump on any apparent point in history where CO_2 apparently increased with no associated rise in temperatures. As more studies arise, they tend to confirm a solid relationship between rising CO_2 and temperatures.

An extinction event 250 million year ago at the end of the Permian era killed 95% of all extant creatures. A large part of Siberia the size of modern Europe, called the Siberian Traps, was host to a continuous volcanic event that lasted a million years and that released huge quantities of CO_2. The oceans warmed and acidified while oxygen declined. Bad bacteria in the ocean could grow and emit hydrogen cyanide, which killed everything in its path. Today's CO_2 generation is 10 times as fast as this historical event and brings us to another potentially disastrous tipping point beyond which an extinction event is the only outcome.

The slowest and largest sink is the geological part of the cycle. This is where most of the Earth's carbon is stored. Much of it is there since the Earth formed, and more of it comes from the millions of years of uninterrupted biomass accumulation, which is then so gradually transformed into sedimentary layers of rock, mostly limestone.

In the distant past of Earth's geology, there was more CO_2 in the atmosphere and the global climate was warmer, more humid and more tropical than today's, resulting in prolific plant life. A paper[198] published in October of 2017 by George Feulner, from the Potsdam Institute for Climate Impact Research, modeled coal formation in the Paleozoic, a period when carbon formation from atmosphere to plants and then to coal and oil sequestered so much carbon that its warming effect, the thing we are experiencing today, was reversed into a cooling effect. Trees, ferns and CO_2- rich atmosphere led directly to absorption of carbon in Pangea, the single supercontinent. This was, suitably, the Carboniferous age some 300 million years ago, named "the Coal Bearer," by the end of which conditions approached a fully developed ice age. Most of the coal ever mined by humans was formed in this age. The huge amounts of stored CO_2 pushed the Earth into a cooling period, almost the situation of the global snowball. The Earth was once again out of balance with the mechanical Milankovitch cycles, overwhelming any climate forcing from these sources even though it was not caused by humankind. Samples of ancient soils and leaves show CO_2 dropping to just 100 parts per million of air at this time. Feulner's models show that CO_2 penetration below 40 parts per million invariably results in global glaciation…but also that if all this carbon is released, 450 parts per million is a threshold where the heat returns once again.

Climate Denial: A War of Words

"The last thing I would want is for Monbiot, Mann, Flannery, Jones, Hansen and the rest of the Climate rogues' gallery to be granted the mercy of quick release. Publicly humiliated? Yes please. Having all their crappy books remaindered? Definitely. Dragged away from their taxpayer funded troughs

and their cushy sinecures, to be replaced by people
who actually know what they're talking about? For
sure. But hanging? Hell no. Hanging is far too good
for such ineffable toerags." —James Delingpole,
English climate change denialist and journalist.

"It is the mark of an educated mind to be able
to entertain a thought without accepting it."
—Aristotle

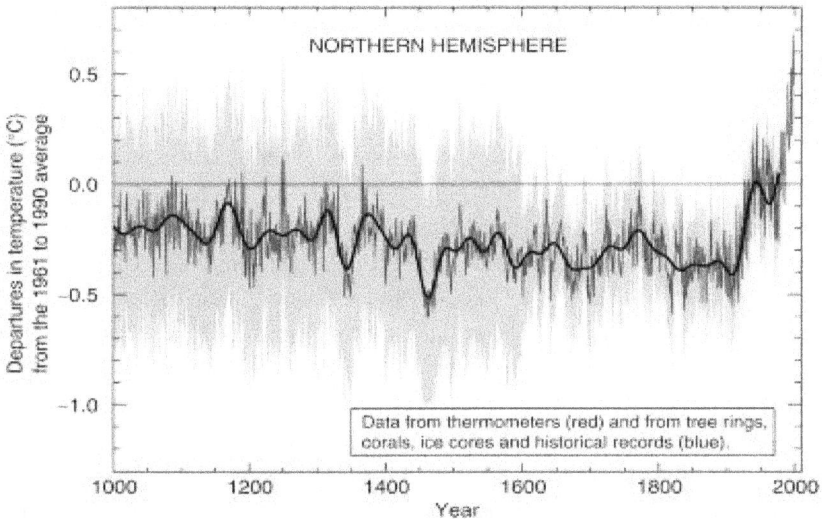

Figure 47: The now famous hockey stick chart showing the increasing global temperature
in the 20th century that was based on dendrochronology and much criticized by those
who wished its implications were not true. Source: IPCC 2001 report.

I want to make a key distinction: Those who deny climate change are
often referred to as climate skeptics.[199] They are not. Skepticism is about
enquiry, questions, observation and the search for truth. It is the willingness to step away from an understanding when that understanding is
flawed.

Nothing in science is relevant without empirical evidence. Feelings,
intuitions and the say-so of someone in authority are among a long list
of logical fallacies that complicate life and that are not appropriate evidence. Scientific evidence is more amenable, based on objective observation and rational analysis and confirmed by other, demanding scientific skeptics in a process called peer review.

Claims are tested and are falsifiable. By its very definition, a scientific claim is based on exhaustive observation and collected data, which is then put to the test by the scientific community. This process results in objective knowledge about things that are true for all human beings. Skepticism is a hard-won moniker, based on objective thinking. Denial is simply denial of something that the denier does not want to be true or that the denier is paid to misinform about. It makes sense that the world is full of things that people don't want to be true, but it's a triumph of the human spirit to elucidate the secrets of nature by proposing how they work and testing those hypotheses.

It is very depressing when denial is done in the face of huge amounts of contrary evidence. Denial has become a resort for those whose understanding of science and the context of history is very weak or who have ulterior motives. This discourse is influenced by politics or economic drivers, and denial positions are often a deliberate policy of some who wish to alter public perceptions.

Holocaust denial, moon landing denial, creationism, Armenian genocide denial and Obama birthers are other examples of distorted opinion that ignores easily accessible evidence. Their effect is amplified when the media provide equal coverage no matter how little actual evidence there may be to support a denial position. The BBC recently invoked editorial power to correct this bias by weighting contrasting views with an analysis of the scientific evidence.[200]

This chart in figure 47 was called the "hockey stick." It was developed by the Intergovernmental Panel on Climate Change (IPCC) Third Assessment Report (TAR) in 2001 and coauthored by a Pennsylvania-based scientist named Michael Mann. It describes the changing temperature over the last thousand years as declining gently and being very stable until the 20th century, when the data shows a sudden increase in temperature, the blade of the hockey stick. When you contemplate its meaning, a feeling of natural alarm might be expected. *We must do something,* you think. *What can be done? Why has this change happened?*

The result was a blizzard of criticism of the methods by which the data was compiled. Michael Mann's team had used proxies for temperatures before 1850, which is when the modern instrumental data was initiated. Prior to this there are many sources of data about annual cycles that can be first correlated to the instrumental data and then used to inform us. Mann and his team analyzed tree rings using a discipline called dendrochronology. Some species of living and dead trees provide a fully anchored chronology dating back as far as 11,000 years, but the team required only 1,000 years of record for this chart. For further corroboration, they also looked at ocean corals and ice cores to determine whether modern temperatures measured by thermometers agreed.

However, it wasn't just criticism that the scientists received. They came under personal attack, suffered death threats and endured the ignominy of defamatory articles from groups like the Competitive Enterprise Institute,[201] a not-for-profit think tank that had employed senior scientists to defend tobacco companies and that were then being funded by such as Exxon Mobil, Philip Morris, the Ford Motor Company, the Koch brothers and many others from this part of the CONG industry. The July 13, 2012, article was titled "The Other Scandal in Unhappy Valley," and it was seized upon by other media groups. The *National Review* was one of these. They called Mann's work fraudulent and bogus. As you might expect of any dignified, sincere person, Mann sued the *National Review* for defamation. The *National Review* complained that free and open debate was being stifled. They had shown their colors and those colors were a thorough ignorance of what basic science was.

They had already condemned Mann's work, via the wrong route, not by the normal method of having it pulled apart in detail by skeptical scientists as scientific peer review but by a simple, misinforming, defaming accusatory article. Not surprisingly, it sat badly with Michael Mann. Science has its own protocol. Peer reviewers analyze results carefully and are frequently able to add value to a hypothesis. A defamatory, opinionated magazine article complaining that free speech is not being respected does not add value.

What really happened was that Mann was a pioneer of methods to discover ancient global annual temperatures and paleoclimates. He and his team, along with many other sober and committed paleoclimatologists, had invested years of time and effort to discover proxies for those ancient temperatures, and they managed to shine an incrementally greater light on a part of our Earth's history. This is what you would expect of a sincere effort to uncover any relationships between the proxies and annual data like temperature or CO_2. This sincere effort was open to criticism through the normal channels of science and, because of its obsession with protocol, "doing it the right way," operated under a considerable consensus.

When their data emerged, it proved unpalatable to the CONG industry, which was funding the various media organs to counter and misinform the public. These groups assumed an indignant position, as if deeply hurt. The truth is that the CONG industries are addressing huge change and opted for opinion beneficial to their strategy. They were, as Shakespeare would have understood, "protesting too much." They were actually not interested in any true analysis of the material or how the temperature data was deduced.

The normal route for dissenters of scientific conclusions is literally to propose a new hypothesis such as the ones by Stephen Hawking on black holes or by the scientists at CERN on Higgs bosons or a million other situations. Experiments are carefully designed to test the new hypotheses, and data is collected. This data is examined by other scientists. Anybody with more to add can then propose to write a paper that explores the data or adds more with a new hypothesis. When a hypothesis is unsupported, it is accepted as wrong and rejected because there was a profound rational examination that cannot relate the data to the phenomenon in question, and new explanations are sought.

When the *National Review* declares that it is defending free speech, it sounds plausible to many, and that is the goal. The *National Review* points out that Mann, accused of being "fraudulent" and "bogus," uses the same terms about his critics, the only manner in which such terms are actually justified. They say he is hypocritical to speak like this. However, this is an asymmetric situation. In the case of Mann and the *National Review*, it is certainly the *National Review* and others who are indeed open to such criticisms. Mann is entitled to throw it back at them since his work is the product of his discipline, where each conclusion is based on exhaustive and reasonable evidence. The scientists are often paid by the public to do careful work in the realms of climate dynamics. The work involves systematic, detailed, annotated record keeping, and any assumptions are laid out in exquisite detail and all of it open to criticism via the appropriate channels.

In 2004, a professor and scientist named Ross McKitrick and a Canadian engineer, Steve McIntyre, attempted to replicate the hockey stick chart, claiming to finally have debunked it, saying that it was full of collation errors, unjustified truncation, extrapolation of source data, obsolete data, geographical mislocations and other serious lapses.[202] They claimed that temperature data attributed to the 15th century was derived from a single tree proxy record, a bristlecone pine, and consequently not valid as a source of information.

In 2007, the National Center for Atmospheric Research carried out an independent review of Michael Mann's methodologies, and although they found slightly different temperatures in the start of the 15th century, they confirmed the main results of the hockey stick and stated that the warming temperatures over the last 115 years were real. Subsequent to this initial study, revisions by the IPCC and many other studies using a variety of proxy sources, as you would expect, have refined the temperature history of the planet.

Other proxies have included stalagmites, borehole temperatures, glacier lengths and lake sediments. In 2008 Mann revised his temperature chart to include all these sources going back 1,700 years resulting in a confirmation that today's temperatures are higher than the highest temperature ranges of this history. We now have other proxies going back almost 50 million years, referenced later in the chapter.

Michael Mann wrote an editorial for the *New York Times,* which he titled "If You See Something, Say Something,"[203] an echo of the explosives caution launched by the new U.S. Department of Homeland Security. His meaning was that, if after careful review, he discovered something that presented a danger to humanity and said nothing, he would be unable to look his daughter in the eye. Thankfully, he and thousands of other scientists have told us what is happening. The wrong reaction to the science is to deny the climate problem without counterevidence.

Attitudes to science in general have caused polarization in the public about some issues, and nothing more than climate change science. It's a complex subject, but scientists themselves are not just under scrutiny but now they are under attack, with personal threats against life and family and climate scientists, in particular. Some 255 scientists wrote a letter[204] to this effect in *Science* magazine in May 2010, since which time the problem has only become worse.

> *We also call for an end to McCarthy-like threats of criminal prosecution against our colleagues based on innuendo and guilt by association, the harassment of scientists by politicians seeking distractions to avoid taking action, and the outright lies being spread about them. Society has two choices: We can ignore the science and hide our heads in the sand and hope we are lucky, or we can act in the public interest to reduce the threat of global climate change quickly and substantively.*

Other climate scientists who know the awful implications of the data have taken to civil disobedience to make the point. The distinguished former director of the NASA Goddard Institute for Space Studies, James Hansen, made a spirited TED talk about the climate predicament. He was arrested in 2009, 2011 and 2013 in association with mountaintop removal coal mining and the Keystone XL oil pipeline, which he described as "game over for the climate." The arrival of activist climate scientists like Dr. Hansen levels the playing field. Anxiety by more moderate scientists about whether this is going too far is an indication of the dignity of this august discipline. The minority of climate scientists who remain unconvinced that climate change is a danger do use scientific methods to prosecute their cases. As we will see, these cases are often rejected for poor science and are often as feeble as the media complaints of the climate deniers. The best way forward is education in objective thinking skills and consistent and patient explanation.

Sometimes the denier community seizes on what looks like a piece of evidence to the layperson's eye. All the time from 1998 to 2012, it looked like global temperatures had peaked in 1998 and plateaued since then. Deniers were claiming this indicated global cooling.

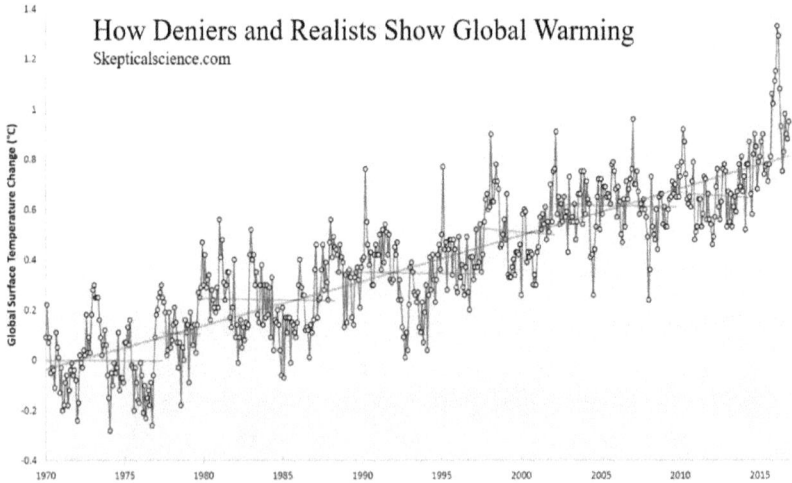

How Deniers and Realists Show Global Warming
Skepticalscience.com

Figure 48: How the denial community shows temperatures as declining (blue lines) and how realists take the wider context into account (red line). This chart is now old and does not show the most recent high temperatures associated with a continuation of the warming, with 2014, 2015 and 2016 being record years. Source: Skepticalscience.com.

Figure 48 shows a chart prepared by Skepticalscience.com[205] illustrating how they demonstrated this. However, when looking at the rest of the data and understanding the full context of the data, it is clear they are "cherry picking," or only using data that describes what they need to show. Laypersons may have little context and may see the incomplete reasoning and gravitate to the deniers' camp.

Temperatures appeared to plateau for a decade after 1998 when the El Niño and La Niña ocean thermal cycles combined to make it a hot year. The year 2013 was hotter despite having no such help, and 2014 was the year with the hottest average temperatures ever recorded to that point, with December 2014 a record high December for the entire planet for the entire period from 1880 to 2014. The year 2016 was the third consecutive year to break a record as the world's hottest.

NOAA is just one of many established science-led organizations that keeps temperature records, which began in 1880. Average global temperatures experienced by the planet can vary if you look at subsets of the data. Temperatures on land and ocean or at different levels of the atmosphere may be very different. In a single year, specific geographical locations may have very different experiences.

The end of 2016 saw world temperatures hit a record high for the third year in a row. It was the hottest year on record. Temperatures over both land and sea in 2016 were on average 58.69°F (14.8°C). This is 2.0°F (1.1°C) above the preindustrial 19th-century average of 56.69°F (13.73°C), according to NOAA. Like the high temperatures of 1998, 2016 was lifted early in the year by heat from the Pacific Ocean also known as the El Niño effect. The COP 21 meeting in Paris in November 2015 saw 194 nations agree to keep global temperatures within 3.6°F (2.0°C) of those levels and aim to keep it to just 2.7°F (1.5°C), meaning that those limits were almost reached in early 2016. Fifteen consecutive monthly temperature records occurred between May 2015 and August 2016, the longest such trend on record.

NASA's data was almost identical, as was the UK Met Office's and the University of East Anglia's, organization that track the same data for the United Nations. As the El Niño effect fades, 2017 became NOAA's third hottest year, although NASA labelled it as the second hottest year. Humankind, led by the United States and China, continues to emit CO_2 from CONG combustion and methane, both of which continue to accumulate in the atmosphere. It's unlikely that the longer-term increase in temperatures will be interrupted unless CO_2 is somehow extracted from the atmosphere.

North America had its second hottest year in 136 years of records in 2017. In the United States alone, 138 deaths and $46 billion in damages can be directly attributed to climate effects. The wildfires in the western USA and Canada were the worst on record. The Great Barrier Reef in Australia suffered a further setback as its coral foundation continued to bleach and die. Phalodi in western India reached a record temperature of 123.8°F (51°C) on May 19, 2016, a national record. Sea ice in the Arctic Ocean and around Antarctica showed record-low levels in mid-January as temperatures were 5.4°F (3°C) above the global mean. The new U.S. administration is almost certain over time to progress from its current position that climate change is a hoax perpetrated by the Chinese. Two hundred nations attending COP 22 in Marrakech declared it was an "urgent duty" to combat climate change. There has not been a cold record set since December 1916. All monthly heat records have been set after 1997. Local records have been set in many different regions such as Tasmania, Western Australia, and northern Europe.

The Hadley Centre, the UK Meteorological Office's institute for climate prediction and research, recently published a paper in the *Bulletin of the American Meteorological Society* titled "Do Global Temperature Trends over the Last Decade Falsify Climate Predictions?" Their conclusion was no.

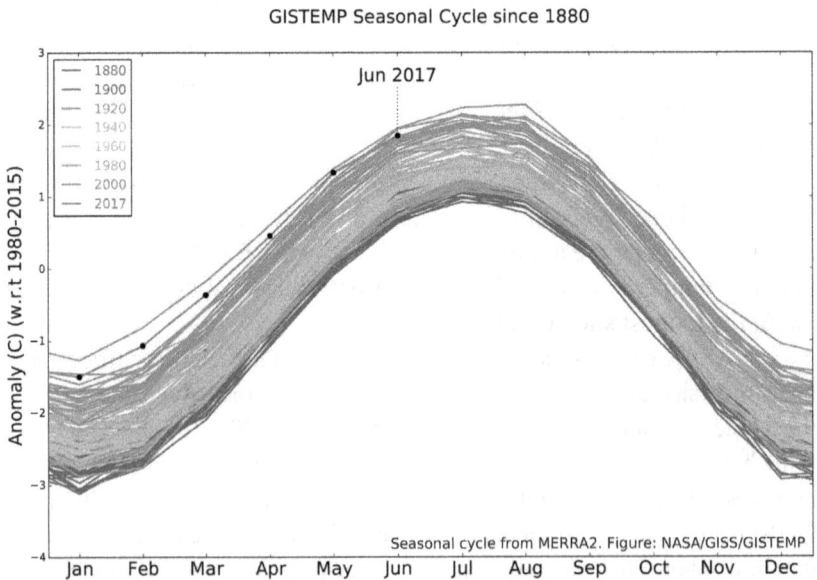

GISTEMP Seasonal Cycle since 1880

Figure 49: NASA data from their GISS satellite data with one of the clearest indications of the warming trend going from blue to red over 137 years. Source: NASA.

"About one in every eight decades has near-zero or negative global temperature trends." Since there had been consistent warming since the 1970s, it was normal to have expected a flat decade. Furthermore, they predict that at least half the years between 2010 and 2019 will be hotter than 1998 as in fact has turned out to be the case with 2010, 2012, 2013, 2014, 2015 and 2016, so far, being in the top 10 hottest years on record. The NOAA released global temperature data up to June 2017, a year that was already the second hottest year in recorded history. This was not expected since El Niño's help in previous years was on the wane, which suggests that these high temperatures no longer require the help of the El Niño phenomenon.

This wonderful chart in figure 49 shows how monthly temperatures differ from the annual figures underlines the rise in temperature of years from 1880 in blue to today in red. The year 2017 is indicated as the second warmest line with monthly black dots. Notably, wildfires in Portugal killed many people and even in the United States caused some airlines to block flights due to high temperatures at the top end of aircraft operational limits. These high temperatures also highlight how little wiggle room remains before we reach the target temperature limit of 2°C.

In figure 48, it's clear that the 1998–2009 decade forms a small part of a wider, longer, significant and accelerating increase in temperatures. It's hard for a denier to suggest that thousands of global PhD scientists have difficulty reading a thermometer.

How worried are you about climate change?		To what extent are humans responsible for climate change?	
Not at all	60%	A little or not at all	53%
Somewhat	21%	Greatly	25%
Very	19%	Somewhat	22%
Total Votes: 7,804		**Total Votes: 7,819**	

Figure 50: A December 2009 spot poll on AOL showing the distribution of responses to questions about climate change.

Dr. Heidi Cullen, chief climate scientist of Climate Central, an independent U.S.-based group of leading scientists and journalists who report on climate facts and their impacts, explained that heat has been taken out of the atmosphere into other systems such as the ocean.

Figure 50 was the result of a spot poll carried out by AOL all the way back in December 2009. In it you can see a distribution of responses on the
subject of whether global warming is anthropogenic or natural. A Pew Research Center study in June 2013, shown in figure 51, found the percentage of residents of each country who said that "Global Climate Change" is a "major threat" to their country.

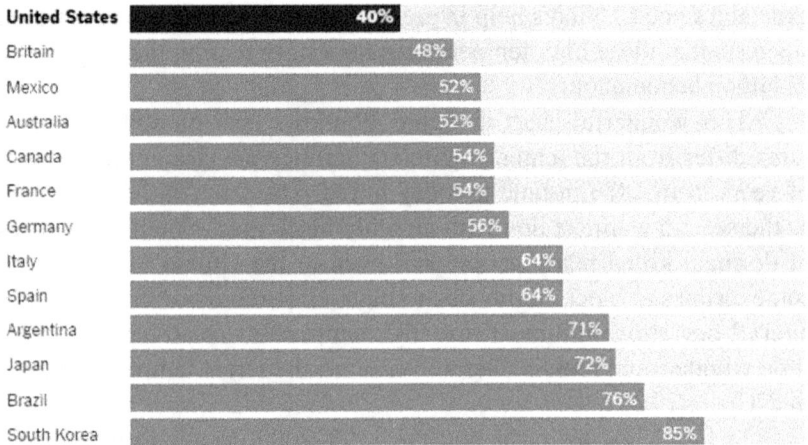

Country	Percentage
United States	40%
Britain	48%
Mexico	52%
Australia	52%
Canada	54%
France	54%
Germany	56%
Italy	64%
Spain	64%
Argentina	71%
Japan	72%
Brazil	76%
South Korea	85%

Figure 51: How the world sees climate change. Percentage of residents of each country who say "global climate change is a major threat" to their country. Source: Pew Research Center, June 2013.

As you can see, the United States is unique in its dismissive attitudes, which can largely be attributed to the politics of fear, where climate change is a very far distant second to jobs, the economy, terrorism, a nuclear North Korea or a nuclear Iran and crime. A Gallup poll conducted in March 2013 confirmed that only one-third of Americans believe that global warming is something major to be addressed.

Today, after Hurricanes Katrina, Irene and Sandy and the horrendous diet of warming and associated hurricanes, floods, tornadoes and wildfires that have impacted the whole planet, there have been remarkable changes in attitudes. People can now accept that anthropogenic climate change is occurring. A Yale University survey on 2014 opinions released in April 2015 showed that 66% of Americans believe the atmosphere is heating up, but only 48% believe that humans are the cause.

There's an improvement in knowing about it and an increase in those who believe it's human caused. This trend is persisting, and the facts of the case have resulted in COP 21, with 194 countries agreeing. The climate is a complex, chaotic issue that plays into the hands of those who are inclined to find fault with the science. It's very difficult for science to be 100% sure of anything in such a system and this is a real gift to the denier. The leak of 13 years of emails from the University of East Anglia's Climate Research Unit (CRU) in November 2009 was a pre-Copenhagen (COP 15) fiesta for the denial community that came to be called "Climategate" in yet another echo of the Nixon-era Watergate scandal. Scientists at the university were coauthors of high-profile scientific papers for the Intergovernmental Panel on Climate Change (IPCC) report describing the impact of climate change, and their integrity had been challenged. The leak was orchestrated in a sophisticated and coordinated attack that yielded 160 megabytes of 1,000 emails and 2,000 documents. Norfolk, UK, police were unable, despite using experts, to trace the perpetrators of the attack.

Two of the emails, taken out of context and edited to select specific words, appeared to be damning, and this inevitably forced a credibility gap to open. In any large quantity of emails, it is highly likely that a naysayer can find something to back a particular cause. The university emails were indeed cherry-picked to substantiate the denial message that climate change was a conspiracy in which scientists manipulated data and suppressed critics. The key scientists at the university who originated those emails were almost immediately the targets of threats, including against their lives and families, but were easily exonerated. Eight committees were formed to investigate the allegations, but none of them found any evidence of fraud or scientific misconduct. Scientists across the world continue to be targeted by anonymous antagonists who threaten scientists and families by saying they know where they lived and had better be careful about how some people might react to their scientific findings.

The investigations did suggest that the scientists do more to repair and regain public confidence, but the science itself was intact. (Remember, independent reviews are frequently carried out and not one has yet found in favor of the denier community, who never have any effective alternative explanations.)

In the United States, the key damning East Anglia emails were read out in Congress by Republican representatives and one of them, Jim Sensenbrenner, a Republican Congressman from Wisconsin, said:

These emails show a pattern of suppression, manip-ulation and secrecy that was inspired by ideology, condescension and profit.

To top this off, Sensenbrenner was once the chairman of the House Science Committee and was on the House Select Committee for Energy Independence and Global Warming from 2007 to 2011. It's hardly pos-sible to conceive a greater misunderstanding unless there is a political agenda. A contemplative and detail-oriented videographer, Peter Sinclair, has produced an excellent series of videos documenting this and many other climate denial issues under the pseudonym "Greenman3160" that are available on YouTube[206]; they put the correct interpretation on both these emails as well as almost every other aspect of climate science.

There is an extreme difference in the way that scientists and deniers deal with "outing" the other side. Peter H. Gleick founded the San Francisco– based Pacific Institute in 1987. He is a prominent hydro-climatologist, or water expert, and member of the National Academy of Science. He was also chairman of the American Geophysical Union's task force on scientific ethics. He has been observing the impacts of climate change on the hydrologic cycle, ocean evaporation, cloud for-mation and movement of water by rain and rivers back to the ocean. Climate change has affected the current deep Californian drought, a major result of the climate impact on the hydrologic cycle. Coal power stations themselves consume almost 50% of the dwindling available fresh water, something that, notwithstanding the issues surrounding coal, is unsustainable in itself. The climate denial machine accuses sci-entists of lying. Scientists are accustomed to handling facts. Allegations such as this poke at the core of their being. This is a major reason why scientists have a deep sense of indignation. The results of such denial frustrate attempts to organize solutions to climate change.

Peter Gleick, like many in the science community, was increas-ingly disturbed by the denialist noise levels and the disproportionate media coverage. On January 27, 2012, Gleick sent an email to the Heartland Institute in Chicago, a major denialist hub and right-wing think tank founded in 1984 that acts as the primary American supporter of climate change denial. It is also listed as a not-for-profit charity. He pretended to be a Heartland Institute board member and asked for institute emails to be forwarded to an additional email address. He succeeded and was copied by the Heartland Institute on their core emails. He then relayed the emails to more than a dozen science activists and bloggers, resulting in a Valentine's Day exposé of the Heartland Institute's financial plans and donor list, which in turn became called "FakeGate," a term coined by denialist James Delingpole,[207] an English author and journalist. The Heartland Institute reacted by saying the emails, which included a revealing strategy document called "Confidential Memo: 2012 Heartland Climate Strategy," were forgeries. This strategy document, however, had the qualities of being just the sort of document a group like the Heartland Institute would produce,[208] and it revealed the extent of the intention to misinform with details about plans to develop teaching materials for American schools.

Unlike the hackers in the University of East Anglia email heist, who are still at large, Peter Gleick did not hide his identity and was forced to step down from being president of the Pacific Institute and many of his other roles. There was a sudden intense discourse within the scientific community about the fairness of a real scientist being punished for dishonesty when he revealed the real intellectual dishonesty of another group when the stakes are deemed to be so high.

Gleick apologized, saying, "My judgment was blinded by my frustration with ongoing efforts, often anonymous, well-funded, and coordinated, to attack climate science and scientists and prevent this debate, and by the lack of transparency of the organizations involved." His transparency and apology speak volumes about the mind-set differences between the two camps. Any potential fixes for combating global warming may not be cheap to deploy. However, it is clear that not deploying any strategy risks far worse economic damage, and economic activity at least will put liquidity into the economy. Many interests are aligned against the use of renewable energy and against the economic impact expected from combating climate change. If truth is the first casualty of war, then we have in fact come to the start of a war. Even the hard-won science that has achieved so much in the last two centuries may be at risk.

Science does have a darker, negative side that's colored with bad things such as concentration camp medicine, forced sterilization, chemical warfare, early and bad medicine, eugenics and many other wrongheaded initiatives. Almost all of this was provoked in the name of science to support a specific political agenda. Even science fiction has its foray into the horrors of science going wrong, starting with Mary Shelley's dark exploration, *Frankenstein*, which is often brought back to life to describe science gone bad. For those who already don't trust government or who don't have much education, there is very little context to tell which side is good and which is bad. This is a fertile ground for those inclined to manipulate information. Earlier, in the technology chapter 1, I make the point that technology is a double-edged sword at once enabling us to have a vision of an improved world but also able to usher in hell on Earth. Our mature choice is to use the positive side of technology. We appear to be far from mature.

Naomi Oreskes, former history of science professor at the University of Southern California at San Diego (USCD) Science Studies Program and now a professor at Harvard, has a video of a presentation called "The American Denial of Global Warming,"[209] available on YouTube, which beautifully illustrates the power of the few to obfuscate the truth for the many. She followed this with *Merchants of Doubt*, which illustrates how sowing doubt about issues is a business like any other. To do it well, you need to have credibility, money and persistence. In this case, there is a tiny coterie of highly qualified scientists whose credibility is high, right along with their greed. Astonishingly, it is the very same team of professional deniers who worked initially to defend Reagan's Strategic Defense Initiative (SDI) and who were later "denial mercenaries" for the tobacco industry and who now are paid to deny global warming.[210]

In this case, it's an entity called the George C. Marshall Institute, which was founded by Professor Robert Jastrow and manned faithfully by Professor Frederick S. Seitz and Dr. William Nierenberg, all individuals with previously impeccable scientific credentials and experience. A gripping documentary,[211] directed by Robert Kenner (*Food Inc.*), with the same name appeared in March 2015 to great acclaim, winning plaudits from the Telluride, New York and Toronto film festivals. The documentary presents "a secretive group of highly charismatic silver-tongued pundits for hire who present themselves in the media as scientific authorities—yet have the contrary aim of spreading maximum confusion about well-studied public threats ranging from toxic chemicals to pharmaceuticals to climate change." In this manner, scientific discourse is split from the public discourse in a divide-and-conquer strategy.

In 1979, the tobacco industry circulated a confidential memo titled "The Smoking and Health Proposal." It had been written in the late 1960s by Brown & Williamson Tobacco to combat growing "anti-tobacco" opinion. In part of the missive, on marketing, it said:

> *Doubt is our product since it is the best means of competing with the "body of fact" that exists in the mind of the general public. It is also a means of establishing a controversy. Unfortunately, we cannot take a position directly opposing the anti-cigarette forces and say that cigarettes are a contributor to good health. No information that we have supports such a claim.*

Given that there is now an industry in misinformation, the media suddenly have a big role to play in presenting opinion on important subjects.

In 1986, Maxwell and Jules Boycoff, of the University of California, Santa Barbara, demonstrated the power of giving a "fair" hearing to wrongheaded viewpoints.[212] They showed that over 50% of all articles in high-end newspapers and magazines on the global warming issue gave equal time to the idea that global warming was not occurring, even though it was difficult to find more than a handful of scientists who were taking this view at the time. From the scientific point of view, climate denial is the equivalent of saying that a cat is larger than an elephant. This metaphor is clear to the layperson, and to the climate scientist the facts discovered about the climate are just as clear. Why must we burden the public with hearing news no different from that that elephants are smaller than cats? What good does it do? Even a casual observation of the degree of polarization in the subject of the climate shows the huge success of the strategy.

The comedian John Oliver created a segment[213] in which he mentions research by James Robert Powell, discussed in the next section, that showed that out of every 100 climate scientists, 97 were part of the mainstream scientific consensus agreeing that anthropocentric climate change was happening, and a solution needed to be found, while only 3 were in the denial camp. Oliver actually invited 96 white-coated "scientists" to accompany Bill Nye the Science Guy, who made 97, and then 3 unfortunate "deniers" to take up their roles in a crowded studio to make a very funny point. Later research illustrates that the deniers are outliers with most often a weak argument and number even fewer than 1%.

The media imperative is to always show the other side of any story. "There are two sides to every story" we hear, and the principle of fairness in the court of public opinion is to air all the information from all sides so that the average listener has a chance to make a fair assessment. When discussing climate, it's common to have a scientist commenting along with someone who may believe the Earth is cooling. However, giving equal, "fair" coverage is tantamount to ignoring and downgrading all the scientific evidence we have accumulated. The important goal is not to give any value to pseudoscience or deniers or the illusion that there is a debate where none exists. The BBC took steps to improve the integrity at least of its science reporting. Steve Jones, a professor of genetics at University College London, was asked to assess the BBC's science content for accuracy and impartiality. His report came out in 2011 and revealed that the BBC gave too much weight to fringe views and recommended "weighting" presentations if arguments have different levels of credibility so that a false equivalence of facts versus opinion could be addressed.[214]

On May 26, 2014, no less prestigious a news organization than the *Wall Street Journal* ran a commentary[215] authored by a director of the Heartland Institute[216] that challenged the view that there was a 97% consensus of scientists that were on the side of climate change being caused by humans. Although there was some level of shock that such an article could come from this source, there was also a sense of a requirement among journalists to air both sides of the story. Denier positions inhibit the possibility that the economy could make a huge boost as it tackles the problem. As I have said before, a significant challenge for humanity is not something to hide from because of fears that the economy will crater if we address it. Instead, it is more likely a stimulus for innovation, creativity and industry that can provide a boost to the economy and all manner of new jobs and industries, not to mention moving us forward to a more sustainable, cleaner, healthier and likely richer basis for humanity from where we can more effectively address other issues such as inequality, poverty, hunger and disease. It is instructive to collect the claims of the deniers and check them out, which I do in the section that follows.

Robert Proctor, while studying the tobacco industry and observing the billions spent obscuring the facts about the health impact of smoking, coined a neologism, a new word, in 1995 with the help of UC Berkeley linguist Iain Boal to describe the deliberate propagation of ignorance about a subject: agnotology. It describes the study or observation and reporting of deliberate efforts to disseminate lies and misinformation to achieve a goal. It works well when a subject matter, such as climate science, is complex and difficult to understand. It is widely used as a marketing tool to create a false vision of the world, by saying there are two sides to a story or that "experts disagree." The tactic is commonly deployed by corporations, politicians, despots and many other highly organized groups such as climate change deniers. A book on the subject, *Their Product Is Doubt*,[217] written by David Michaels, says that although knowledge is widely available, we live in a world of "radical ignorance" caused by faith, tradition or propaganda, and in this context, it's amazing any truth actually gets through. This also complements the work of Naomi Oreskes with her book *Merchants of Doubt*.

The chart showing global average temperatures is no longer useful to any group claiming there has been no warming for a long time. Gavin A. Schmidt, head of NASA's climate science unit, the Goddard Institute for Space Studies, and cofounder of the award-winning climate science blog *RealClimate*, explained that there is no evidence for a pause in the long-term global warming rate and that the current warming is unprecedented over the last 1,000 years. Once, Dr. Schmidt was invited onto denier journalist John Stossel's show to debate with Dr. Roy Spencer, mentioned in the next section as one of the 3% of climate scientists who disagree that humans have caused global warming. Many actual scientists are reluctant to debate deniers because it authenticates discredited positions. The media show us a person, the denier, who effectively says that cats are bigger than elephants. Needless to say, the denier community accused Dr. Schmidt of shying away from Spencer's difficult questions and not wanting to show himself up in the face of the more powerful denier arguments.[218]

Denier Positions

> *"It ain't what you don't know that gets you into trouble. It's what you know for sure that just ain't so."* —Mark Twain.

Sigmund Freud's youngest child, Anna, followed her father into his newly minted profession of psychoanalysis. She spent time studying how human beings practice denial to stay in their comfort zone. After working in the field of child psychoanalysis, she classified denial as a mechanism of the immature mind[219] because it conflicts with the ability to learn from and cope with reality. A person who denies there is global warming may parse the issue into several subjects. They may deny there is warming or that the warming is anthropogenic or that it will have serious consequences. They may say that there is nothing in global warming to worry about and even persuade listeners that there are actually advantages. They won't accept the conclusions based on the conscientious efforts and long-term established hard evidence of the scientists, who are simply trying to discover the truth, while being human.

There are many interested parties for whom combating global warming is likely a poor business plan. There are the CONG companies that, though now admitting their activities and products are sources of greenhouse gases, nonetheless regard the adoption of renewable energy as a market grab on their territory. When push comes to shove, however, we can eventually expect tactics that are every bit as malevolent as those practiced by individuals in the tobacco industry in defense of smoking and highlighted at length by Naomi Oreskes in her book *Merchants of Doubt*. The media are ranged broadly across the opinion bandwidth. Some media are also owned and controlled by groups that are contrary to any attempts to prevent global warming.

A very good friend of mine worked for CNN for 17 years. In 2006, he aspired to be the roving green reporter, but they were nonplussed by his insistence. He was sensitive to the issues and knew what he could believe and what not. As many of us do, he felt a need to act. He approached CNBC as well. They held the same uninformed disregard for the greatest transition humanity has ever faced and didn't see a need to cover it. CNBC, the TV station to Wall Street and the global financial sector, maintains a solid attitude that sustainable technologies are too expensive and can be written off as unimportant or a market bubble, as though the unsustainable condition were something to preserve and protect.

It is definitely true to say that solar and wind are intermittent, which makes it impossible to guarantee power at a particular time. Making solar and wind base load, the high quality we have come to expect from CONG, is today to virtually double their price by adding energy storage systems, although this is improving fast. The point of this book is that the actual cost of internalizing the CONG externalities, which we are now talking about in detail, makes renewable energy look very cheap in comparison, and as the equipment plummets in price, there is no real reason to look to CONG for new energy capacity anymore.

There is an opportunity cost. The vision is of the developed world making the most of its profound capacities for innovation, learning, experience and creativity, all powerful forces that can first transform the economies that adopt this approach into sustainable powerhouses, then implement export businesses in sustainability technology to those economies that are still struggling. Very few observers seem to appreciate that this new sustainable world operates without significant externalities. Obviously, it's not something that happens without a huge struggle, or inevitable, unexpected curveballs that delay and complicate the progress, but it's also a great and worthy vision of a future and provides a reason and justification for our prowess and humanity. Few ideas can so powerfully trap an author into such romantic, ethereal verbal reveries as this one!

Instead of a positive, inspiring vision, we suffer a debilitating polemic about whether established science is true or not. We are following a deeply frustrating path that leads closer to a future of depletion and geopolitical stress locked into a destructive circle. Clearly, the impact of the former vision is only present in an inadequate quorum of minds. Efforts so far to mitigate climate change must rely on direct approaches such as implementing the ever more efficient and economic renewable energy technologies as well as using the stick and establishing carbon taxes, or carbon markets to internalize the externalities of fossil fuel burning. The Intergovernmental Panel on Climate Change (IPCC) says that we need to leave at least two-thirds of the Earth's remaining fossil fuels underground if we are to keep temperature rises to 3.6°F (2°C) above those of the start of the industrial revolution. Unfortunately, this comes at the same time as a palpable sense of pride in America's ability to outproduce Saudi Arabia in oil production and the attractions of cheaper U.S.-based energy production. The political and economic status quo in the United States reveals a class that believes it is unconscionable to address climate change when there are other economic challenges such as jobs, growth and the missing middle class. They should realize that the fastest way to rekindle all those things is to embrace the infrastructure renewal offered by replacing CONG energy with renewable energy. It's the biggest economic opportunity in the world!

There is a long line of people who, for various reasons, find it useful to deny things, including climate change. If you reason carefully with them, they won't listen. When they don't listen to you, you get frustrated. Your efforts to enlighten are not wanted. It is not the rational argument that matters; it's the upset that matters...and the votes. Climate science is a complex enough subject and too few have much conviction about it. It's easy to tilt opinion the wrong way with a bit of artifice and to paint scientists with deception and fraud. A bit of debating sophistry and they can run rings around the doddering, nerdy, complex, incomprehensible scientists and spread dissension in a community that ought to benefit from rational discourse. One telling way of presenting this issue is to ask, "Is it more likely that the scientists have excellent marketing skills and are highly practiced in spinning the story their way or that the large fossil fuel corporations do?"

Humanity's CONG morality has fired a potentially suicidal bullet at our civilization. Despite observations about how long that bullet will take to hit us and details about its speed and direction, all deduced with the accumulated training of centuries of logical, rational thought, the denial community make every effort to protect their interests, sociopathically, without empathy or concern for people or the quality of evidence to the contrary. The best way to correct this is to educate people to look for themselves and install the objective, critical thinking skills needed to make up their own minds. Another good way is to shake off all frustration, knowing the goal is to frustrate you, and to instead sympathize by explaining that scientist have a lack of communication skills to get their complex points across, but that they are correct in the last analysis.

Here are two dozen points offered up by the climate deniers in defense of their position. None of them hold water on closer inspection, but that is to be expected since it is not the conscientious investigator they are hoping to convince:

- **Scientists who speak out against the view that global warming is man-made risk persecution, death threats, loss of funding, personal attacks and damage to their reputations.**

    This is precisely the other way around. Real scientists do a lot of work to establish the facts about climate change. If it's in consensus, they will be attacked by the denier community, which is backed by CONG interests. All the threats are made to individual scientists, such as Michael Mann, for doing their jobs properly. The actual scientists, also, are a world apart from the denial camp because they carry on the "business" of science in peer- reviewed journals and discuss the evidence available that would shed light on a phenomenon. The effort to rebuff this deliberate misinformation takes time away from real science and its benefits to argue a mindless case that is not backed up with a carefully written and peer-reviewed report. If the denial comes as an attack in a magazine or newspaper, it's not science. A notable documentary film[220] that complained that marginal scientists were being penalized was Ben Stein's *Expelled: No Intelligence Allowed*. Ben Stein's scientists were being marginalized for a very good reason: They were expressing the equivalent that the number 10 is smaller than 6.

- **Relax, CO_2 is good for you! (And plants love it too.)**
    In September 2009, the *Washington Post* reported that an oil industry executive, H. Leighton Steward, joined with Corbin J. Robertson Jr., CEO and main shareholder of Natural Resource Partners, a Houston-based owner of coal resources. They wanted to tell the world that higher CO_2 levels are good for us. They invested $1 million in a company called CO2isgreen[221] to lobby and educate the public. Their goal at the time was to defeat the Waxman Markey cap and trade bill then percolating through Congress. Co2isgreen ran an advertisement[222] in New Mexico and Montana that began: "Congress is considering a law that would classify carbon dioxide as pollution. This will cost us jobs." The advertisement finished with a bracing reminder to go to the CO2isgreen website "because we ALL need CO_2." This is a perfect example of deep pockets being used to perpetrate falsehoods against the interests of an easily confused public.

There is no mystery that plants really do thrive with more CO_2. There is also evidence that plants have an upper limit for the gas. We can see the changes in atmospheric CO_2 throughout history because we can look at the tiny bubbles of ancient atmosphere trapped in ice layers, formed millennia ago in falling snow. No matter how much more CO_2 may help plants, though, a drought has a pretty bad effect on them. Many plants such as rice slow down their growth when the ambient temperature increases above 90°F. In a paper mentioned earlier, written in October 2009, Aradhna K. Tripati and her team convincingly linked CO_2 and climate change over the long term. They demonstrated that atmospheric CO_2 has varied between 180 and 280 parts per million for the last 800,000 years. They looked at carbon held in the shells of tiny creatures over the last 20 million years. At about 15 million years ago, when the parts per million of CO_2 were similar to the level today, the ocean was 25 to 40 meters higher and temperatures were approximately 5°F–9°F (3°C–6°C) warmer. The key point they noticed was the rate of change of the parts per million of carbon. Over the natural cycles, the rate of time it took to add 100 ppm of CO_2 to the atmosphere was normally 5,000 to 20,000 years. Humans did it in just 120 years at exactly the time of the start of the industrial revolution and our huge appetite for carbon-generating energy sources.

Since the discovery of CO_2, its heat-absorbing properties have been confirmed. Previous changes of CO_2 took thousands of years, but today we are emptying the millions of years of geological CO_2 accumulation by burning all of it as fossil fuels within two centuries. The Earth has experienced a sort of calm with CO_2 levels for 15 million years, and if conditions that were present the last time levels were this high were to revisit us, humanity will have a very tough time.

- **Human breathing releases CO_2 and causes global warming.**
  One interesting subject deniers often raise is that human breath is a big source of CO_2 and therefore must contribute to global warming. In fact, we breathe in air with 400 ppm (hardly registering as a percentage) of CO_2 and breathe out air with approximately 5% CO_2. The extra carbon in exhaled air comes from the food we eat. The CO_2 is collected from our food via the process of the Krebs cycle and finally emerges as a gas in the blood that is put into the air in the lungs. That means that the carbon we absorb comes from plants and animals every year and therefore is carbon neutral. It's not new or extra carbon from fossil sources.

Seven billion people breathing out 5% CO_2 sounds like a lot, but because it comes from fresh food, it's all carbon neutral.[223] The question of how much CO_2 is represented by human breathing alone, leaving out the CO_2 produced by other flora and fauna around the world, often comes up in the greenhouse gas discussion. The average volume of either male or female human breathing is half a liter and represents what is termed "tidal volume." The actual lung volume is greater than this, but there are on average 900 breaths per hour.[224] As of June 20, 2015, there are 7.323[225] billion people on the planet. A single collective human breath holds 3.66 billion liters of exhaled air and so 5% of this represents 183 million liters of pure CO_2. Ignoring for the moment that this is carbon neutral due to the carbon being derived from food that was grown this year, this still represents only 0.0000000132% of all the existing CO_2 in the atmosphere.

If we go further and collect all the CO_2 breathed out by humans during an entire year, we come up with 336 cubic miles of pure CO_2. That's the 5% of exhaled breath that is CO_2. Even when you collect the entire CO_2 exhaled in an entire year by all humans and compare it to the atmospheric content of CO_2 of over 331,066 cubic miles, it amounts to only 0.105%. Therefore, human CO_2 exhaled gas is an irrelevant part of the carbon cycle. Much more CO_2 is exhaled or transpired by our animal and plant cousins, but this, too, hardly scratches the surface.

- **Arctic ice is growing, or staying the same, not shrinking.**
  Every winter arctic ice cover grows again as the cold returns and freezes the water surface. At the very coldest latitudes, the ice was once a permanent fixture and it got deeper as winter snow piled up on it and over time was crushed into a new layer of ice. Hundreds of years resulted in meters of ice depth. As global warming has affected the poles more than it has the temperate part of the planet, ice there is being melted faster. Today the ice is now much thinner. The geographic extent of arctic ice can be large in winter, but very little of it is as thick now, so in volume terms, the critical issue, it is disappearing.

It is frequently possible now to sail to the North Pole, which is expected to be completely ice free by 2030. Also, now you can sail all the way from the North Atlantic to the Pacific Ocean along the top of

Canada. This was the original fabled Northwest Passage that the ill-fated 1845 Franklin expedition failed to discover.

Instead, the expedition disappeared into the Arctic ice and only recently was it discovered that the entire crew of 24 officers and 110 men had abandoned their ice-bound ships and all hands had died of starvation, lead poisoning, hypothermia, tuberculosis and scurvy.

Franklin was a liberal rear admiral much admired by his men, but with a reward of £20,000 for anybody discovering the expedition, more men were to die trying to win this prize than died in the original expedition. Over 25 expeditions were to try in the four decades after his death.[226]

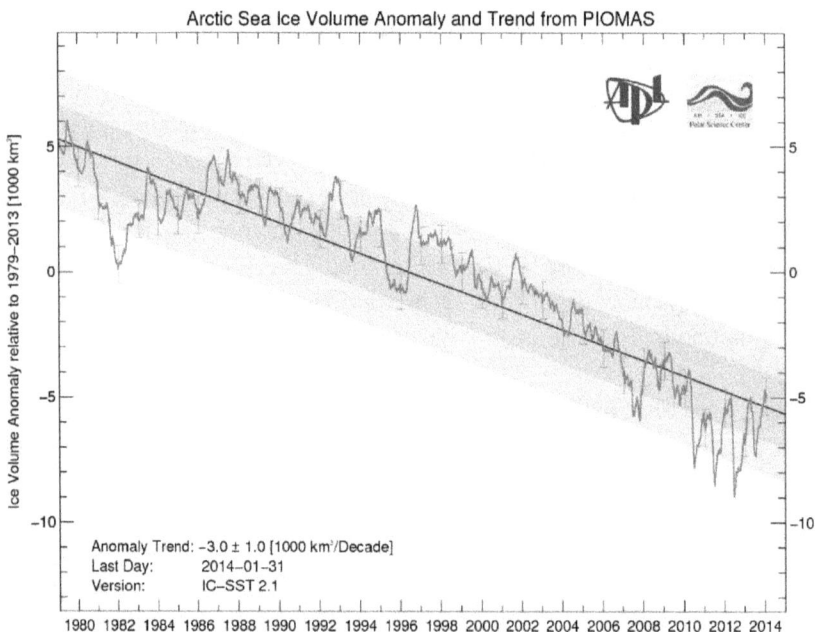

Figure 52: Arctic sea ice volumes and trends showing a definitive decline over the period 1980 to 2014. Source: Polar Science Center, University of Washington.

As ice is replaced by dark water, the Earth's albedo[227] (its reflectiveness) changes the most it can, from reflective white to absorbing black, making the whole issue worse. This is called a tipping point, the idea that if all the ice melts, then the chance for new ice may have to wait thousands of years.

It's a negative feedback loop, where warmer temperatures lead to less ice albedo, which leads to more retention of heat. Arctic ice is like ice in a glass because it is not held above the water on land. When it melts, the level of the ocean does not rise. However, melting Antarctic or Greenland ice is like a new lump of ice in your drink and it pushes the liquid surface over the rim. There is a chance that faster warming at the poles will precipitate a slippage of the Ross Ice Shelf or other land-borne ice formations into the Antarctic Ocean with the result that oceans could rise by 20 to 50 feet.

Iceberg calving, the splitting off of chunks of ice from a glacier or larger ice formation, is now so common that it has become a cruise liner or adventure spectator sport. YouTube is full of exciting ice collapses that in truth are majestic, large-scale heaving events, as though the Earth itself is turning in its grave. There is much footage of the amazing capability of ice to powerfully but gently overturn once it breaks loose, causing mini- tsunamis and enormous blue and white spectacles.

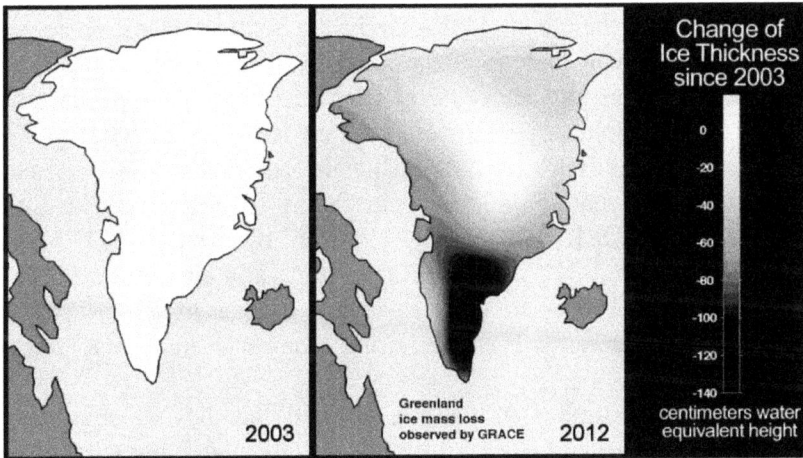

Figure 53: NASA's Gravity Recovery and Climate Experiment (GRACE) reveals the extent of ice loss and gain and shows strikingly the huge amounts of water melting into the ocean from surface ice, glaciers and black snow effects. Source: NASA.

The Greenland ice sheet is melting fast now, and satellites measure its decline. Every drop is increasing the level of the ocean. Happily, even thin ice has a high albedo, but the data tells us the volume is declining. Measuring the volumes of the ice is pretty accurate today, although deniers try to claim that there are inaccuracies. Volume of the Arctic ice is gauged by a satellite that measures its height above the water line. Data on ice volumes from the Pan-Arctic Ice Ocean Modeling and Assimilation System (PIOMAS)[228] model, aircraft and submarine observations and the satellite data all match pretty well.

The NOAA National Snow and Ice Data Center (NSIDC) collects data from weather satellites to measure sea ice extent with passive microwave instruments. There are two measurements. One is sea ice extent and the satellite takes a two-dimensional picture of ocean where at least 15% of the surface is covered in ice. First-year ice is thin, but older ice has accumulated thickness or volume, the second measurement. This form of ice makes up most of the ice at the North Pole. Volume is a more important indicator of the impact of global warming than extent is. This does not inform us about the thickness of the ice. NSIDC data tells us that the decadal trend for Arctic sea ice extent is a retreat of –2.24%.

The May 2015 measurement of 12.65 million square kilometers is the third-smallest May measurement on record after 2005 and 2006. May is the start of the melt season, and the Bering Sea especially so with no ice by the end of the month. Interestingly, the Southern Hemisphere sea ice has been growing and May's figures showed the largest southern sea ice extent on record and that it is increasing by a rate of 2.8% per decade, offsetting the shrinking Northern Hemisphere ice cover. In 2002, NASA launched the Gravity Recovery and Climate Experiment (GRACE) satellites in a joint venture with the Deutsche Forschungsanstalt fur Luft und Raumfahrt (DLR) in Germany. These are two satellites 137 miles apart in polar orbit 310 miles high that map variations in the Earth's gravitational field. They are very sensitive to changes in the surface and other variations in mass, even measuring changes as small as a human hair.

During this time, figure 53 was created and shows a significant change of over 1 meter of ice over much of Greenland, but hardly any of Greenland has gone without some melting during the period. Such observations are confirmed by parallel studies of ice melting such as the InSAR, GPS and photographic records of receding glaciers. A recent documentary called *Chasing Ice* by James Balog showed the glaciers breaking up at the bottom end where they touch the ocean. Glacier calving events have reduced the size of glaciers globally. Greenland is currently experiencing loss of ice of 230 gigatons annually. All the water in Chesapeake Bay is about 80 gigatons. If all of Greenland's ice were to melt, there would be the equivalent of 7 meters of extra global ocean depth.

For longer than five years, NASA has been examining the Arctic and Antarctic ice sheets with its six-year Operation IceBridge,[229] the largest polar ice survey ever undertaken. They have a C-130 Hercules and a DC-8 aircraft fitted with radar and laser equipment that produces incredible images of what is under the ice. In the Arctic that's mainly Greenland, and of course in the Antarctic it has its work cut out mapping the entire continent, looking for changes. We have to be aware that we are leading inexorably toward this tipping point that has so many consequences. A TED talk given by Lewis Gordon Pugh[230] focuses on a 1-kilometer swim that he made across the North Pole to help attract attention to the melting ice. The video clearly shows a lot of open water that he used to complete the swim in water that was minus –1.5°. It froze his hands so badly it took months for the feeling to return.

If you add this tipping point to others, such as the release of large amounts of methane from melting tundra, then it exacerbates the issue further. Interestingly and counterintuitively, Antarctic ice is growing and in June of 2014 recorded its largest extent ever with ice covering over 15 million square kilometers beating the previous 2010 record. The reason is not that there is colder down there. It's actually due to the fact that warmer air contains more moisture, which leads to more snow falling on the ocean which makes the ocean less salty, which means it can freeze at higher temperatures. Also, important to know about Antarctic ice is that the ocean ice holds back the land ice, forming dams that effectively slow down the flow of land ice into the ocean. Melt ponds like those in Greenland have been covering the Antarctic's Larsen B ice shelf, which broke off from the main body of ice in just weeks in 2002 when it was the size of Delaware. Now glaciologist Erin Pettit, professor of geoscience at the University of Alaska in Fairbanks, has shown that the Scar Inlet ice shelf, the size of Rhode Island and 1,000 feet thick fragmented in the Northern Spring, the end of the Antarctic summer of 2017. The location is well named; Cape Disappointment.

- **The science of climatology is new and untested.**

The science of climate change is hundreds of years old, as we have seen. We add new information to our knowledge base constantly. Just as the design of a car has changed in the last century, the knowledge of our climate has also advanced, improved and allowed us finally to begin to predict events with more accuracy. Models that were awkward and incorrect in the late 20th century have now become advanced and accurate. New technology in satellites and instruments are testing new hypotheses and accelerating our learning. The data has been retrieved from 20 million years and, like all scientific evidence, stands in its own right and combats uncertainty with observation and peer-reviewed theory. Deniers, notably, rarely put forward effective alternative explanations or evidence. The science, however, is solid, undeniable, transparent and available.

- **Global temperatures are declining.**

Occasionally they do, but as we have seen, the whole picture is very alarming, and temperatures are rising over time. Figure 48 shows how a selected, shorter period, showing a decline, offers the cherry-picker deniers the possibility of selecting the part of the global warming chart that shows a decline in temperature and offering that as the total story. A look at the wider picture by checking out longer timelines makes this effort very silly and pointless as the clear trends become evident. Sixteen of the 17 years since 2000 have been the hottest years on record for the entire planet. A *New York Times* Climate Desk analysis showed that global warming may make it impossible for 9 of the 21 cities that have hosted the Winter Olympics to do so today.[231]

- **It was hotter in medieval times, but polar bears survived then.**

Now we know that the last time temperatures got this hot was about 4 million years ago at a different stage of the evolution of polar bears. They survived then because, later research revealed, it did not get hot enough after all to melt the ice. It was originally the University of Washington that mentioned the plight of the polar bears in a report that launched the unfortunate animal into the global consciousness. A 1990 IPCC report showed a chart (figure 43, page 191) that indicated a medieval warming period. The data for this chart were based on dendrochronology and historical reports alone. Dendrochronology is the counting and examination of tree rings, which can take us back 1,000 years or more. The initial tree ring data was subsequently improved with data from many other sources, such as ice cores going back 800,000 years and geological evidence. It is not today's understanding that there was a Medieval Warm Period and in fact the last time global temperatures were this warm is indicated by the shells of minute sea creatures to be over 15 million years ago.

• **Solar cycles explain global warming.**

Observations of the sun have shown that every 11 years there are solar cycles. The way we know this is by counting sunspots that appear to increase until there is a phase of maximum sunspots and maximum energy output, which is called the solar maximum. These sunspots were first seen and recorded by Galileo. All the sunspot evidence points the act that we should be in a global cooling phase.

We are currently in a very weak solar maximum, the weakest since 1913, when sunspot counts are minimal along with the lower energy readings being picked up by NASA's SORCE satellite. This would normally be a reason for cooling because less energy is coming in from the Sun in the last two or three cycles. Add to this that the Milankovitch orbital cycles which explain ice age cycles which suggest that we ought to be heading for a new ice age and cooling gently, and both systems are working against the observed warming. Human (anthropogenic) warming comes from our fossil fuel emissions of the known, heat catching gas, CO_2, along with methane and nitrous oxides, into our atmosphere over the last 200 years. We are currently in a very week solar phase with very few sunspots.

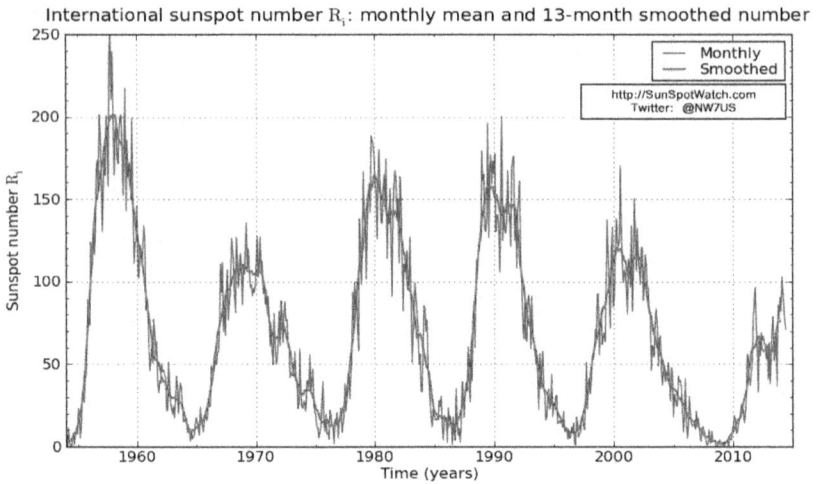

International sunspot number R_i: monthly mean and 13-month smoothed number

SILSO graphics (http://sidc.be) Royal Observatory of Belgium 01/07/2014

Figure 54: This chart shows the sunspot cycle since 1955. We are currently in the top of the sixth such cycle, and it is the lowest peak in this period. High sunspot activity is correlated with higher energy levels, so rising atmospheric temperatures do not correlate with solar activity. Source: The Royal Observatory of Belgium.

- **Humankind's role in greenhouse gas production is negligible.**

Human activities release 135 times more CO_2 than volcanic activity (U.S. Geological Survey).[232] We are releasing a large part of the CO2 laid down by nature over hundreds of millions of years. About 345 million years ago, rising sea levels and swamps submerged large areas of vegetation in the early northern land masses. This material was impacted by high temperature and pressurized by the constantly increasing overburden of heavy sedimentary material. It became coal, oil and gas. About 160 million years ago, phytoplankton deposits were slowly transformed to oil in a similar manner. There might be 4 to 5 trillion barrels of oil on the planet and we have already burned 1.3 trillion of that (not to mention coal and gas). In geological terms, the last 200 years is very sudden! BP show proven reserves remaining amount to 1.6966 trillion barrels.[233] The carbon in this oil has also decayed radioactively to become the stable isotope ^{12}C, revealing that all the radioactive isotopes had decayed over long periods of time, proving their fossil origin.

- **There is an 800-year lag between rising temperature and CO_2 levels.**

This item is the product of a documentary called *The Great Global Warming Swindle* from the BBC that became a central policy item for the denial camp.[234] It was produced to represent the views of the tiny minority of climate scientists who disagree with the consensus and believe that global warming is not the product of human production of CO2 emissions. Later, Ofcom, the British broadcasting regulatory agency, formally criticized the program for misrepresenting the known data as presented by the Intergovernmental Panel on Climate Change. In one section, the film highlights the results of a paper by Nicholas Caillon,[235] showing the temperature and CO2 rising after a glacial period 240,000 years ago.

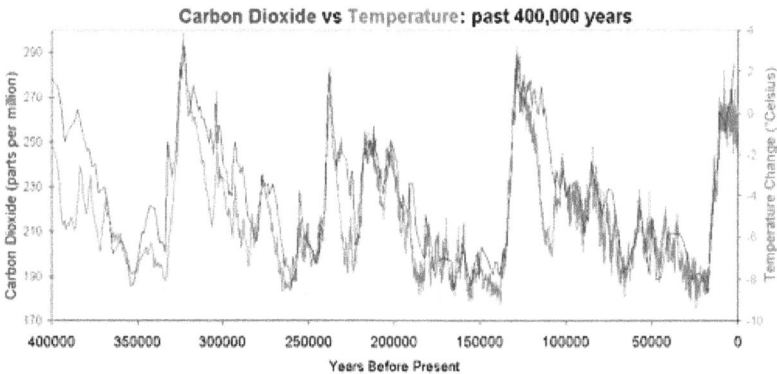

Figure 55: Temperature and CO2 relationship over the last 400,000 years from the Vostok ice core records. Source: Science Magazine

The CO_2 has an 800-year lag on the temperature. The documentary says this shows that CO_2 cannot, therefore, be responsible for warming if it's a product of warming. The actual science, though, says both can happen. CO_2 warms the atmosphere wherever it comes from. Without humans, the Earth goes through its orbital cycles, getting closer to the sun and hotter and then getting farther away and cooling.

As the Earth warms up, CO_2 comes out of the ocean and into the atmosphere. As it does so, the CO_2 adds its own extra heat retention, causing an acceleration of the warming. This happens in a normal warming phase of the Earth's orbit of the sun.

Over the last 400,000 years, CO_2 and temperatures have been in lockstep. Changes in CO_2 follow changes in temperature by about 600–1,000 years. This leads to the inevitable claim by deniers that the current phase of warming cannot be due to CO_2.

CO_2 didn't initiate warming in the past because it didn't come first. Another study done in 2012 by Shakun et al. discovered that initial orbital warming was only 10% of the total warming and that the remainder was down to the release of CO_2, which then retained the sun's heat in the atmosphere until the orbital cycle took the Earth farther away from the sun, allowing cooling to cause the oceans to absorb more CO_2 and accelerate the cooling further. Today, human release of CO_2 is accelerating warming in a time when we would normally be cooling thanks to the Milankovitch cycle. An article in *Science* magazine illustrated the lag of CO2 over temperature. The University of Queensland in Australia has collected a key group of climate scientists to make a series of videos all about climate denial and the known explanations for most phenomena.[236]

- **There is no scientific consensus for warming.**

The Petition Project,[237] launched in 1998, is a listing of 31,487 "American scientists" who signed a petition claiming there was no evidence for human- caused global warming and that CO_2 was beneficial. Its purpose was to counter the claim of scientific consensus and prevent government action, but it tripped up by including very spurious names and inadequate qualifications for the so-called listed scientists of whom only 9,029 had PhDs. It transparently aimed to support the fossil fuel energy industry.

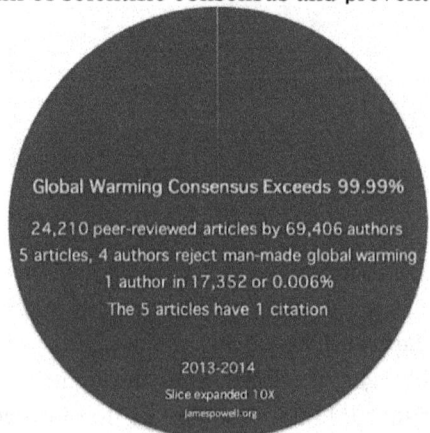

Global Warming Consensus Exceeds 99.99%

24,210 peer-reviewed articles by 69,406 authors
5 articles, 4 authors reject man-made global warming
1 author in 17,352 or 0.006%
The 5 articles have 1 citation

2013-2014
Slice expanded 10X
jamespowell.org

On the NASA website,[238] there is a statement that 97% of climate scientists are part of a consensus that says that warming trends identified over the past century are very likely due to human activity. The list includes most of the leading scientific

Figure 56: James Lawrence Powell searched 24,210 scientific papers on climate change between 2013 and 2014 and discovered only 5 which disagreed, or 0.006%. Source: http://www.jamespowell.org.

organizations worldwide that have issued public statements endorsing this position. This consensus does not have the hallmarks of a conspiracy. In addition, there is a long list of scientists and contributing organizations that help to author the IPCC's Assessment Reports on Climate Change. On the IPCC website, there is also a document describing how authors of the report are selected. In 2013, research[239] by Cook et al. looked at the summaries of 12,000 scientific papers on the subject and concluded that 97.1% indicated humans were to blame. In 2009, Doran and Zimmerman found 97.4% support among those most active in climate research. Anderegg et al. in 2010 found 97–98% support.

Despite all this support from those in the know, the general public in most Western nations remains confused. In a U.S. survey, Dew, in 2012, discovered that 55% of those polled either disagreed that there was a scientific consensus or did not know there was one. Any public policy enacted to address global warming generally will require democratic support. The main source of this anomaly in understanding is a common practice in the media of allowing "equal time" to explore different opinions under a doctrine of fairness that distorts scientific fact. Balancing the positions skews the message, creates the impression that there is a debate and creates a false equivalence. This means that the denier community have, for the moment, an incredibly powerful ally in the media.

This raises the question about who are the 3% of climate scientists who still buck the trend and disagree with the consensus. In an article in the *Guardian* newspaper in the UK, mention is made of Roy Spencer, whose research suggests that humans have a minimal role in global warming. He is often asked to testify before the U.S. Congress and is interviewed by media outlets that have a fairness regime where any points of view are basically given equal time even though discredited by consensus. Spencer wrote a blog, "Top Ten Good Skeptical Arguments," arguing against the consensus, similar to these questions you are currently reading. Some of these positions held by deniers are represented by his questions. One of them, for instance, is, "If global warming science is so 'settled,' why did global warming stop 15 years ago, contrary to all 'consensus' predictions?" My "Global temperatures are declining" entry covers the answer to Spencer's simplistic question, which obviously makes him not very serious and one of the cherry pickers. This piece is the answer to the consensus question.

Science magazine published an article by history of science professor Naomi Oreskes in 2005. She had searched for articles published between 1993 and 2003 using the keywords "global climate change" and found 928. She classified them according to their abstracts and found that none rejected human-caused global warming. Of course, she was attacked for her work, but her conclusions still stand.

The search for scientific disagreement must come from peer-reviewed literature, where evidence and not opinion is the basis for any differences. Using the search terms "global warming" and "global climate change," James Lawrence Powell, a science author and geochemist who was appointed to the National Science Board by President Reagan and then by President Bush, searched for and found 13,950 scientific papers written by 33,690 individual authors and published between January 1991 and November 2012 that discussed climate. In scientific research, often a whole group of authors write papers that are published.

Powell used the scientific literature because peer-reviewed literature is the highest level of scientific dialogue and an analysis would shed light on the subject. He searched for articles that rejected human-caused global warming by looking for explicit terms of rejection that stated that human-caused warming is "false" or that something else better explains it. Articles that claimed just a minor flaw or discrepancy were not deemed to be rejections. In 2013–2014, Powell did the research again,[240] this time with climate scientist authors as his focus. He now had 24,210 articles of which only 5, or 0.006%, clearly rejected human-caused climate warming or supported a cause other than CO_2 emissions. He found that only 4 out of 69,406 authors of peer-reviewed literature disagreed with the anthropogenic angle. Only two rejected anthropogenic global warming.

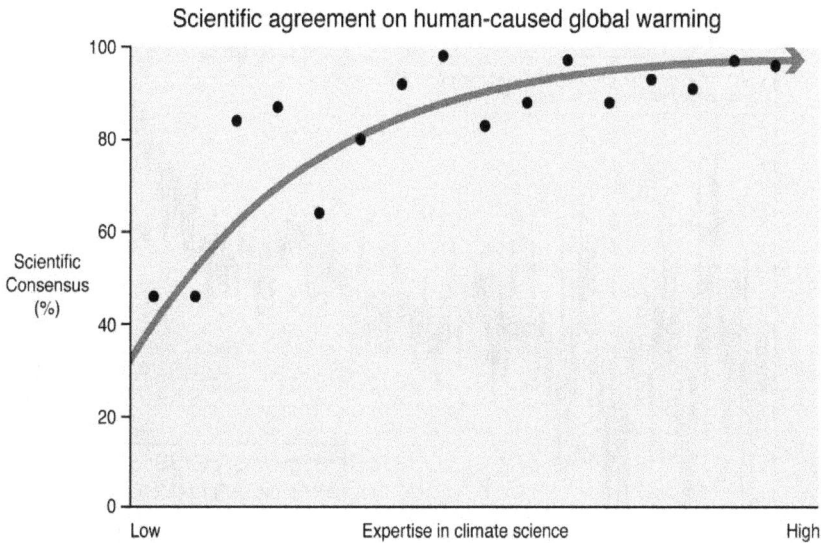

Scientific agreement on human-caused global warming

Figure 57: *The greater the climate expertise of the scientist authors of climate studies, the greater the consensus that humans cause global climate change. Source: Cook et al. 2016.*

This says that 99.94% of climate scientists have a consensus and not the still impressive but more widely spoken about 97% that resulted from the NASA studies. The outliers have no influence on the science and the word *consensus* ought really to be *unanimity*.

In 2013, Cook et al. published a study[241] that aimed to generate an accurate perception of the scientific literature and the degree of scientific consensus to inform public opinion. They also concluded that consensus was 99.9%, as though 97% was somehow a bad number. Even trying to find a negative article on human-induced global warming is very demanding. Virtually none of the most vocal global warming deniers ever wrote a peer-reviewed scientific article, and the reason is because they don't have any evidence to put forward that would stand up to scrutiny. This means that the so-called debate over climate change doesn't exist and instead is the product of apostate scientists working for fossil fuel companies and right-wing foundations.

Annual Land-Surface Average Temperature

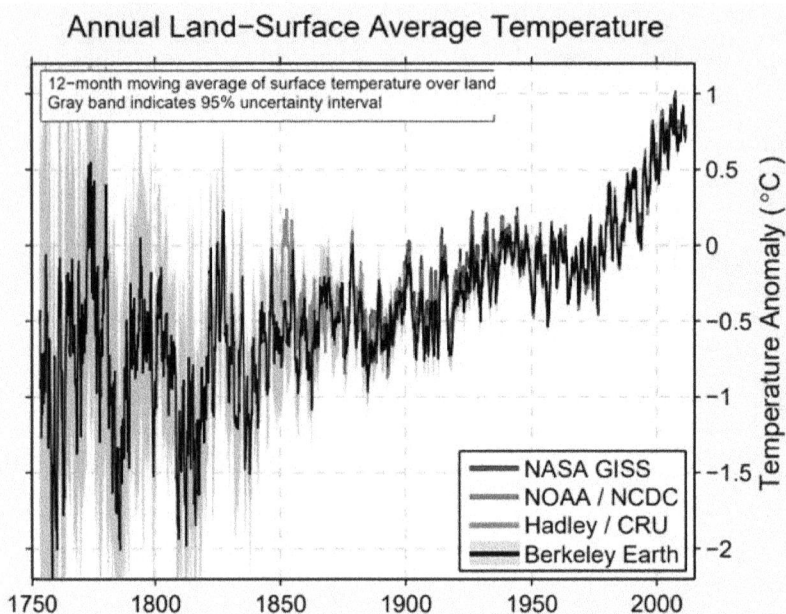

Figure 58: Professor Richard Muller's compounded chart of temperature changes since 1750 shows the agreement between the NASA, NOAA, UK's Hadley/CRU study and the Berkeley Earth Surface Temperature project study (BEST) in black. Source: Berkeley, archives.

The Pew Research Center polled members of the American Association for the Advancement of Science. They found that 87% of these scientists agreed that the Earth is becoming warmer because of human activity. This number was open to critique because only 18.1% of the scientists responded and half of them were from a different scientific discipline.

In 2016, Cook et al. followed up their research by combining seven climate consensus studies,[242] including those of Naomi Oreskes, Peter Doran, William Anderegg, Bart Verheggen, Ed Maibach, J. Stuart Carlton and John Cook.

They all coauthored a paper they felt would settle the consensus question once and for all. It made two key conclusions: (1) however you measure the consensus, between 90% and 100% of scientists agree humans are responsible for climate change; and (2) the greater the expertise in climate science, the greater the consensus.

Occasionally, a climate change denier changes his spots when, as a scientist, he has a closer look and has to agree with the conclusions of all the other scientists. Charles G. Koch, who runs the Heartland Institute, the climate denial group funded by Exxon Mobil among many others, also runs a charitable foundation that, along with others, spent $150,000 in 2010 on a program called the Berkeley Earth Surface Temperature project (BEST). This project, founded by UC Berkeley professor Dr. Richard Muller and his daughter, Elizabeth, employed senior scientists such as Robert Rohde to verify claims made by mainstream science that humankind was responsible for global warming.

Muller had previously been openly in denial that the climate was changing, and his original denial position was fueled when he, in his words, "identified problems in previous climate studies that, in my mind, threw doubt on the very existence of global warming." In July of 2012, Muller wrote an op-ed piece in the *New York Times*[243] where he described how, in the final analysis, the data matched the mainstream scientific conclusions already reached by many of his peers. His team used algorithms to analyze many more data sets from temperature stations around the world, 1.6 billion data points from 16 preexisting archives, that increased the statistical importance of the conclusions. Once the study was concluded, he embraced the view that Earth's surface temperatures have indeed risen by 2.5°F (1.4°C) in 250 years and 1.5°F (0.8°C) over the last 50 years alone, entirely because of human activities.

Muller looked into one of the false claims of the denialists, that the sun was responsible for Earth's heating experience, and discovered that the long sunspot data record was reliable and that it indeed pointed to lack of changes in the sun's output. Even the United Nation's IPCC 2007 report allowed for the possibility that the sun might have contributed to the end of the Little Ice Age, a period of cooling from the 14th century to about 1850, and that humankind was only responsible for "most of" the last 50 years. Muller's data showed that this rise in temperature was not explained by the stable output of the sun and, most excitingly, he concluded that the apparent stability of temperatures in the last 20 years was due to the flattening offset effects of the El Niño and other ocean currents. He said by far the best match by any variable to the shape of the temperature curve was not solar activity, or world population, but the record of atmospheric CO_2 consistent with the greenhouse effect. He concluded that although this did not prove that CO_2 was necessarily causal in explaining the temperature rise, any alternative hypothesis had the very hard job of matching the data at least as well as the CO_2 data did. This result was deeply embarrassing to the Koch brothers, who were searching for supporting evidence for a denialist position useful to their conservative viewpoint and protection of use of fossil fuel resources.

The deniers aim to disseminate the thought that climate scientists disagree about global warming to make it appear that there is a serious or fundamental issue that casts doubt. On the contrary, human-caused global warming is the main platform of climate change on which every-one who knows anything—the climate scientists—agree. We know that human emissions of greenhouse gases are responsible for climate change. These are established facts about which virtually all the perti-nent scientists agree.

- **I can't comment, I'm not a scientist.**

Time and time again, when politicians are asked their opinion about climate they defer making a statement, declaring that they aren't scien-tists themselves and that they are not qualified to respond. Inevitably, if the science is saying something you don't want to be true, this is the sort of thing you'd say. There is, of course, an easy way to cope with this question and that is to refer to the actual real scientists. Instead of calling it alarmism, we can be grateful to have an early warning on which we can act from people we pay to be experts at the subject.

It's interesting that in every other branch of science we trust the scientists implicitly. For example, crafting the materials and the hun-dreds of man- years of accumulated science work that, say, go into air-craft design has today resulted in over one million people in the world flying every hour, trusting their lives to the human understanding of aerodynamic forces and materials.

Instead of not flying and making the excuse "I'm not an aircraft designer," we trust the scientists to do their jobs and we get on the plane. The same can be said of scientists who design multi-billion-dollar weap-ons systems or who work tirelessly to develop treatments for diseases such as cancer or the ones who developed the miracle of communica-tions with the smartphone or the ones who brought us computer chips or who explore the solar system and have recently facilitated the sending back of stunning images for the first time in human history of the far-thest planet in our system, Pluto. We trust these scientists without question to get on with what they are doing. We do not say when asked about Pluto, "Well, I'm not a scientist." Instead, we share how amazing it is.

Climate experts are scientists, too, and are trained in the same disciplines of observation, hypothesis, experimentation and testing as all these other scientific disciplines.

- **The Earth has always had cycles of cooling and warming.**

Yes. They are called Milankovitch cycles, as mentioned previously. The distance from the sun changes according to the Earth's tilt, spin, speed, precession and the phase of the Earth's orbit cycle between circular and oval shape over about 100,000 years. Extra CO_2, such as that generated now by humans, forces extra heat to be retained and currently works against the long-term natural cycles, which were originally cooling the planet. The temperature cycles have been shown to be in lockstep with the concentration of CO_2 in the atmosphere, so as we throw more CO_2 into the system, the temperature rises, even if the other cycles normally would bring us cooling.

- **It's cold outside, so where's the warming?**

No particular weather event can prove or disprove climate change. Instead, it is indicated by long-term patterns of activity, recorded patiently by thousands of trained observers with instrumentation. In the warming world, the Arctic is warming twice as fast. This weakens the polar vortex, the strong winds which whistle around the Arctic. The winds go less fast, and as the vortex weakens, its edges become less defined and wavier, and parts of it make southerly excursions, bringing bitter arctic air to southern climes. It also means warm air goes north. This means that now the bitterly cold winter experience is a product of global warming but at the same time is not a local indication of warming.

Global warming has been criticized as an insufficient nickname for climate change. The term has difficulty accommodating colder winters and extreme floods or droughts, heat waves and blizzards. "Global weirding," a phrase that Thomas Friedman of the *New York Times* said was coined by Hunter Lovins, the cofounder of the Rocky Mountain Institute, is perhaps a better if slightly nonserious expression. It addresses the issue of extreme weather caused by the disturbance of GHGs. The jet stream is a wind that whistles around the planet in a counterclockwise direction (looking from the North Pole) at a height of 5 kilometers.

Warm air from the equator moves north and south. In the north, it is swung eastwards by the Coriolis force, making it look like a fast-moving, fluctuating river of air. Normally, it didn't fluctuate from north to south as much as it is beginning to do. The strength of the winds has diminished by about 10% since 1985 as the temperature gradient between the poles and the equator has reduced. Remember, the Arctic and Antarctic are warming faster than the temperate and tropical latitudes. The strength of the vortex has diminished because arctic air that is no longer as cold because of the sublimation (evaporation of solid ice directly to water vapor) of ice there becomes less linear and more turbulent. The arctic vortex's instability is a major source of weirding, with colder air coming south and warmer air going north. Often it manifests as cooler days in winter or snow in Texas, when the Arctic jet stream billows south more than usual.

Don't confuse the weather with climate. Winters normally get cold. You might say that if there was an abnormally warm winter, the scientists would be eager to say this was an example of global warming and …you'd be right. It would be.

- **The science community is a conspiracy designed to preserve prestigious science jobs. The IPCC is a political body telling us what scientists believe. When did scientists' beliefs constitute "proof" and when was scientific proof determined by a vote, especially when those concerned are from the Global Warming Believers Party?**

While new jobs would be a great thing, few groups are expert enough to promulgate a strategy of falsehoods designed to create science jobs on this scale. However, accusing the scientists of manipulating data is very easy. This argument is used by the fossil fuel companies as a means to preserve workforce jobs and appeal for votes. Ironically, when scientists do break the rules and "out" the deniers by, for example, releasing stolen emails from the deniers' camp, invariably they are obliged to resign their job for conduct unbecoming of a scientist, whereas this is not the case on the deniers' side.

The IPCC organizes the world's top climate scientists every 5 to 7 years to summarize the up-to-date status of independent research on climate. The IPCC report and its 99% expert consensus are not inherently "proof" of anything. They merely summarize the evidence. It is the evidence that generates the consensus. There is, however, evidence of another kind, which is that the "consensus" can have a very severe effect on the lives and careers of those who don't conform. Occasionally, this is reflected in an article or a movie describing how otherwise worthy scientists can't get papers written with their colleagues or must leave their job. This, of course, is overkill, but it clearly doesn't pay to believe in fairies when there is earnest work to be done.

- **The Earth reflects excess heat into space anyway, so any extra warmth will not have disastrous effects or need expensive offsets.**

It does, but not enough escapes the dragnet of CO_2 to make the difference. There is always equilibrium between the heat coming in and the heat going out. Greenhouse gases like CO_2, methane and others keep more heat in.

- **Water vapor is really responsible for the increase in temperatures.**

Water vapor increases solely with temperature. At a higher temperature, more water evaporates. When extra CO_2 causes warmer air that in turn produces water vapor that makes the air warmer still as more heat from the sun is retained. It's a positive feedback loop that roughly doubles the effect of CO_2. If CO_2 alone were to cause a 1.0°F (0.55°C) change, water vapor will make it 2.0°F (1.11°C).

When these feedback loops are calculated, you can get a 3 times multiplication in the temperature by having the CO_2 there in the first place. Water also falls as rain, so the amount of vapor in the air changes dynamically and water vapor changes its concentration rapidly, unlike CO_2, which can stay in the air for centuries. When the air is humid, it often feels hotter, and this is called the heat index. With overall global warming, such heat waves are the equivalent magnitude of other natural disasters such as hurricanes or tornadoes. The possibility is that these heat waves could increase in intensity and come as waves of several days, reducing a locality to zero activity while its population do everything possible to escape the heat. Infrastructure and technology are essential, in this case, for avoiding the worst effects of this weather phenomenon, but the best way is just to get rid of CONG.

- **Cosmic rays cause clouds that can cool the planet.**

Cosmic rays are extremely high-energy, or fast-moving, radiation that manifest as atomic particles that originate from outside our solar system. They ionize atmospheric gases and also do things like transmute nitrogen into carbon-14 (^{14}C). Ninety-nine percent of these particles consist of the nuclei of common atoms, and 1% are electrons. ninety percent of the nuclei are protons, 9% are alpha particles (2 protons and 2 electrons bound together, or a helium nucleus) and 1% are nuclei of heavier elements. Little is known of their origins, but one possibility is that they originate from exploding stars or massive supernovae or galactic events.

Cosmic rays can damage electronics, and outside the atmosphere are much more damaging to life. Astronauts know the effect well and, during the Apollo missions, reported seeing flashes of light every 2.9 minutes. A special helmet worn to detect cosmic rays demonstrated that the flashes coincided with the arrival of a cosmic particle. The energy in a cosmic ray proton can approach tens of millions of times as much as protons that travel around the Large Hadron Collider, which is at about the same kinetic energy as a baseball traveling at 56 mph. The average energy content of a cosmic ray is far lower, but still very powerful in its potential for damage.

When the cosmic particles ionize atmospheric molecules and clump them together sufficiently to seed clouds, these clouds reflect sunlight directly back into space. Magnetic fields from the sun can deflect cosmic rays away from Earth, resulting in fewer clouds and less reflected energy and a warmer planet. More solar activity results in a warmer Earth. Less solar activity results in more cosmic ray–seeded cloud cover. The ionizing molecules create aerosols that permit water vapor to condense.

In 1997, Danish physicist Henrik Svensmark[244] showed correlations between distribution of clouds and cosmic ray flux around the world. Later he built on this by showing how gamma rays can cause molecules in the atmosphere to clump together and form aerosols that are also called cloud condensation nuclei (CCN). And then a group called the CLOUD collaboration at CERN in Geneva discovered that rays can boost aerosol production in a small artificial atmosphere by an order of magnitude, adding credence to the idea. Aerosols need to be 50 nanometers in size before they will permit water vapor to condense.

The idea is potentially very important. Jasper Kirkby, a CERN physicist,[245] was able to show that nucleation of cloud water vapor comes from existing aerosols of sulfuric acid, ammonia, amines and biogenic vapors from trees. Using an ionizing particle beam from the CERN synchrotron, he simulated the effect of cosmic rays on clouds. Clouds reflect sunlight at an estimated rate of 30 w/m², while anthropogenic warming is only responsible for 1.5 w/m² over the entire twentieth century. This means only a very small effect can negate the impact of CO_2, or radiative forcing (in plain English, the driving of a warming atmosphere by increasing the amount of CO_2, which in turn captures more heat). Two things are needed for "cloud seeds": one is a direct aerosol or dust seeding, the other is condensable vapor.

CERN particle beams control ionization by cosmic rays which could accomplish the former, but the latter has not been shown. Peter Laut[246] checked and corrected Svensmark's report and his new conclusions did not support any correlation between clouds and cosmic rays. Since records of cosmic rays have only been made for a few decades, the use of proxies for cosmic rays established a much longer timeline going back over 60,000 years. At about 39,000 years ago in the geological records there is something called the Laschamp Anomaly. This describes a few years when there was a very high cosmic ray flux and near zero magnetic fields, but there is no apparent impact of this on the climate record. Svensmark claimed a relationship which made a certain sense, but a closer look at the data refuted it. The jury is still out, and the idea is still controversial. Needless to say, for those who wish to find a way to show that CO_2 is not the evil force the scientists say it is, Svensmark's theory is a powerful tool.

- **Volcanoes generate much more CO_2 than humans do.**

> *"Such a sore snowe and a frost that men myght goo with carttes over the Temse and horses, and it las-tyd tylle Candelmas."* — Chronicles of the Grey Friars of London, 1506.

It turns out, volcanoes are a reason for global cooling, not warming. Studies[247] at Kilauea in Hawaii show that an eruption discharges between 8,000 and 30,000 metric tons of CO_2 into the atmosphere each day. Self- evidently, active erupting volcanoes release much more CO_2 than sleeping ones do.

Gas studies at volcanoes worldwide have helped volcanologists tally up a global volcanic CO_2 budget in the same way that nations around the globe have cooperated to determine how much CO_2 is released by human activity through the burning of fossil fuels. These studies show that, globally, volcanoes release a total of about 200 million tons of CO_2 annually. This seems like a huge amount of CO_2, but a visit to the U.S. Department of

Energy's Carbon Dioxide Information Analysis Center (CDIAC) website[248] helps to put the volcanic CO2 tally into perspective. The human-caused, global fossil fuel CO2 emissions for 2014 tipped the scales at 9.9 billion tons; 200 million tons of volcanic CO2 is just over 2% of that value.

Looking back through the comparatively short duration of human history, volcanic activity has, with a few notable disturbances, remained relatively steady. Spectacular eruptions like that of Mount Pinatubo on June 15, 1991, actually increase the Earth's albedo, cooling the atmosphere by injecting solar-energy-reflecting ash and other small particles into the atmosphere.

Figure 59: A Frost Fair on the River Thames in London in 1683, by Thomas Wyke. Source: Licensed under Public Domain by Wikimedia Commons.

At the end of the 13th and through to the 15th century there was a period called the Little Ice Age (LIA). Elements of this period lasted well into the 19th century. Advancing glaciers crashed into northern European towns. The River Thames was frozen over, and fairgrounds were common in winter on the river. Between 1309 and 1850, the Thames froze at least 23 times. Canals in the Netherlands were iced over, and other continents also show signs that this occurred. Having the Little Ice Age occur right when the medieval warming period shown earlier in figure 43 is supposed to have happened is a delightful thing to savor, because it too flies in the face of the thin arguments the deniers put in place. The medieval warming chart was the first chart produced for temperatures of the last one thousand years by the IPCC and indicated a period warmer than today, derived inaccurately from tree rings and other proxies, the science of which has inevitably improved over time. If it was so warm then, they said, why are we worried now? The polar bears clearly survived that event. Why is this any different?

On April 10, 1815, Mount Tambora in Indonesia erupted with the biggest blast in recorded history. Although there had been rumblings and earthquakes on the mountain since 1812, it had been dormant for a thousand years. The volcano's lush slopes were covered with villages and nobody evacuated, leaving 10,000 killed by pyroclastic flows and tsunamis. This enormous eruption two hundred years ago changed the global climate, causing a "year without a summer" in the Northern Hemisphere. Sulfur dioxide from Mount Tambora lingered in the atmosphere for several years, cooling the planet and triggering crop failures, famine and disease pandemics in America, Europe and Asia.

The answer is that there was no Medieval Warm Period. A study[249] by Gifford Miller, a geological sciences professor at the University of Colorado at Boulder shows that the centuries of cold in the medieval period coincided directly with a period of intense volcanic activity. Precise data from ice records in Arctic Canada and Iceland show the cold period starting abruptly between AD 1275 and 1300, followed by a ramp up between 1430 and 1455. Explosive volcanism has been shown to coincide with abrupt summer cooling, and long after the volcanic aerosols are gone the sea ice or ocean feedbacks can keep the cold period going. Four large, sulfur-rich eruptions took place, each able to lace the upper atmosphere with reflective sulfur aerosols and throw the solar irradiation back into space, freezing the planet.

Perhaps we can look at imminent volcanic activity and hope that it might offset global warming for as much as a couple of decades and allow us to pacify the fossil fuel lobbies and arrange a compromise that halts emissions of CO_2 and find a way to sequester the huge quantities of CO_2 already in the atmosphere.

Mount Pinatubo erupted in 1991, but since 1998 there has also been much volcanic activity from volcanoes such as the Augustie Volcano in Alaska. Ryan Neely, a lead author on the University of Colorado, Boulder, study, explained that observations suggest that increases in stratospheric aerosols since 2000 have prevented as much as 25% of the human greenhouse gas warming. Another study from 2011 by Solomon also suggests that greenhouse gas effect was offset 25% by stratospheric volcanic aerosols, with observations that the optical depth in the atmosphere of the aerosols had increased by 4–7% since 2000, making the atmosphere more opaque than in previous years. In short, volcanic eruptions will not counter the greenhouse effect. When they are active, they cool the planet while human greenhouse gas emissions just continue to rise.

- **Why should we believe the predictions of models when they can't even explain the past?**

Norman Philips[250] was a theoretical meteorologist and former director of the National Meteorological Center of the NOAA National Weather Service (NWS). He produced the first realistic mathematical model of the Earth's atmosphere, called the "General Circulation Model," in 1956. His goal was to increase the accuracy of our understanding and capability of predicting the weather. Since this time, models have embraced computing and have become much more accurate and sophisticated. Inevitably, and in the tradition of computing, where we say, "garbage in, garbage out," models are only as good as the observations and rules of interaction of the various inputs. Nevertheless, it has been virtually impossible to understand complex systems without manipulating a model first to secure a limited amount of predictive capability. Today's models are much more accurate, and after decades of tweaking have a much-improved predictive capability and are responsible for much of science's new confidence about its understanding of the complexity of the weather. A model is not a perfect representation of reality and does not echo nature exactly but is a human construct of the weather arising from our growing understanding of it.

Interaction between the atmosphere, oceans, land surface, ice and the sun are very complex, and so models are only able to predict with estimates of what the future holds. Events predicted far off into the future are less likely because much can change that is not predictable when more time is considered. Since we can't wait 30 years to see whether a model is accurate, models are often tested against what we know happed in the past (hindcasting) by selecting a historic starting point and operating the model as if we were back then. If models are good at predicting what actually happened, then there is more confidence in their ability to look at the future.

Models have shown us that CO_2 in the atmosphere does explain the warming we are currently experiencing. The eruption of Mount Pinatubo gave modelers a chance to test this by feeding computers with the data about the eruption. They were successful in predicting the effects of the Pinatubo volcano and now have been successful also in predicting the greater warming in the Arctic along with cooling in the stratosphere. Many models tend to be conservative rather than alarmist, with the actual, real-world outcomes often being more extreme. With improving data sources, models become more reliable and are an essential and now reliable component as a guide to climate change.

There is a great documentary[251] by Sir Paul Nurse, the head of Britain's Royal Society and a Nobel Prize winner himself. In order to say something meaningful about the increasing level of animosity generated against scientists, he made the documentary with the BBC and visited NASA, where Dr. Robert Bindschadler showed him two large screens. The top screen is a picture of Earth from space in real time. The lower screen is the model of Earth, running off the enormous collection of accumulated data. Movements of clouds and wind and false color indications of temperatures are all very much in synchrony. Modeling has become very accurate now that many decades of work have gone into the issue.

- **All the best-known scientists in history, Aristotle, Archimedes, Galileo, Tesla, Faraday, Newton, Pasteur, Edison, Einstein, never mentioned climate change, so that proves it's only a recent politically based hoax.**

Yes, this actually was said in public. On his August 30 broadcast,[252] Mark Levin, a U.S. national radio host, listed these famous historical science figures: "Aristotle, Archimedes, Galileo, Tesla, Faraday, Newton, Pasteur, Einstein and Edison." He then asked his question in a voice that says to the listener that he has a seriously big reveal to make: "What do they all have in common?" His answer? "Not a single one of them ever wrote about man- made climate change." Levin repeats himself as if to rub in this fabulous proof that climate change is actually just a modern political effort by the Left to adopt an idea that crept in from Europe, like Marxism.

Instead, of course, just like the design of chips, planes, cars or anything, humankind is improving our knowledge of what climate change is and how it operates. We have only just really learned about climate change recently. What we know about climate is, in fact, due to the work of historical scientists who specialized in what they were doing, as discussed in the climate chapter. Most of the scientists mentioned by Levin worked on other aspects of the natural world and not climate. All the scientists who worked on climate discovered aspects of CO_2 that added to the accruing knowledge of the situation, resulting in what we now know. You might as well have said that Galileo had the recent pictures of Pluto!

- **Warming is actually good. Why is a little warming a bad thing?**

It is true to say that the frozen parts of the planet that are effectively barren would "wake up" if global warming happens. Global warming will make huge areas of land habitable and able to grow crops. Large areas of currently almost empty territory in Russia, Canada and Alaska, which have been frozen, stand to become habitable and arable centers of civilization in a warmer Earth, and previously ice-trapped ocean cities can become thriving ports.

If the cold areas become pleasant, though, it means the areas we call hot today will likely be hellish, and those places currently have huge populations. This is already leading to the start of mass migrations, which are difficult to manage in the best of circumstances. Tropical diseases have already begun to spread their active zones north. Malaria and dengue fever are two examples that move with their delivery vectors, the mosquito. If we make a negative impact and any of an ice albedo, methane or clathrate tipping point occurs, all bets are off. Parts of the tropics, the Middle East, India, China will be uninhabitable for much of the year. Already long-term financing vehicles such as mortgages are being refused in Florida because of existing ocean flooding levels. The ocean is expected to be approximately 3 feet higher[253] by the end of the 21st century and 6 feet if nothing is done. Tipping points threaten to accelerate sea level rise.

Already, ice and bogs that used to be solid permafrost have melted and there are amazing videos of pond ice being broken and the escaping gases being ignited.[254] An index of this process is the "drunken" forest phenomenon, where newly warmed topsoil, no longer frozen solid, actually moves and the previously solid support for tree root systems becomes soft and allows trees to adopt any angle but vertical. This is a spectacular demonstration of warming tundra in arctic regions. Further tipping point events for sea level rise include ice shelves slipping off the land or melting faster into the ocean. Of course, the consequences are severe. Whatever happens, there is almost certainty that sea level rise will also continue for centuries beyond 2100.

Climate scientists at Climate Analytics tell us in a new study[255] that unless coal disappears from human energy production by 2050, millions of Pacific Islanders and inhabitants of low-lying land worldwide will be forced to migrate as waters rise. A strict carbon budget leads to half a meter (1.6 feet) of sea level rise by 2100. Viewing this as optimistic, however, leads us to the catastrophic business-as-usual version, where a greater than 1.9°C warming would cause a runaway melting effect on the West Antarctic Ice Sheet.

Researchers used high-resolution seafloor imaging and new modeling techniques to discover that likely sea level rise by 2100 will be 1.32 meters or 4.3 feet if the world continues to burn fossil fuels at current rates. This is 50% greater than the IPCC's AR5 report. The report underlines the idea that as ice cracks and breaks off, it is surrounded by warmer water and air and this accelerates the melting

Zillow, a leading real estate information company, recently added a story[256] to its blog summarizing the impact of sea-level rise and coastal flooding data from the National Oceanic and Atmospheric Administration. In 2016 scientists said that if CO_2 continued in a business-as-usual fashion, the year 2100 would usher in ocean levels 6 feet higher than today's level, effectively flooding 1.87 million homes in the United States valued at $882 billion, half of which are in Florida. Fort Lauderdale alone would have 38,000 flooded homes, and 37,500 in Miami. The city, which already has difficulty when daily high tides flood streets, is spending half a billion dollars to install pumps and raise road levels. If southern Florida was built on granite like New York, instead of spongy, porous limestone, they'd be able to contemplate a barrier system similar to "the Big U" planned by New York,[257] which consists of a 10-foot-high berm or sea wall running 2 miles along the East River to protect lower Manhattan and starting in 2018. In Florida, such a wall would merely be bypassed by water flowing underneath it in the porous limestone rock.

Miami and New York would do well to examine a phase of Chicago history when buildings were added but at different levels, causing streets to be a series of steps to go up and down. Added to this, in 1854 there was an outbreak of typhoid, dysentery and then cholera that killed 6% of the city's population. This was a disaster of the first order, and on examination it was found to be due to the fact that Chicago buildings were all built only marginally higher than the level of Lake Michigan, resulting in a lack of drainage of sewage and allowing bacteria and pathogens to pool and infect the vulnerable. By 1856, the city got organized and a plan for a sewage system was implemented, part of which was to raise all buildings in the city to a higher level, making both the streets level for pedestrians as well as allowing the sewage system to drain away from the city.

Bostonian engineer James Brown lifted a 750-ton brick building by 6 feet 2 inches, the first of more than 50 buildings raised just that year. Soon they were lifting buildings that were 200 feet long. Six hundred men spent five days raising half a city block on Lake Street that was 320 feet long, constructed of brick and stone, four and five stories high, occupying an acre and weighing in at 35,000 tons. They used 6,000 jack screws, built new foundations and drainage underneath and raised it by 4 feet 8 inches. The Tremont House hotel, also on an acre of land, was similarly raised by 6 feet while guests, including a senator and VIPs, continued to stay there and business continued as normal. Wooden buildings were moved to the outskirts of the city while new stone buildings were erected in the city center at the new higher levels. Businesses and hotels remained open while the works moved forward. This is likely to be an engineering trick that returns to use as relentless sea level rise impacts coastal communities.

We normally talk about temperature in terms of how warm it will be today, as in its 82°F (27.8°C), with a high of 83°F (28.3°C) and a low of 68°F (20°C). When climate scientists calculate the rises in global temperatures, they come up with numbers like 1.6°F (0.91°C), and its little surprise this doesn't fire up the public imagination. Areas of the planet can become too hot and this extreme weather will simply fry every living thing on the surface of the ground. If average temperatures were to rise by only 12.6°F (7°C), it would be too hot and humid to even live in some regions. With a rise of 21.6°F (12°C), most of the human population would have to move or rely on continuous air-conditioning to survive. Today we have AC for comfort, not survival. In these stifling conditions without AC you… cook.

If our skin gets hotter than 95°F (35°C) for a few hours, we cannot survive. Although many people live and work in 113°F (45°C) conditions, normal in New Delhi, for example, they keep cool by sweating in the relatively dry heat and by living in air-conditioned spaces.

Increase in Earth's average Temperature	Approximate Maximum Fahrenheit Heat Wave Temperature	Heat Index	
21.6 °F (+12 °C)	147 °F (63.9 °C)	200 °F (93.3 °C)	
10.8 °F (+ 6 °C)	136 °F (57.8 °C)	170 °F (76.7 °C)	
7.2 °F (+ 4 °C)	132 °F (55.6 °C)	155 °F (68.3 °C)	< Happening Today!
3.6 °F (+ 2 °C)	129 °F (53.9 °C)	142 °F (61.1 °C)	

Figure 60: Heat waves and the high heat index temperatures that are likely, given the expected rise in Earth's temperature. Source: Steven Sherwood. University of New South Wales.

To make this difference, scientists talk about a type of thermometer called a "wet bulb," which is kept in a well-ventilated wet cloth. It measures the temperature of saturated humid air. Human survival depends on wet bulb temperatures of less than 95°F (35°C).

We are releasing carbon at huge rates. Half of the human fossil fuel emissions have occurred since 1980. One hundred ninety-four countries committed to preventing 3.6°F (2.0°C) in temperature rise, responsible for 87% of emissions, but this means restricting ourselves to just 565 gigatons of CO_2 emissions by 2050, well below the business-as-usual case.

At current rates of about 31 gigatons of emissions per year, with a growth rate of about 3%, this is likely to happen by 2030,[258] and that's hoping that there are no tipping points like those already mentioned, which threaten to make it a one-way journey and accelerate the process. The established fossil fuel reserves identify 2,795 gigatons[259] of CONG in proven reserves, not including shale gas or all coal reserves, and valued at $27 trillion.[260] This is the value we are asking the CONG companies to give up. How difficult it will be. Also, pension funds investing in carbon assets can sell and not strand their carbon if they do it in time. These numbers were quoted by Bill McKibben in his *Rolling Stone* magazine article and were sourced from a group called the Carbon Tracker Initiative in London, made up of a group of professional financial analysts concerned not just with carbon as a financial risk associated with trillions in stranded assets but also the real climate impact and the much more serious effect that will have on global markets. This is 5 times the target burnable carbon, and all of it is currently allocated for production and burning.

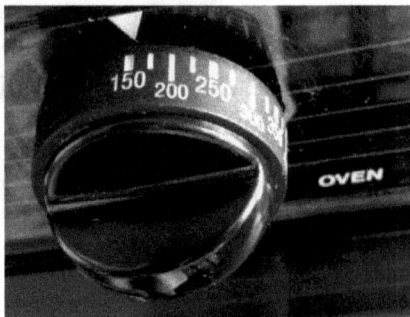

Figure 61: Even the temperature knob on an everyday oven starts at 150°F (65.6°C). Source: NEF Advisors, LLC.

Droughts, heat waves and wildfires are all at record levels as I write this. If we are seeing all of this with average temperatures up 1.3°F (0.8°C), then 3.6°F (2.0°C) will be disastrous. If hot countries suffer like this today, the target numbers make them effectively unlivable. The land area in the Middle East, India and Pakistan, South East Asia, Africa, Central America and Australia that would no longer be habitable would dwarf the area of land newly submerged under the ocean from melting polar ice. Of course, the same impact is true for animals, and perhaps even the hardy cockroach will meet its extreme conditions as well. Thermal maximums of the Paleocene–Eocene periods saw significant species extinctions 55 million years ago.

Heat stress would not be there all year-round initially, but the conditions we'd have to survive would be like heat waves not too different from those hitting the Middle East today. A study[261] by Steven Sherwood and Matthew Huber explored human and mammal hyperthermia and concluded that it's just going to be too hot to live in many areas.

We could expect climate deniers to say this is too extreme to be believed, but these temperatures are not waiting for the average rise to be 3.6°F (2°C) or even 10.8°F (6°C). Heat waves in the summer of 2015, 2016 and 2017 happened all over the Middle East and Asia. A world record was set in Furnace Creek in Death Valley in the United States, where the hottest air temperature ever recorded anywhere on Earth came in at 134°F (56.7°C) on July 10, 1913. For five consecutive days, the heat wave reached 129°F (54°C) or above. The hottest surface temperature was recorded in the Lut Desert in the provinces of Kerman and Sistan in Baluchistan, eastern Iran, where the surface of its sand reached 159°F (70°C).

The heat index in Bandar Mahshahr, an Iranian town, in July 2015 was the second highest heat index temperature ever recorded, the first being in 2003 in Saudi Arabia, where the heat index reached 178°F (81°C), and in the same year on July 7 the heat index reached 155°F to 160°F (68.3°C to 71.1°C) in Dharan, with the actual temperature at 108°F (42.2°C) and a dewpoint of 95°F. These temperatures are horrific.

°C (°F)	LOCATION	COUNTRY	DATE
56.7 °C (134.1 °F)	Furnace Creek Ranch, Death Valley, California	United States	10-Jul-1913
55.0 °C (131.0 °F)	Kebili	Tunisia	7-Jul-1931
54.0 °C (129.2 °F)	Tirat Zvi	Israel	21-Jun-1942
54.0 °C (129.2 °F)	Mitribah	Kuwait	21-Jul-2016
53.9 °C (129.0 °F)	Basra	Iraq	22-Jul-2016
53.5 °C (128.3 °F)	Mohenjo-daro, Sindh	Pakistan	26-May-2010
53.0 °C (127.4 °F)	Dehloran	Iran	22-Jul-2016
52.0 °C (125.6 °F)	Jeddah	Saudi Arabia	23-Jun-2010
51.3 °C (124.3 °F)	El Bayadh, El Bayadh Province	Algeria	2-Sep-1979
51.0 °C (124.0 °F)	Phalodi, Rajasthan	India	19-May-2016
50.7 °C (123.3 °F)	Semara	Western Sahara	13-Jul-1931
50.7 °C (123.3 °F)	Oodnadatta, South Australia	Australia	2-Jan-1960
50.4 °C (122.7 °F)	Doha	Qatar	14-Jul-2010
50.3 °C (122.5 °F)	Ayding Lake, Turpan, Xinjiang	China	24-Jul-2015
50.1 °C (122.2 °F)	Repetek Biosphere State Reserve, Karakum Desert	Turkmenistan	7-Jul-2015
50.0 °C (122.0 °F)	Dunbrody, Eastern Cape	South Africa	1-Apr-1905

Figure 62. Highest recorded actual temperatures worldwide. These are only the recorded records above 122ºF (50.0ºC), and it's telling that more than half of these records (in green) occurred since 2010.

In 2003 in Europe, heat waves with temperatures far lower than these examples, killed 70,000 old and vulnerable people in a region where air conditioners do not traditionally get installed. In the summer of 2015, a heat wave in Pakistan killed 1,200 people, and now they dig anticipatory summer graves in Karachi before the heat strikes.

In the Middle East, authorities are mandating holidays to keep people out of the intense heat. In Baghdad, Iraq, the heat index temperature reached 126°F (71.1°C). This is difficult enough in peacetime, but with millions of displaced persons, keeping cool is a vital but less available possibility. The Guinness Book of Records does not record the highest temperatures, but they do record deaths from heat waves. In Russia in 2010, "freak" heat waves, droughts, forest fires and smog killed up to 56,000 people. The dead included thousands who drowned after taking to waterways, rivers and pools to cool down. Reliable electricity supply can mean cooling for those with an air conditioner, but in the developing and often hotter world, electricity supplies are far from reliable, with brownouts and blackouts sending thousands to "shelter" in air-conditioned shopping malls and even fountains, rivers and irrigation canals.

Australia, the sunshine continent, was in the top 10 list for hottest recorded temperatures ever, just once, but a study from the Australian National University led by Sophie Lewis working on models for the Intergovernmental Panel on Climate Change found that if 3.6°F (2°C) was reached, the actual temperatures in New South Wales would increase by 6.88°F (3.8°C). This means that Sydney and Melbourne will see days over 122°F (50°C) even if the 3.6°F (2°C) limit is achieved. Current records for the two cities stand at 114.44°F (45.8°C) for Sydney in January 2013 and 115.52°F (46.4°C) for Melbourne on February 7, 2009.

Business is very good for electrical generators, which represent an effective insurance policy against the heat so long as the fuel supply and machinery can be relied upon. Those who can leave the hot regions entirely and wait out the heat in cooler geographies such as Western Europe.

Canada was a signatory in the 1990 Kyoto Treaty, but when the tar sands in Alberta, which contain half the quantity of carbon that is the 3.6°F (2.0°C) limit quantity, became economic because of the rising oil price, the country's commitment to cut carbon intensity vanished and indeed they withdrew from the treaty before facing fines for failing to cut the promised carbon emissions.

The implication is that the external costs of unmitigated climate change are too high to calculate, making fossil fuels something we need to stop today and then do our best to get rid of the excess CO_2 over the coming decades as quickly as possible. Given extant industry and jobs, though, this presents a tough challenge.

- **Why is the Medieval Warm Period a fake but the European heat wave summer of 2003 an indication of global significance?**

The Medieval Warm Period (MWP) was an early assessment of historical temperatures in the first IPCC report in 1990, which was based on historical records and dendrochronology alone. Much of the information gathering was the work of Hubert H. Lamb,[262] a British climatologist who worked in the United Kingdom's Meteorological Office before founding the Climate Research Unit. He was convinced that dendrochronological data that extended back into the Holocene (11,000 years) after the last ice age suggested that there had been a period about 3.0°F (2.0°C) warmer than today. He cited northern latitudes of vineyards, higher tree lines, central Norwegian farming locations, Viking colonies in Greenland and in northern Mexico and Canada, where tree remains of species that grew at lower latitudes were found well north of their current ranges.

Subsequent reports became (predictably) more accurate as different sources of information bearing on global temperatures were recorded. The European heat wave in 2003 was significant because it was part of a warming phenomenon that made that year into one of the warmest years in modern times. I have spoken about heat waves in the Middle East, India and Pakistan, already hot countries, that are killing thousands of people. The Medieval Warm Period was a product of less direct measurement techniques, which inevitably were improved upon to reveal that warming is greatest today and not as originally inferred.

- **The urban heat island effect (UHI) explains more about global warming than CO_2.**

An early criticism of reading the temperature data was that many thermometers, enough to change the outcome, were situated in urban settings where the phenomenon of the urban heat island (UHI) effect prevailed. Human activity in cities generates higher temperatures than in rural locations due to dark surfaces such as roads, roofs, active vehicles, hot emissions, greenhouse heating behind windows and a lack of green or light-colored surfaces. This phenomenon was first identified in 1810 by Luke Howard, a British chemist and amateur meteorologist who introduced the first nomenclature for clouds, which improved on Frenchman Jean Baptiste Lamarck's names because they were in universal Latin (cumulus, cirrus and stratus are examples). He noticed that the city of London was 6.6°F (3.7°C) warmer than temperatures in the countryside.[263]

UHI effect also shows up in low wind conditions and is especially noticeable in the summer. As an urban center grows, there is more heat in the center. The effect can be seen downwind of cities, where more rain falls and also influences growing seasons, spring emergence of flowers and leaves on trees. The UHI decreases air quality and increases ozone and can be mitigated by reflective surface colors and green roofs. Despite its reality, the effect has been demonstrated to have no influence on global temperatures. Temperature recordings took UHI into account long ago.

What the Military Thinks

No government agency in the United States values renewable energy more than the military. The U.S. military is the largest single customer for energy in the world. Renewable energy has all the hallmarks of resilience they demand. It's also the case that the need for fossil fuels has sparked many of the conflicts they have been involved in. These drivers have sharpened minds on this subject. The Pentagon consumes 110 million barrels of oil annually, or 1.5% of total U.S. consumption. In 2007, Congress passed the National Defense Authorization Act (NDAA) to ensure that the Defense Department obtains 25% of its energy from renewables by 2025 to improve "mission readiness and national security."

The Pentagon lists climate change as one of the major risks to U.S. security. On October 13, 2014, it released a report that directly implicated climate change, saying it poses an immediate threat to U.S. national security, along with terrorism, infectious disease, global poverty and food shortages, which also top the list of ills that will cause global lawlessness.

Climate change will contribute to food and water scarcity, will increase the spread of disease, and may spur or exacerbate mass migration.

The conflict in Syria and the Tuareg rebellion in northern Mali are both partly to do with droughts that were the most severe since the beginning of the instrumental records. This caused resource stresses within the countries.[264] Farmers make a beeline for the city if their farms don't work and a beeline for rebellion if politics don't work. One of the big effects of climate change is its threat to agriculture, and already parts of the world are showing this is true.

In March of 2015, scientists at Columbia and Santa Barbara Universities published a paper[265] directly linking the Syrian conflict with climate change. "For Syria, a country marked by poor governance and unsustainable agricultural and environmental policies, the drought had a catalytic effect, contributing to political unrest." Up to 1.5 million Syrian farmers and their families were directly the victims of severe water shortages made much worse by farm and fuel policy mismanagement. It's very difficult to establish true causality in such circumstances, but certainly there was a threat multiplication caused by the existing political instability.

Many U.S. military bases are following recommendations to install renewable energy capacity to provide military resilience, the capacity to act despite a possible blackout of the local grid. These projects also diversify the country's electricity supply. Few institutions value a distributed power supply more than the military, which is aiming for full power access if the grid should go down for whatever reason.

It has become apocryphal that by the time a gallon of military diesel fuel reaches its intended point of use on the battlefield that it costs well above $100 per gallon. Fuel needs are a major vulnerability in theaters such as Iraq and Afghanistan, where convoys were frequent targets for roadside bombs. In 2006, commanders in Iraq voiced frustrations by asking the Defense Department to free soldiers from the "tether" of fuel. The Defense Logistics Agency purchases fuel at its market cost, but that cost goes up five times if its transported by ground to a forward location in peacetime. In-flight refueling boosts the price by 20 times, to more than $40 per gallon, and if the fuel is delivered "in theater" during hostilities, the price climbs to between $100 and $600. Shipping fuel by helicopter is $400 or more.[266]

Impact of Climate Change on Global Security

"I don't want to believe. I want to know"—Carl Sagan.

The American Department of Defense (DOD) is the world's largest single consumer of energy, with a $20 billion annual bill. The Navy, Air Force and Army each has an ambitious schedule for implementing what amounts to a 25% replacement of energy sources with renewable energy by 2025. More importantly, though, the other externalities of a world pushing into the extremes of a fossil fuel nightmare include the likelihood that climate change constitutes, in itself, the largest single risk to national security and is expected to act as a catalyst for global conflict, the Syrian conflict being the current example of this effect. Refugees from low-lying areas like large river deltas such as the Mekong River in Vietnam and the Ganges River in Bangladesh represent a new potential wave of refugee humanity needing resources.

Naval bases are under threat from rising ocean levels. Increasing storm frequency and amplitude tie up valuable military assets helping people survive. Water, food and energy are resources that I believe can be sustainably plentiful. Increasingly, the word *resilience* is being used to describe a community that can withstand disruptions in global energy, food or water supplies. Resilience is clearly a desirable quality and a fundamental part of global peace depends on the spreading use of such resources. The U.S. military fully expects flooding, droughts, extreme storms, food and water shortages and infrastructure damage will all continue. The Quadrennial Defense Review links global terrorism and extreme weather patterns. These problems multiply threats by intensifying environmental degradation, increasing poverty and destabilizing political and social situations, which in turn are ideal conditions to foment and enable terrorism and violence.

In 2007, the military authored a report citing the instabilities climate change would cause. In the ensuing years, the scientific certainty has moved climate change from a multiplier to a direct driver of conflict. The military interest is primarily strategic and security, whereas the civilian issues revolve around cost, environment and politics. The politics in the USA is a particularly harrowing situation where the dynamic between those who just don't understand, such as Senator James M. Inhofe, who remains a vocal foe of preparing for climate change, and the general pro–fossil fuel contingent in the Republican caucus. Political opposition remains merely a set of opinions, however, ranged against careful observation and facts.

On a domestic level, the U.S. military are already saving money by installing many solar plants. A 16.4-MW solar photovoltaic project at Davis- Monthan Air Force Base in Tucson, Arizona, will save $500,000 annually for 25 years and beyond and provide 35% of the base's power needs. Nellis Air Force Base is another location where the previously largest U.S. solar PV installation had been installed. The Navy has been working closely with algae companies to generate liquid fuels that can run ships and planes and has had successes in both areas.

Climate Impact on Business

"Climate change forces investors in the 21st Century to reconsider our understanding of economic and investment risk. This study provides the New York Common Retirement Fund with valuable insights that will inform our efforts to manage climate risk and build out our portfolio in ways that protect and enhance investment returns." —New York State Comptroller Thomas P. DiNapoli, trustee of the New York State Common Retirement Fund.

In 1973, Munich Re, the world's largest reinsurance company whose job is to sell insurance to insurance companies, first linked extreme weather events to climate change with a study called "Munich Re Flood Risk Brochure." The kind of risks posed by weather-related events caused the company to keep a huge database of weather-related losses going back to 1980. The records show increases in worldwide droughts and flooding.

Weather-related claims are now $50 billion annually, which has been doubling every decade since the 1980s. About half of the losses associated with natural catastrophic events are covered by the reinsurance sector. Climate change is very real for this industry and is a critical risk factor for insurance rates. In the United States, many vulnerable areas that continue to incur increasing damage have been selected for development. Drought facilitates wildfires, which destroy homes and businesses, which are covered by either insurance or federal disaster assistance. Better land use planning and building codes would go a long way to avoiding the unsustainable cycle of damage and funding recovery with taxpayer dollars in a continuous cycle.

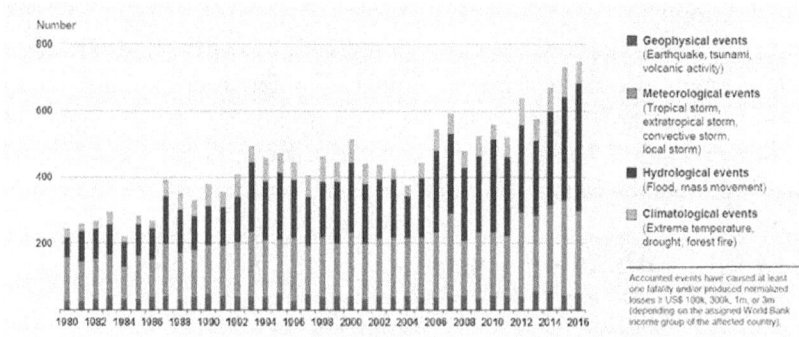

Figure 63: Munich Re's summary of worldwide catastrophe events since 1980. Red is Geophysical such as earthquakes and tsunamis and shows a stable rate of incidence, green is meteorological events such as tropical storms, hurricanes and tornadoes, blue is hydrological events such as floods and orange is climatological events such as extreme temperatures, drought and wildfires. Source: Munich Re's magazine on global catastrophes called "Topics GEO", 2016 edition.

In the United States, the government acts as the reinsurer in general, one huge reason why there is less climate activism in U.S. insurance companies than there is in Europe. The U.S. FEMA National Flood Insurance Program is an example of an unsustainable policy and a paradigm for private sector. It has not priced its policies for catastrophic risk and in average years' experiences mostly a revenue/cost-neutral basis. Then it has years such as those where Hurricanes Katrina and Sandy boosted losses to $24 billion.

Insurance companies cannot do what government does and are compelled to search for risk-appropriate rates or leave the industry. Munich Re's annual industry review is called "Topics GEO,"[267] which in 2018 celebrated its 23rd year of acting as a detailed diary of world climate events, much like BP's annual energy survey celebrating its 67th edition in 2018.

This chart shows the increase in insurance losses covering four different causes: geophysical, meteorological, climatological and hydrological. After the 2016 U.S. presidential election, we know the Republican

Party is never going to acknowledge climate change as a top issue. America—otherwise a world leader in science—elected a president who ignores established facts in favor of political ideology that places its citizens at considerable risk. It is dramatically clear that the U.S. administration is chewing on some version of denialism that's fueled by falsehoods, like the world faces a conspiracy by fraudulent scientists, that the numbers are wrong, that there is no need to panic about climate change, that it would be prohibitively expensive to address and that there are more important near-term policy issues.

The year 2017 exacted huge losses, catapulting itself to top the list as the year with the highest insured claims in history. Around the world, there were 710 natural catastrophes, 16 of which caused more than $1 billion in damages each in just the United States, which for the first time accounted for more than 50% of Munich Re's losses. The bulk of the damage was from hurricanes. Hurricane Harvey caused flooding in Houston, Texas, that was a one-in-a-thousand-year event. Then Hurricanes Irma and Maria, both category 5s, hit the Caribbean with withering force, completely disintegrating the electric grid in the disaster in Puerto Rico and amounting to $90 billion in losses alone. Even after 4 months, half the island was still without power or running water. Finally, the wildfires in California raged until January 2018, the longest and most damaging ever. For the first time it became clear as a bell that increased ocean heat was contributing to the intensity of the storms and that a warmer atmosphere can hold more moisture, contributing to floods and mudslides.

The polar vortex, the now tortuously twisting normally arctic winds, visits the Sahara and Texas, at will, with snowy conditions, while Alaska continues to warm faster than anywhere else; NOAA tells us the average temperature there in December 2017 smashed the old record by 2°F (1.11°C). California and southern Europe suffered intense dryness, causing wildfires, while northern Europe suffered torrential rainfall. In Asia more monsoon conditions caused extensive flooding along with record high temperatures.

The insurance industry is on the line for $135 billion in insured losses against total damages of $330 billion, double the 10-year average, from hurricanes and non-climate-related events such as earthquakes in Mexico. The 2011 Tōhoku earthquake and the floods in Thailand of that year put overall losses at $354 billion inflation-adjusted. The 2011 losses were mostly from the earthquake, while subsequent years have seen a profile like 2017, where the impact of natural catastrophes to do with climate becomes the norm. The inevitable rise in insurance premiums is also related to the higher yields experienced by the capital markets, supplying significant cash to the insurance industry and making reinsurance an attractive investment vehicle.

Faced with increasing claims from storms, tornadoes, floods, droughts and climate change damage, the insurance industry is scrambling to switch sides. In 2011, the Heartland Institute put out an ill-considered advertisement on a Chicago billboard that compared climate alarmists to Ted Kaczynski, the Unabomber, who had killed innocent people with letter bombs. Two dozen insurance companies, including State Farm, the Association of Bermuda Insurers and Re-Insurers, representing 22 companies that financed the Heartland Institute, immediately withdrew their support.

Weather patterns are the key issue for insurance companies. If climate change is happening and its influencing storms, tornadoes, hurricanes, floods, droughts, landslides and more, increases in damage results in increased insurance claims and litigation about claims. Property and liability insurers seek to mitigate risk, and climate change threatens them with losses on a huge scale, so some of them are paying attention. Insurance companies have actively researched the situation and are leaders in awareness of the impacts of climate change. This results in mitigation efforts via stronger building codes, better planning and design, better materials and innovative efforts to help developing countries invest in renewable energy, such as the Desertec initiative to illustrate how relatively little desert land could replace our CONG energy resources with solar with no climate impact. They have also enhanced their liability exposures by helping clients with climate risk and the environment, threatening lawsuits for failure to protect or disclose such harm. In 2009, the insurance industry's largest body, the Geneva Association, which represents the largest insurers, resolved all these concerns into a Climate Change and Economic Impact project (CC+I) and produced a report called "The Insurance Industry and Climate Change— Contribution to the Global Debate."

The failure of the COP meetings (until 2015) to discipline governments into a comprehensive global policy to reduce CO_2 emissions, identified as the major cause of climate change, culminated in the 2015 UN meeting organized in Paris. Binding agreements were being sought but not delivered. In 2014, failure of communities in Chicago to prepare for the heavy rains and flooding were the subject of lawsuits by subsidiaries of U.S. insurer Farmers Insurance. They filed class action suits citing that global warming had triggered plenty of warnings that the floods would occur, and their liabilities could have been reduced by planning.

In 2013, the Geneva Association explained in a report that the strong association between insurance liabilities and climate was leading to rising oceans. The report outlined that the technique of looking into the past to establish insurance risk was no longer useful with climate change. They now need models to analyze various scenarios but believe certain catastrophic risks may be uninsurable.

In February 2014, California, Connecticut, Minnesota, New York and Washington all agreed to participate in a National Association of Insurance Commissioners climate risk disclosure survey initiative, which was originally launched in 2009. California needs all insurers, writing more than

$100 million in premiums to respond to the 8-question survey while the other four states required insurance companies writing premiums of $300 million or more to answer. Responses are available to the public.[268]

The insurance sector has been active in the "alarmist" camp for climate change for years, informing policymakers and the public about the threat of climate change. They were the first industry to adopt public statements on the environment and form business coalitions calling on federal and state governments to enact legislation to reduce greenhouse gases. It's the reinsurance industry that bears the final responsibility and has been most active. They proactively sponsor research and work to find ways to adapt to extreme weather, especially in developing countries. Paperless billing is just one of many initiatives that have grown out of greener corporate and personal policies. Boston-based Karen Clark & Co., a hurricane modeling consultant, said that property insurance losses will increase by 30–40% in the next 20 years if hurricane wind speeds increase by only 2–5%.

After Katrina, rating agencies evaluated the capital adequacy of property insurers and raised thresholds. The possibility of a mega-disaster, the 250- year threat that has become a one-in-100-year threat, raises its head. Hurricane Katrina was considered a one-in-400-year storm, with a 0.25% chance of occurring in any one year. Poor countries are disproportionately at risk, due to their locations in tropical zones, and they have fewer resources. They have traditionally relied on emergency donations from wealthier countries during catastrophes. Although such catastrophes include earthquakes, in general we do not consider these to be caused by climate change. These sources of disaster relief may not be available from a world coping with its own disasters, and so there are increasingly initiatives addressing better climate risk management, such as micro-insurance, livelihood protection policies, loss of income policies and disaster recovery bonds as well as multinational government insurance pools using private- public partnerships. The Munich Climate Insurance Initiative was launched in 2005 in response to the idea that insurance-related solutions can play a role in adapting to climate change. A balance between carbon emitters and developing country needs in the face of climate change is sought, but the pace of change is frustratingly slow.

One form of liability that can now be insured against is that a company can be sued for harm to the environment caused by its actions or inactions. Management of a company can also be sued for failure to properly manage the company's global warming liability exposure. As yet untested legal concepts are being dusted down for action in "sustainability" cases. *Sustainability* is broadly defined for the purposes of such legal actions as "meeting the needs of the present generation, without compromising the ability of future generations to meet their own needs." To insulate themselves from such liabilities, it is in the interests of companies to ensure they cause no harm, have an emissions reduction program and plan energy conservation or renewable energy projects. The insurance industry is in a creative buzz about new opportunities and risks to insure.

Emissions of carbon can come back to haunt a company. Carbon project risk management consultancy services are now more common. Curbing global warming has inspired policies that incentivize policyholders to contribute to such efforts, including automobile discounts for driving electric or hydrogen cars or limiting mileage. Pay as You Drive (PAYD) options reward vehicle owners for driving less and combat the fact that 25% of all U.S. greenhouse gases are released by automobiles. Property policies now exist for homeowners and property managers who generate their own renewable energy and sell surplus back to the grid, cutting power outages and increasing the amount of energy that comes without pollution.

According to a climate change research organization named Ceres, buildings account for more than a third of greenhouse gas emissions.

Green commercial building is experiencing a boom, and property insurance mentions increasingly ask about emissions, allowing insurers to offer policies that enforce the greenness of a property. They also allow building owners to replace damaged buildings with green alternatives, including equipment that brings energy efficiency, heating, ventilation, air- conditioning, building recertification, inspections, low-volatility paints, ventilation, debris recycling, replacement of green roofs and replacement of renewable energy equipment. I am aware that air-sourced heat pumps can keep a building warm in winter and cool in the summer, and biofuels sourced from cellulose can answer, if necessary, almost all of the liquid fuel emissions. Existing technology can do so much to reduce these emissions right now and at hardly any greater cost if not more economically.

In a normal world, proven CONG reserves in the ground are worth trillions of dollars. CONG stock prices represent those reserves, and there is a very big risk that over the next few years, it will no longer be possible to extract that carbon; in this case, those assets would become stranded. This would be useful *if* the planet did not have a limit to the amount of CO_2 it injects into its ecosystem. A major commentator from the financial community is Jeremy Grantham[269] of Grantham Mayo van Otterloo, a Boston-based fund group whose stranded assets argument we visit in the next chapter. Insurance companies are the canary in the coal mine. They are sensing the changes because they see the losses from climate change damage claims climbing. They are businesses that are signaling a change and yet this subject is passionately rejected by the denial community to protect extraction energy businesses that are damaging the Earth's climate.

Mercer, the global financial consultancy, highlighted the risks of CO_2 emissions in a 2009 report called *Through the Looking Glass*. It has also written a report called *Investing in a Time of Climate Change*.[270] They make the point that climate issues will have an impact on investment returns and that climate change is a new "return variable" to be considered when making an investment. The report estimates that in scenarios where the temperature climbs by up to 4°C, annual returns from the coal subsector could fall between 18% and 74% over the coming 35 years.

Balancing Risks: Economic Stress Delays Action

"It is difficult to get a man to understand something when his salary depends on his not understanding it." —Upton Sinclair, American author.

During the market crash of 2008, fear gripped investors because the impact on the economy of the housing mortgage bust was unclear. There were fears that one of the principal pillars of economic stability, owning your own house, might have no meaning if the value of that house was suddenly below its cost. This threatened a fundamental assumption about our economy and civilization. In the uncertainty of 2008, the market crashed in September and October and remained weak until the spring of 2009. This was short-term risk. The best thing to do was simply not hold assets that held such risk, and so anywhere possible where there was liquidity, investors were selling and shorting the markets. The paradox for us in our fund was that we were holding a portfolio of companies whose products and technologies offered solutions to the oncoming threats of depleting CONG and climate change and addressed all the remaining externalities of CONG also. There is a huge risk humankind takes in not addressing these wider challenges.

Market timing of course was to blame, and there will be plenty of time for investors to appreciate and benefit from improving economics in the technologies that combat climate change. The market crash brought antipathy to innovation. Even today, seven years later, it is still difficult to find equity investors for early-stage companies. We can absorb short-term dips and their effect on our ability to solve long-term problems, but we cannot afford the risk of ignoring our creative, innovative capabilities for long, because they offer us a route to a higher quality of life, convenience and a chance to use our time in a way that makes us happy. We have a limited amount of time to fix our problem, and the market crash impacted our ability to do something fast. The relative success of the Paris climate talks in late 2015 resulted in an inspiring agreement where 194 countries decided they would do something about emissions and renewable energy and significant immediate changes. Along with the news that 2015 was the hottest year in recorded history and that 2016 would be even hotter came news from Standard & Poor's suggesting that at least $16.5 trillion would be spent tackling climate change as a result of the COP 21 accord, virtually doubling the world's wind and solar capacity over the next 15 years. They said that there was increasing sensitivity in the investment world about stranded carbon and that institutional investors and national pension funds were already looking closely at the fossil fuel corporations they own in their portfolios. S&P is among a growing group calling for a price on carbon to help control emissions.

In the United States, the Obama administration set a moratorium for coal leases on public lands, and New York State announced closure of its last three coal plants, the very reverse of the Trump approach. Canada banned oil tanker traffic on its western coast, and China raised its coal consumption fee 500% and will increase its renewable energy by 21%, to double by 2020. India announced tighter emissions standards for coal-fired power stations and accelerated car pollution standards. Australia resumed its volatile support for wind power but in the same breath also permitted massive coal extraction. Morocco announced a $40 billion investment in renewable energy, but the UK pressed ahead with its intentions to reduce incentives for renewable energy. Many more public and private initiatives will help support the goals of freeing the Earth from emissions over the coming years.

Professor Brian Cox has made the link between pre-hominids and the expansion of brain size from 400 cubic centimeters 3 million years ago, when *Australopithecus* was around, to our current 1,600 cubic centimeters in Irhoud 1, a 300,000-year-old fossil that basically represents modern humans.[271] This fourfold increase in the volume of our brains coincided with some extremely choppy climate changes in the parts of the world where humankind was growing up. These were not caused by humans, but we did have to react to them to survive. Professor Cox makes no bones about linking the size of the brain to the scale of the challenges of the changing environment. We are the only product of evolution so far than can rationally think up a response to a change in our environment. Let us hope that the new mental capacity we fought for is effectively brought to bear on what certainly promises to be a new reason to expand our brains in the coming decades.

Nuclear Winter

"A great many reputable scientists are telling us that such a war could just end up in no victory for

anyone because we would wipe out the Earth as we know it." —Ronald Reagan, 1985.

Nuclear impacts show up in both the climate and geopolitical externality chapters 3 and 5. The nuclear climate impact is significant. Russian and American nuclear scientists together created a dystopic view of a post nuclear engagement world in the 1980s. It was a nightmarish world of smoke-obliterated sunlight that would deprive us of food harvests and turn the world dark for at least a year, resulting in starvation and refugees. While the media has made quite a living on this postapocalyptic meme, luckily it never has yet come to be a reality.

An analysis of a war with 100 Hiroshima-sized bombs, less than half of the Indian and Pakistani arsenal, would go much further. Nine countries still have 15,695 nuclear weapons, according to the Ploughshares Fund, 94% of which are held by the United States and Russia. Any one participant can alter the global environment. There are over 1,000 times too many nuclear warheads to be safe from this outcome.

More advanced climate modeling shows that the effects of such an international conflagration would likely last for more than a decade because the smoke would rise above rainclouds 25 miles into the upper atmosphere. Yields of wheat, rice, corn and soybeans would decline by 10–40% for 5 years. The ozone layer would also be depleted, resulting in much more UV reaching the Earth's surface and the resulting impact on plants and animals.

These initial calculations were made when the world peaked at 70,000 nuclear warheads, a number now vastly reduced and still under pressure to come down, not least due to this awful vision. George H. W. Bush unilaterally reduced the U.S. arsenal as the Soviet Union collapsed and plans exist to continue the process.

How We Know It's a Man-Made Problem

"Intelligence is the ability to adapt to change." — Stephen Hawking.

I may be biased and have it all wrong, so if we are to believe anything, it should be based on good, solid scientific evidence. So much public money has been spent in the pursuit of the scientific truth of this matter, and thousands of scientists from hundreds of institutions in hundreds of countries have reviewed the evidence in the spirit of skepticism, which is what science is at heart, that we can simply quote the evidence and make a clear conclusion. The stuff that one scientist observes is checked by many others. It took decades before scientific evidence could prove that smoking caused cancer. Today we have even better evidence that climate is caused by humans. Much of the issue can be divided into the natural factors that have affected the Earth's climate, like volcanoes, sun and orbital mechanics versus the carbon emissions of human activities forcing warming. One really great chart describing this was put together by Bloomberg New Energy Finance using NASA data,[272] but even the Environmental Protection Agency had the same thought expressed in the chart in figure 64, where the various natural and human forces are graphically demonstrated.

Normally speaking, the Earth's current orbital cycle is 16,000 years ahead of a new ice age, and the solar cycle, as illustrated by declining sunspots over the last 20 years, has resulted in a smaller energy output from the sun. Both factors normally result in cooling. Instead, there has been a sudden release of CO_2 from the acceleration of human activities. We are producing emissions from burning fossil fuels, burning forests, putting waste into landfills, animal husbandry, agricultural fertilizer and industrial activity. Over the last 200 years, this has caused the atmosphere to heat up rapidly.

The Earth is retaining more heat at a time when orbital mechanics and solar activity both point to cooling. The reason is the extra greenhouse gases emitted by us. Multiple, evidence-based scientific studies have concluded this is so, and there is almost total consensus on this point. Basic accounting for the extra CO_2 is all it needs.

Separating Human and Natural Influences on Climate

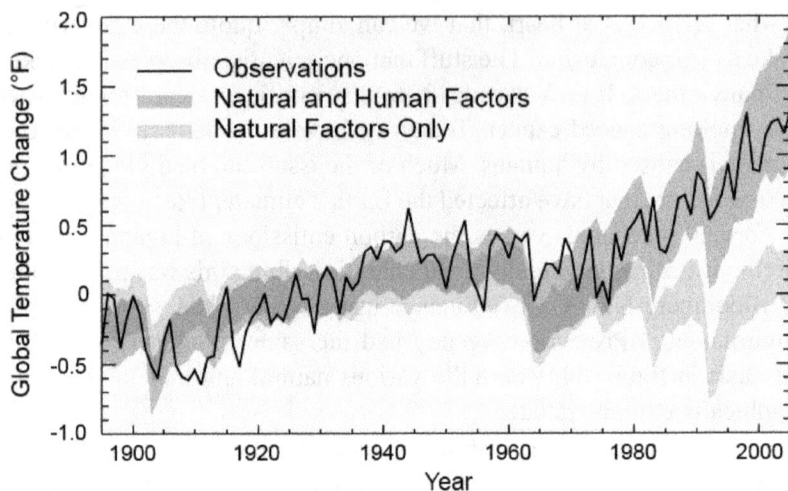

Figure 64: EPA graphic discriminating between the impact on rising global temperatures between natural (green band) and human factors (blue band). Source: U.S. EPA (prior to EPA website material being removed by the Trump administration in January 2017).

The quantities are enormous. In the last 200 years, human activities have suddenly released the carbon from well over a trillion barrels of oil, with accompanying huge quantities of peat, gas and coal, into the Earth's oceans, atmosphere, soil and geological systems. In 1900, there were 1.65 billion humans alive and only a fraction of them drove cars or worked in a way that released CO_2. Luckily, along with the increase in population also went an increase in our ability to observe and collect information, temperatures, measurements of distance, quantities of energy, with ever-increasing accuracy. Satellites and cameras, gas sensors, infrared detectors, analysis of ice cores, wood, lake bottoms, ancient animals and many other approaches have produced large amounts of data that have been peer analyzed by thousands of scientists, operating with skepticism and on meager grants.

Today we have 7.4 billion people and almost everybody in the developed world and much of the developing world who can drives a vehicle, works in a factory, eats foods that has traveled thousands of miles or flies frequently. In 2014, there were 109,000 flights a day worldwide! The Air Transport Action Group (ATAG) released a document called "Aviation Benefits Beyond Borders"[273] in 2014 and explained that 65 billion people had paid to fly in the last 100 years and that currently 3 billion people a year climb onboard an airplane that takes them from A to B with considerable release of CO_2. There are a million passengers in the air at any given moment today! Our use of fossil fuel energy has accelerated, and in 2016 we extracted 96.6 million barrels of oil per day, or 35 billion barrels a year, from the ground globally.[274] It's all the more satisfying that carbon-neutral transportation fuels, both ground and aviation, made from cellulose are able to replace fossil fuels in the near future.[275]

This has caused the atmospheric proportion of CO2 to climb from 280 parts per million in 1860 to over 405 parts per million in 2016. This is highly likely to continue to accelerate, as shown by the Keeling Curve. We are today witnessing increased temperatures, weather volatility, wildfires, droughts, spreading disease geography, flooding, reduced groundwater availability, threatened agriculture and food production, human and animal migration and species extinction at a rate unparalleled by any historical events, natural or human caused. The volumes of CO_2 released by human activities are easy to track because we know how much carbon we are releasing. Increasing temperature is now something that has been observed and measured for a long time. The long history of climate science puts understanding and measuring all the variables technically well within our capability. A 2016 study[276] found that without the human emissions from coal and oil, there is very little possibility that 13 out of the 14 hottest years on record would have happened. We have become skilled observers.

The evolution of the conclusions of the Intergovernmental Panel on Climate Change (IPCC) reports shows an inexorable increase in the certainty of evidence of human complicity. Here we look at the evolving conclusions:

1990—The report **did not quantify** the human contribution to global warming.
1995—"The balance of evidence **suggests** a discernible human influence over time."
2001—Human-emitted greenhouse gases **are likely** (67–90% chance) responsible for more than half of Earth's temperature increase since 1951.

2007—Human-emitted greenhouse gases **are very likely** (at least 90% chance) responsible for more than half of Earth's temperature increase since 1951.

2013—Human-emitted greenhouse gases **are extremely likely** (at least 95% chance) responsible for more than half of Earth's temperature increase since 1951.

We have used proxy indicators to show us the relative concentrations of CO_2 and temperatures going back 20 million years that illustrate that today is hotter than at any other time in that span.

Oxygen Levels are Decreasing

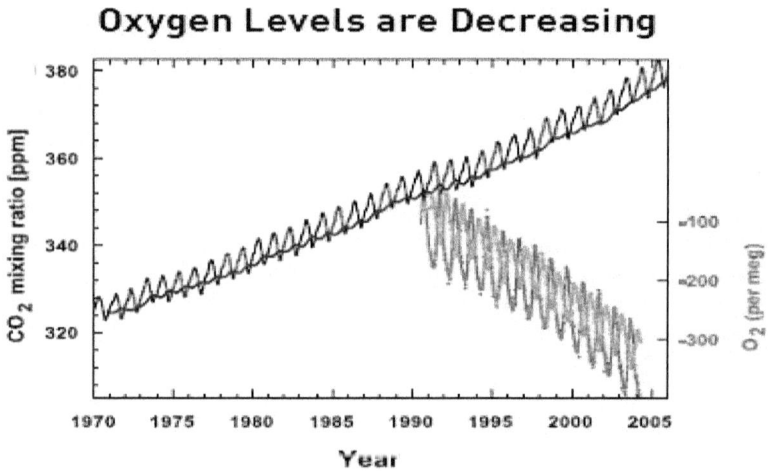

Figure 65: Decreasing oxygen levels measured from the late 1980s. Source: Figure 2.3(a) in Chapter 2 [PDF] of the IPCC report.

Our research has revealed that nature alone does change CO_2 concentrations, but it takes thousands of years. Humanity has done this in 200 years, without much indication of a slowdown. If we have been burning all the fossil fuels, then the chemistry of our atmosphere will reflect unmistakable changes caused by this human fingerprint.

Anything that burns consumes oxygen. Is the concentration of oxygen in the air decreasing? Lack of suitable instruments to measure these changes meant that we were not able to answer this question until the late 1980s. Then we checked, and *voila!* Figure 65 from the IPCC report shows decreasing oxygen levels in blue and red alongside the increasing CO_2 levels in black.

The CO_2 we release into the atmosphere from burning fossil fuels is carbon-12 (^{12}C) and not the radioactive isotope carbon-14 (^{14}C), meaning that, while it was cooking deep underground for millions of years, it had already decayed to its stable isotope form (^{12}C). This is a smoking gun and a marker that inculpates humanity directly.

Another study[277] looking at anthropogenic forcing found that almost two- thirds of the atmospheric and ocean temperature impacts can be confidently attributed to humans. Mathematical models matching the natural and anthropogenic components of climate change now accurately show an unequivocal relationship to reality when the human forcing is included. Despite the criticism that climate models receive from the denial community, they are now, as you might expect after millions of rational man-years of development, much more accurate and detailed. Figure 66 echoes the EPA chart above.

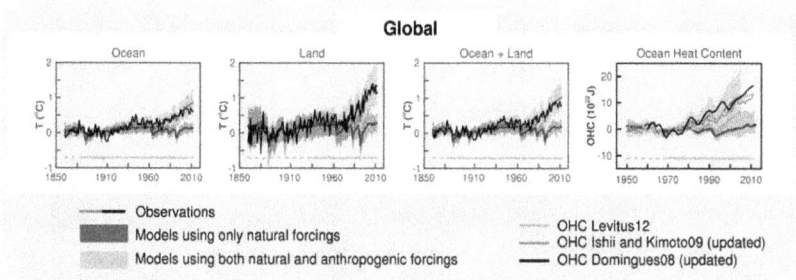

Figure 66: Models showing the impact of natural and human drivers. When they are combined they match the observed reality best. Source: IPCC.

Even if this is not an issue that is as black and white as it is, there would be no sense in taking the risks we are taking, and plenty of sense in welcoming the innovation cycle and its boost to the economy that human beings bring to these challenges. Even if these changes were not human caused, we would do well to avoid the impacts of the effects.

Chapter 4: Depletion

CONG Externalities

"Valuations of the oil and gas sector still assume that they will be able to take all proven and probable reserves out of the ground and burn them. Based on credible data we cannot be allowed to do that." —Jeremy Grantham, "Unburnable Carbon"

Burning wood or dung over the millennia did not entrain any externalities because volumes of wood grew so much faster than humans could burn it and the volumes burned were tiny, so the CO_2 emitted was therefore minimal. The externalities of fossil fuels and nuclear power have only come about as the volumes of fossil fuels (for which I use the CONG initialism: coal, oil, nuclear and gas, and sometimes, coal, oil, natural gas) combusted have "exploded" in the last 200 years. Media attention to the evils of CONG energy only focuses on one or two of the extensive insults we submit to from these sources of energy while, in fact, and obviously, we are hit hard by them in several dimensions. As we have discussed in earlier chapters, we expose ourselves, needlessly, to eight different and significant sources of damage. We've just had a hard look at one of them, the climate issue, which revealed that CONG has led us into potentially irreversible danger. This fourth chapter concerns CONG's seven other egregious impacts that we still have not yet covered. We will also demonstrate that renewable energy in any of its configurations does not hold a candle to the damage we are experiencing from CONG.

The idea that fossil fuels have externalities has been gnawing in our common consciousness for a long time. The early climate scientists recognized the dangers but there was no real environmental movement until 1960 after the publication of Rachel Carson's book *Silent Spring*,[278] which was an alarm bell against the use of pesticides and chemicals in our environment. This use of pesticide was well intentioned and killed mosquitoes, but also killed birds. The book had a huge impact, polarizing a nascent environmentalist movement against the chemical industry, which tried to fight back fiercely.

DDT was initially successful as an insecticide and the industry blamed the environmentalists for having effectively sentenced tens of millions of people in developing countries to death by malaria. Rachel Carson herself developed breast cancer and DDT was later identified as a chemical carcinogen, although a direct causality could not be determined. Also, insect populations developed resistance to DDT. Chemical companies had a profitable business (otherwise they surely would not be sensitive) that they fought to defend even though it caused cancer and was not an effective long- term strategy against insects anyway. It was dangerous and ineffective. This was a very similar situation to the tobacco industry.

Today, the idea of a sixth extinction event, with indicators like the plight of bees due to insecticides used for agriculture, is an echo of this initial awakening of the environmental movement. Rachel Carson's book stirred public concern about the effects humanity was having on the environment and led to the establishment of the U.S. Environmental Protection Agency (EPA). President Richard Nixon signed the agency into law on January 1, 1970. In turn, it also promptly succeeded in banning DDT, a common pesticide at the time.

Congress requested a report on the cost of the energy system in 2005. In 2009, the National Research Council (NRC), a branch of the U.S. National Academies, finally produced the study called "Hidden Costs of Energy: Unpriced Consequences of Energy Production and Use."[279] It analyzed the hidden costs that we are calling external costs or externalities of producing energy, and it used the most recent year that full data was available, 2005. The NRC report did not assess the effect of energy production on climate, the environment, inflation or national security (geopolitics), but it did at least dwell on the effects of energy production on health. These major gaps, of course, mean that its final tally of the load on the economy that CONG represents was significantly understated, but it is useful as a guide to our own efforts here.

I am making a more "thorough" multifaceted review of the external costs. The big number was that $120 billion was paid by the government to cover health costs, imposed mostly by the coal industry, which accounted understandably for almost exactly 50%, or $62 billion, in hidden damages due to pollutants such as sulfur dioxide, nitrogen oxide, particulate matter, mercury pollution and other emissions. Given that coal at that time accounted for slightly more than 50% of electricity generation, that makes sense. The report also found that $56 billion was the sum of health damages arising from the use of gasoline and diesel engines in transportation. Cars accounting for $36 billion, and heavy-duty vehicles, the other $20 billion.

While it's no surprise that congressional curiosity conspicuously skirted the many other consequences of carbon dioxide such as the climate concerns, it was nonetheless a considerable conclusion attracting further attention to the fact that our energy paradigm was hurting us badly and that we needed to replace it with benign, drop-in solutions. Luckily, there are some, and their impact is being delayed by social forces unwilling to transition from the status quo. Where have we heard that before! (Hint—the tobacco industry.)

Depletion

"We use 1 million years of fossil fuel in 1 year!"
—Professor Sir Harry Kroto.

"I was in New York in the 30s. I had a box seat at the Depression. I can assure you it was a very educational experience. We shut the country down because of monetary reasons. We had manpower and abundant raw materials. Yet we shut the country down. We're doing the same kind of thing now but with a different material outlook. We are not in the position we were in 1929–30 with regard to the future. Then the physical system was ready to roll. This time it's not. We are in a crisis in the evolution of human society. It's unique to both human and geologic history. It has never happened before and it can't possibly happen again. You can only use oil once. You can only use metals once. Soon all the oil is going to be burned and all the metals mined and scattered." —Marion King Hubbert, Shell geophysicist and professor and discoverer of Hubbert's Curve.

On March 8, 1956, a geophysicist named M. King Hubbert presented a paper,[280] "Nuclear Energy and the Fossil Fuels," at the spring meeting of the Southern District of the American Petroleum Institute, which was holding a three-day conference at the Plaza Hotel in San Antonio, Texas. He had been asked to fill the role of a presenter "who could give them a broad-brush picture of the overall world energy situation." He stated that he expected the American production of oil would reach a peak between 1965 and 1970.

By their very nature, CONG resources are finite. They run out. They were produced agonizingly slowly in nature from the accumulation of organic material, dying animals and plants, in rock strata over hundreds of millions of years of geological time. It's called the Carboniferous period for a reason. We have effectively burned millions of years' worth of this slowly accumulated energy source in the space of 150 years, placing much of the CO_2 from the period into the atmosphere from where it has been absorbed by the oceans, land and plants as best they can. This is on top of any CO_2 that was already in the system.

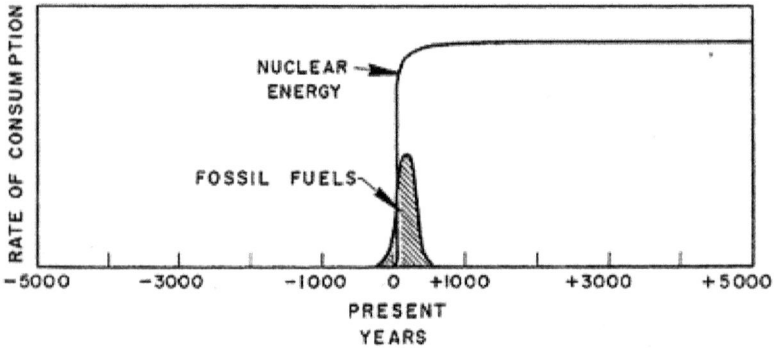

Figure 67: Relative magnitudes of possible fossil fuel and nuclear energy consumption seen in time perspective of minus to plus 5,000 years. Written by M. King Hubbert about the impact of nuclear power. The chart demonstrates a then common sense of optimism about the way in which nuclear power was a complete energy solution. Source: M. King Hubbert, 1956.

	2017 ENERGY SOURCE	MTOE	2017 GLOBAL RESERVES	2017 CONSUMPTION	%	YEARS REMAINING	10 YEAR GROWTH RATE %	% FROM 2016/2017	YEARS REMAINING	FINAL YEAR
C	Coal		584,441	3,732	28%	157	0.74%	-0.01%	69	2083
O	Oil		239,300	4,622	34%	52	1.05%	1.42%	45	2059
N	Nuclear		Huge	596	4.4%	160,000	-0.37%	0.88%		Never
G	Natural Gas		170,793	2,840	21%	60	2.2%	2.69%	39	2053
Total Fossil Fuel Energy Consumption		994,533		11,790	87%					
Total World Primary Energy Consumption				13,511	100%					

Figure 68: Global reserves of CONG in millions of tonnes of oil equivalent (MTOE) and the changing consumption over a decade. Source BP Statistical Review 2018.

These figures from the BP Statistical Review, 2018 (figure 68), show that 11,790 million tons of oil equivalent (MTOE) of fossil fuels, representing 90% of all energy consumed, were burned in 2017 alone. It's also worth stating that all this fossil fuel originally came from the sun anyway. Millions of years' worth of seasonal plant biomass, growing and dying, was slowly compressed as a sedimentary rock formation and then increasingly cooked by heat and pressure as the strata it lay in got deeper and deeper. The plants were made of the H, C and O available in the Earth's water and atmosphere, via photosynthesis, which clips the C from CO_2 and adds it to the H and the O in H_2O to make carbohydrates, which then, over geological time, became hydrocarbons.

If you simply divide the known reserves by the current consumption figure, you obtain a number that is a linear path to the number of years of this resource that remain, as in the fifth column of figure 68. However, more realistically, we need to take into account that consumption is growing, driven by the vision of a higher quality of life, demand for air travel doubling every 15 years, and so forth. In this case the time remaining for these proven reserves shortens dramatically, as shown in columns 8 and 9. If nothing changes, at about halfway through the 21st century, CONG resources will be severely curtailed, and the climate is already objecting strongly.

That linear figure of dividing the last year's consumption into the proven reserves number has no real bearing on how many years of reserves actually remain. Here, I have taken the last decade's growth rate and calculated how much coal would be burned at that growth rate over time and checked the year at which the proven reserves will be consumed. At current usage rates, oil will be exhausted by 2059 and gas by 2053, without new discovery. Coal will effectively be gone by 2083. Coal consumption has been declining since 2014. If this remains the case, then we can see potentially much of the carbon left mercifully unburned, in the ground.

These numbers are alarming and require that alternative sources of energy be put in place not just to replace the CONG energy source but far sooner to offset the already burned carbon. These numbers show that there is still significantly less time remaining than suggested by the linear figures. Over the last decade, coal use has grown but recently its share of the energy market has declined. It only grew by 0.53% from 2013 to 2014 and declined by −3.25% over 2014–15. Its 10-year average growth rate is 0.74%. At this level of growth of coal consumption, there would be 69 years of coal remaining, oil would only have 45 years and gas only 39 years remaining. This means that the 21st century is definitively the century in which new energy sources and preferably sustainable ones have to be in place.

CONG RESERVES (MTOE)	1997	2007	%	2017	%
COAL	534,668	474,906	-11.18%	500,256	5.34%
OIL	162,694	199,794	22.80%	237,524	18.88%
NATURAL GAS	113,048	144,289	27.63%	170,764	18.35%

Figure 69: Proven reserves numbers over two decades, showing an increase as we locate further supplies of fossil fuels, reflecting new methods of extraction. For coal, however, there are no new reserves discovered. Source: BP Statistical Review 2018.

Coal reserves have dropped by over 6.44% in the last 20 years. In the same time period, oil reserves have expanded as more deep water and polar reserves have been confirmed along with significant finds in Latin America. Oil sands and shale oil have further increased oil reserves in non-OPEC countries. During this same period, gas reserves have also steadily built up all over the world as shale gas has been released by new technology. Both oil and gas reserves have climbed by about 25% per decade as these discoveries are confirmed. It's possible that these continual new gains might extend the life of fossil fuels, but I'd dare to say that it's only new technology releasing reserves locked in previously impenetrable rock. You can't say never, but no such equivalent new resources are likely to be found in the future, so such CONG growth is most likely now over.

The estimated recoverable reserves (ERR) of coal, reported by the U.S. Energy Information Administration (EIA), are often said to represent 200 years of supply. The EIA has also acknowledged that these reserves have not been analyzed for profitability of extraction.[281] The years of availability figure is therefore deemed to be far lower because of constraints like coal mine life span; economic recoverability; and geological, legal and transportation factors that affect existing coal production and expansion resources. These constraints end up reducing that number to as little as 20– 30 years.

Nuclear reserves[282] are literally subject to $E = mc^2$. The amount of energy available is so enormous for just a small amount of matter. It would be nice to be able to obtain that energy from any form of matter, but some elements make it easier than others. Actinide metals such as thorium and uranium offer us access to isotopes of these elements that split apart when hit by a neutron, releasing fission products that are minutely lighter than the original unstable source element. This difference in mass is energy captured as thermal energy, which is used to make electricity by common steam turbine.

This is even more important given that the proven CONG carbon reserves are still in the ground and must not be burned, since they will tip the climate into an unstoppable disaster for humanity. Luckily, solutions exist and are plentiful. The question remains, will we deploy them in time? Will governments be able to work in isolation of lobbies whose goals are to resist necessary changes in search of continuation of the unsustainable status quo?

Chapter 5: Geopolitics

Geopolitics

"...the United States has been occupying the lands of Islam in the holiest of places, the Arabian Peninsula, plundering its riches, dictating to its rulers, humiliating its people, terrorizing its neighbors, and turning its bases in the Peninsula into a spearhead through which to fight the neighboring Muslim peoples." —Osama Bin Laden, 2005.

After the industrial revolution, coal use transitioned to oil use. Oil is not distributed evenly around the planet and so the fate of billions of people and dollars depended on whether a country was a buyer or a seller. In this chapter I explore some events triggered by geopolitics, starting with the unique technological energy edge that the Germans developed in the Second World War and the background to the Middle Eastern conflicts with Israel and Saudi Arabia and the two U.S.-led Iraqi wars. Since the Middle East has been traditionally associated with oil, it's no surprise that this region is also the focus of significant struggles frequently driven by energy. I also describe the Falklands Conflict in which I was involved as an active combatant and reveal the energy side of the story, which is not often told. Also, I explore the different kinds of issue that countries with large reserves of energy run into with their balance of payments, as is the case with Venezuela. Finally, it is useful to explore the impacts of both peaceful and warlike nuclear geopolitics, which offers more than enough material with the situations of India and Pakistan, Iran and North Korea. I'm not going to retell all the events, merely highlight the energy part of the issues.

The Middle East Becomes the Mecca of Oil

In 1931, The United States signed a significant treaty with Saudi Arabia, even though at the time it was managing its interests in the Middle East via its embassy in Egypt. A wealthy Chicago businessman and Arab specialist, Charles Richard Crane, was originally working on waterworks in the Saudi desert when he detected oil in 1931. He was instrumental in the formation of the Californian Arabian Standard Oil Company (CASOC). The British had been protecting Saudi Arabia from the Turks and had a favored status with the Saudis. The kingdom became an official country in 1932 under King Abdulaziz Al-Saud.

More oil reserves were discovered at Al Khobar in 1938. This led to the discovery of Ghawar, the world's largest onshore oilfield, and Safaniya, the world's largest offshore oilfield and the world's largest oil reservoir. The Saudi need for economic gain and the U.S. need for oil made for an ideal partnership in respect of which the United States installed an embassy in Jeddah.

Even distraction by the Second World War and support for Israel failed to derail this mutual favored status. An event in Dharan in 1940 reminded the United States of the reality of energy in a contentious world. Italy, fighting with the Axis powers, bombed an oil installation in Saudi Arabia. The bombing mission[283] of October 18, 1940, was remarkable. It was still the opening months of the war and the design of bomber aircraft had not developed much since the First World War. The four Italian planes were Savoia-Marchetti SM.82s, with fabric-covered fuselages, wooden wings and three, 9-cylinder, 880-horsepower Alfa Romeo propeller engines. They each carried 3,310 lbs. of bombs and 1,300 gallons of fuel. The mission was the brainchild of test pilot Captain Paolo Moci, who participated under the leadership of Lieutenant Colonel Ettore Muti.

Their mission was to bomb the refinery at Manama, Bahrain, a British protectorate, to cut fuel supplies to the Royal Navy. They took off from the island of Rhodes, flew across the Eastern Mediterranean over Cyprus into Lebanon, into Vichy French Syria, then south through British-controlled Iraq to the Arabian Gulf to target Bahrain. One aircraft got lost in nighttime clouds after they passed Damascus and its pilot, staying focused on the mission but geographically challenged, mistakenly attacked the CASOC oil facilities in Dharan in Saudi Arabia instead of the oil refineries on the island of Bahrain. He did more damage to those facilities than his colleagues did to the ones in Bahrain. The other three planes remained focused on their original target, however. Unsuspecting, the British even turned on the Bahrain airport runway lights when they heard the planes' engines, not conceiving that there could be an attack so far from enemy territory.

Fortunately, the bombs did very little damage despite the targets being well lit. The pilots were met with zero resistance and now, with lighter aircraft, they had to fly west across the Arabian Peninsula, cross the Red Sea and find Eritrea, at the time in Italian hands. Miraculously, the missing pilot was found, and the four planes found their main home airport, Massawa, which happened to be under attack by the RAF that same day, and so Captain Moci was forced to divert the flight 35 miles south to Zul'a. Finally, after 2,610 miles, the equivalent of flying across the Atlantic Ocean, which took 15 hours and 30 minutes at 170 miles per hour, all four planes touched down safely in Africa. Each plane had less than 40 gallons of fuel remaining. This was a remarkable feat of airmanship and a record for a bombing mission. It caused significant ripples among the Allies, even though the bombs did little physical damage.

Necessity Was the Mother of Invention for Germans in WWII

German ingenuity in the Second World War was brought to bear on the issue of energy. Hitler's advance into Russia and North Africa were partly motivated by a need to obtain control of energy supplies to supplement the meager supplies of oil at home. Perhaps a better strategy would have been to use all its might to control a significant source of oil first and then prosecute the war. Instead, Germany had to fall back on its prodigious domestic supply of coal alone. This meant that its scientists were working hard at technologies that have survived into the modern age.

German sensitivity to the importance of energy derived from the First World War, when shortages of foreign exchange meant they were unable to import energy and were instead dependent on their own peacetime resources. By 1933, when Hitler first became Chancellor, the Germans were pioneering a technique for the generation of liquid fuel from coal. In 1938, Germany used only 44 million barrels[284] of oil annually, 60% of which was imported. For comparison, total annual consumption of oil was 76 million barrels per year in the UK, 183 million in Russia and 1 billion barrels in the United States (compared with a global 95 million barrels per day, or 34.6 billion barrels of oil annually, today). These imports to Germany ceased the moment war was declared except for controlled areas such as Romania and Russia, which was still observing the "friendship" treaty that the two countries had signed in 1939.

In June 1941, Germany invaded Russia in Operation Barbarossa, completely ignoring that treaty, partly to secure energy. It targeted the oil fields at Maikop in the Caucuses as well as the larger wells at Grozny and Baku. The wells pumped 19 million barrels, 32 million barrels and 170 million barrels, respectively, per year. In its invasion, Germany failed to take the latter two and Maikop was destroyed by the Russians themselves. The Germans only managed to extract 4.7 million barrels from Russia during the war, the same as they would have received if they had not invaded.

The Fischer-Tropsch method of extracting usable liquid fuels from coal was still in the R&D phase when the war began and produced quantities of diesel-like fuel. To make gasoline from coal, it required to be hydrogenated. Coal is a hydrocarbon with a small amount of hydrogen, whereas gasoline is a hydrocarbon with a lot of hydrogen. The challenge was to add hydrogen (hydrogenation) to the coal molecules. It turns out that with pressure and a catalyst it works. If you add tetraethyl lead to gasoline, it increases the octane required for aircraft engines, so this method became the most relied upon to generate the liquid fuels required. The license for this additive was acquired in 1935 from the American holder of the patents.

Fischer-Tropsch reactors generated a maximum of 2.8 million barrels per year, but the hydrogenation factories generated another 34 million barrels and synthetics provided 50% of all liquid fuel. By 1943, total production from all sources amounted to 71 million barrels of which synthetic fuels amounted to 36 million barrels. Fuel rationing cut supply to all but essential services. To cope with this, usage of solid fuels in transportation grew very fast and wood chips, lignite, coal, gas and coke and peat moss were used to effect significant savings in gasoline. Thousands of cars and trucks were converted and equipped with a device shaped like a water heater that extracted combustible gas from the solid fuels.

The octane rating of a fuel controls the compression ratio which in turn speaks to the power output of the engine. Power, important for aircraft, requires higher octane, which resulted in a 15% increase in desperately needed speed, longer ranges and increased flying altitudes. Before the war, two hydrogenation plants were in production. By the start of the war, seven such plants were in operation and two others were under construction. Four more high-octane aviation plants were built. Plans to add even more by 1942 were finally given the go-ahead, but Allied bombing raids on the existing hydrogenation plants managed to destroy production capacity, ushering in the end of the German war effort. Today, the entire peak German wartime production is consumed in just four and a half days. Lack of energy in Germany was the driver that caused the invention of new resources. Human beings are capable of adapting to tough circumstances, but they often need to be pressurized first.

Shortly after the war, another nation whose politics resulted in it being isolated and constantly under economic embargo was also gasping for fuel. It was a country with lots of coal, but not much oil. South Africa was to become the champion of solids-to-liquids technology by operating three Fischer-Tropsch plants, which even today still supply that economy with liquid transportation fuels.

Post–Second World War Middle East

The United States opened a consulate in Dharan in Saudi Arabia in 1944. In 1945, Roosevelt and King Ibn Saud signed a treaty onboard the USS *Quincy* after the ship had sailed from the Yalta Conference, where it had hosted Churchill and Stalin. At the start of the Cold War, the Saudis were inclined to be friendlier with the Egyptians, who were getting closer to the Russians. In 1956, however, Eisenhower's opposition to the British, French and Israeli plan to seize the Suez Canal was stimulated by his need to resist the encroachment of the Russians in the Middle East. The new Saudi king admired this and decided to start cooperating with the United States, which were eager to keep Saudi as a bulwark against communism. U.S. military bases were established. In 1964, Faisal became king and cooperated with the United States until the Yom Kippur War in 1973, when he was politically unable to remain aligned with the U.S. position of unconditional support for Israel. The result was that Saudi Arabia joined the rest of the Arab nations in the oil embargo. The United States worked to recover the relationship by selling military technology, which the Saudis were able to buy with their oil wealth.

The great powers, Britain and the United States, have long influenced the affairs of other nations, seeking leaders of their stripe, able to, among other things, maintain the flow of energy at almost total moral cost and at the expense of the Golden Rule, "do unto others as you would have them do unto you."

After the Second World War, President Truman founded the CIA with a budget of $82 million to collect and analyze global intelligence. Dwight Eisenhower increased the budget 10-fold, creating a veritable army for the U.S. administration's executive branch. There was a series of secret events coordinated by the United States from the overthrow of democratically elected Mossadegh in Iran, to the change of the Arbenz government in Guatemala, the installation of a pro-Western government in Egypt in 1954, with the follow-up dissuasion of the British intervention there in 1956. Further interference in China and the Congo brought attention from Congress, and so a report was commissioned and led by Robert Lovett, former secretary of defense, and David Bruce, a distinguished diplomat. The report,[285] which was referred to briefly and in some detail was delivered to Eisenhower in 1956. It then disappeared, but not before it identified that the CIA, in particular, was making mischief globally and answering to nobody. This is an excerpt:

> *Should not someone, somewhere in an authoritative position in our government, on a continuing basis, be . . . calculating . . . the long- range wisdom of activities which have entailed a virtual abandonment of the international "golden rule," and which, if successful to the degree claimed for them, are responsible in a great measure for stirring up the turmoil and raising the doubts about us that exist in many countries of the world today? . . . Where will we be tomorrow?*

Unequal Distribution Creates Vulnerable Chokepoints

As a parachute lieutenant, as mentioned in the introduction, I was privileged to spend some time in the Oman on the southeast side of Saudi Arabia and just north of Yemen. Our training exercise involved moving soldiers through the mountains, culminating in a live firing attack on an unsuspecting desert mound. On the way, we visited small communities to offer medical services to anybody in need. This exercise was about moving over difficult terrain. At one stage,

STRAIT	BPD	%
Turkish	17.00	35.64%
Malacca	15.20	31.87%
Suez Canal	4.50	9.43%
Bab El Mandeb	3.80	7.97%
Danish	3.30	6.92%
Hormuz	2.90	6.08%
Panama Canal	1.00	2.10%

Figure 70: Oil trade maritime distribution chokepoints. Source: EIA.

I announced a five-minute rest stop for my platoon near the top of the Jebel Akhdar mountains overlooking the Straits of Hormuz on the Persian Gulf. We saw the lines of supertankers carrying crude oil to their markets and the empty tankers returning for a refill. The navies of several countries patrol there—at a cost not reflected in the price of a gallon of gasoline—to ensure that these essential tankers are able to get their strategic cargoes to their destinations. This location in global geography has long since been seen to be vulnerable and yet it is one of the most essential parts of the oil energy regime.

The supply of oil is essential in today's economy. There are seven major oil shipping chokepoints in the world that control 84%, or 56.5 million barrels per day, of the world's oil that is moved on the ocean. This is the focus of global oil security. Even a temporary blockage at one of these points can result in a supply and price shock for the nations concerned with that particular chokepoint. Adapting routes to allow for any number of issues, such as piracy, terrorism, accidents or spills, can change the economics for oil-trading nations.

OPEC and the First Oil Shock

The U.S. relationships with both Israel and Saudi Arabia led to massive geopolitical energy complexity. The Third Arab-Israeli War in 1967, also known as the "Six Days War," was the result of turbulent relations that had never really recovered from the 1948 Arab-Israeli War, and I mention it here because it's an amazing story and hardly ever told as being one of the reasons for the later Yom Kippur War, which resulted in the big 1971 oil shock.

Israel's neighboring Arab countries had been involved in many previous border skirmishes and there was a United Nations Emergency Force (UNEF) in the Sinai to keep combatants apart. Egyptian President Abdel Nasser, on bad Russian information, was told that the Israelis were preparing an attack on the Syrian border. He dismissed the UNEF forces and took their positions. Iraqi, Egyptian and Jordanian troops took up positions in Jordan. Egyptian military forces were deployed along the Israeli border east of the Sinai Peninsula, and tensions came to a head.

On June 4, the Israeli cabinet voted to go to war, and on June 5 initiated a fast attack that took the Egyptian forces by surprise. The Israeli air force succeeded in destroying almost the entire Egyptian air force in a series of preemptive attacks and simultaneously launched a ground offensive into the Gaza Strip and then into the Sinai Peninsula. President Nasser ordered the evacuation of the Sinai Peninsula, and the Israelis chased the Egyptians across the Peninsula all the way to the Suez Canal, inflicting significant casualties on the desperate retreating Egyptians as they did. On June 8, the Israelis reached the western coast of the Sinai Peninsula.

Both Jordan and Syria had been induced to attack in the north of Israel during the first Israeli airstrike, convinced that Egypt had repulsed the Israelis. On June 5, the Israelis fought back against the Jordanians, turning the direction of battle and seizing East Jerusalem, Bethlehem and Nablus. King Hussein of Jordan ordered his troops to fall back, and they moved to the east of the River Jordan, leaving the Israelis in control of the West Bank. Also, in the evening of June 5, the Israelis launched another air strike against the Syrians and destroyed two-thirds of their air force, and by June 9, Defense Minister Moshe Dayan ordered his troops to occupy the Golan Heights, a strategic and valuable piece of high ground that rises to a 1,700- foot-high plateau above the Sea of Galilee to the north of Israel. By June 10, the Israelis succeeded in breaking through the extensive fortifications, forcing the Syrians to retreat eastward. On June 11, a ceasefire was signed. A thousand Israelis had been killed, and they had tripled the land under their control, vastly improving the defensibility of the small country.

There were over 20,000 Arab casualties. The Israeli success resulted from surprise, initiative and good execution. The results were that Arab populations in the seized territories became refugees: 300,000 West Bank Palestinians fled to Jordan, and about 100,000 Syrians left the Golan. Remaining Jewish minorities in Arab countries were forcibly expelled.

Humiliated by the loss of territory, Egypt, now under President Anwar Sadat,[286] was determined to recover the Sinai. They formed another alliance with Syria to do so. Israel, Russia and the United States did not believe that the Egyptians would actually attack, believing their deliberate misinformation about lack of spare parts, poor training and equipment. The Israelis nevertheless reinforced their positions on the Golan Heights, which was later to prove critical.

This time around, the Jordanian king decided not to participate but met with Sadat and Assad in Alexandria. They needed to know where he stood. On the night of September 25, King Hussein then flew to Tel

Aviv to tell Israeli prime minister Golda Meir of the impending Syrian attack. The Israelis did not believe there was anything the Egyptians were capable of doing and ignored this and 10 other warnings about an impending attack they received in September. The Israeli success in the Six-Day War had cemented the effectiveness of a preemptive attack, but the leading members of the intelligence community did not believe the Egyptians had it in them to fight back this time.

On October 6, the same day as the first attacks were launched, Henry Kissinger sent a telegram to Golda Meir saying, "Don't Pre-empt." Most OPEC customers were afraid of supplying or helping Israel prior to any fighting because they did not want to have their oil supplies interrupted, and that included the United States.

Egypt and Syria both launched the 1973 Arab-Israeli War on October 6 to recover pride and lost territory. All the action took place on Arab lands. The destruction of Israel was not in the minds of the Arab leaders at this time, although the Israelis could not be certain of that. Egyptian and Syrian forces crossed their ceasefire lines and entered into the Golan Heights and the Sinai Peninsula, respectively. The attacks happened during Ramadan, which coincided that year with the Jewish holy day of Yom Kippur. Egypt crossed the Suez Canal with massive force and penetrated deeply into the Sinai unopposed. After three days, Israel managed to stop the Egyptian advance. In the Golan Heights, the Israelis mounted a counterattack that pushed the Syrians back to their previous ceasefire lines and then punched harder, pushing into Syria with a four-day offensive, eventually firing artillery on the outskirts of Damascus. Russia and the United States were connected to opposing combatants. Sadat knew he'd do better in subsequent peace negotia-tions if he was in possession of two particular mountain passes in the Sinai, so he ordered another push to occupy them. Now the Israelis were ready and quickly repulsed this attack and counterattacked, pushing the Egyptian army apart and penetrating all the way across the peninsula and the Suez Canal into Egypt, advancing southward toward Suez.

On October 22, a UN-brokered ceasefire unraveled, allowing the Israelis to completely surround the Egyptian Third Army and the city of Suez. The Soviet Union and the United States helped arrange a ceasefire on October 25. The Egyptians fought well enough to make the Israelis realize the cost. The Egyptians were now willing to discuss peace. The main fighting stopped on October 28, although clashes continued all the way until January 18, 1974. The Israelis had occupied territory within

Egypt west of the Suez Canal and were only 150 km from an unde-fended Cairo. They also surrounded the Egyptian Third Army in the Sinai but were themselves expensively stretched to achieve this posi-tion. They had also advanced within 40 km of Damascus. The Syrians needed to have all their tanks replaced by the Soviet Union and needed support from the Jordanians and the Iraqis. This time the Israelis were not embarrassed by any claim that they had acted preemptively. Instead, they had contained both attacks, counterattacked and put both enemy capital cities at risk of occupation. The Israeli navy had also decisively beaten the Syrian and Egyptian navies.

Nixon and Kissinger decided to resupply the Israelis, which wasn't surprising given the immediate help from the Soviets for the Syrians. The Israelis had gone as far as ordering nuclear warheads to be prepared for delivery by plane into Syria and Egypt, but in the event the ceasefire and termination of meaningful hostilities put the nuclear option back to bed. Kissinger is said to have explained the Israeli resupply to Sadat by saying that the Israelis were contemplating the nuclear option, but in the end the decision was based on other factors. Russian nuclear weapons may have been on ships in the port of Alexandria after their reconnais-sance had noticed the Israeli nuclear weapons being prepared. European countries (except Portugal and the Netherlands) refused airbases to the U.S. planes resupplying Israel because they were fearful of an Arab oil embargo.

U.S. Starlifter and Galaxy aircraft flew 567 resupply missions with 22,395 tons of materiel. More was delivered by sea. Lost aircraft were replaced, and lots of the latest military hardware and ammunition were delivered, amounting to about $4.25 billion worth today. The Russian Antonov heavy lift aircraft flew 900 sorties to resupply the Syrians, Egyptians and Iraqis. Other Arab states joined in with Egypt and Syria, notably Algeria, which sent $200 million to the Soviets for resupply of materiel and also supplied squadrons of fighter jets. The Iraqis sent a squadron of Hawker Hunters to the Egyptians, who used them success-fully in antiarmor operations. Cuba sent 4,000 troops. Pakistan air force pilots flew missions in Syrian aircraft and shot down an Israeli aircraft. Even North Korea sent 20 pilots and 19 noncombat personnel to Egypt. The Israelis engaged and shot down one Korean MIG, but Egyptian anti-aircraft defenses shot down another in an example of "blue on blue," when you have lots of confused allies. Libya provided an armored bri-gade and two squadrons of Mirage fighters. Saudi Arabia sent 3,000 soldiers to Syria with armored cars. Kuwait sent 3,000 soldiers to Syria and Egypt.

U.S. Crude Oil Production and Imports

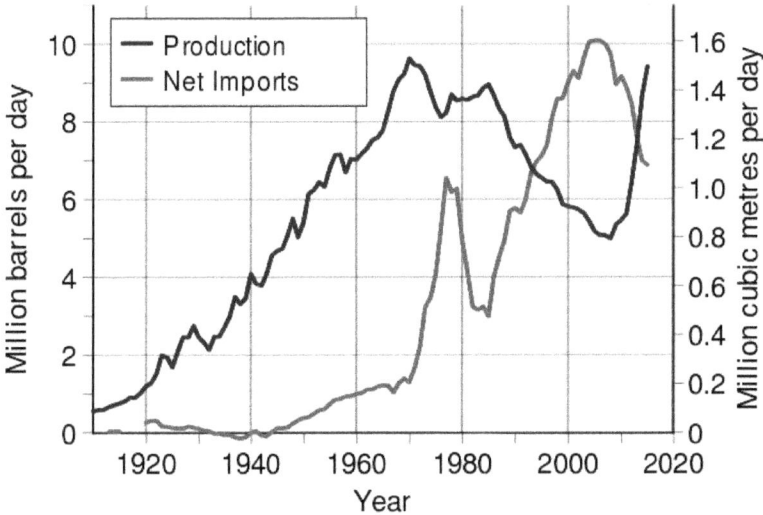

Figure 71: U.S. oil production and imports 1920 to 2005. Here you can clearly see the peak in U.S. domestic oil supply in 1970 coinciding with the price rises that resulted from the West's support of Israel during the Yom Kippur War. Source: EIA.

Morocco sent an infantry brigade to Egypt, where they were deployed to defend Port Said. Lebanon sent radar antiaircraft units, and the Sudan sent an infantry brigade just in time for the end of the war. There was also a Palestinian brigade in Egypt prior to the war breaking out. Not since the Spanish Civil War had so many nationalities taken sides in a struggle that could have been a prelude to World War Three.

The Israelis became complacent after their stunning victory in the 6 Day War and the sudden reversals and losses woke them back up to reality. An independent enquiry recommended dismissal of many of the Israeli military leaders and also the resignation of the Golda Meir government in April 1974 along with defense minister Moshe Dayan.

Ultimately, the situation led to the successful peace negotiations between Sadat and Menachem Begin at Camp David during the Carter presidency. For this, Egypt was thrown out of the Arab League and Anwar Sadat was later assassinated by officers of his own army for making peace with the Israelis.

On October 17, 1973, OPEC reacted to Nixon's arms resupply to the Israelis by increasing the price of oil by 70% to $5.11 per barrel and imposing an embargo on supplies to the United States, the UK, Japan, Canada and the Netherlands, which led directly to the 1973 energy crisis. They cut production by 5% and threatened to repeat this on a monthly basis. The crisis caused a rift within NATO as some members sought to distance themselves from the actions of Israel and the United States. The need for stable energy supplies was even stronger than a need to stay on good terms with the United States.

Algeria	Canada	Gabon	Kuwait	Oman	Sweden
Albania	Chad	Georgia	Kyrgyzstan	Pakistan	Switzerland
Angola	Chile	Germany	Latvia	Panama	Syria
Argentina	China	Ghana	Liberia	Papua NG	Taiwan
Aruba	Columbia	Gibraltar	Libya	Peru	Thailand
Australia	Congo	Greece	Lithuania	Philippines	Togo
Austria	Congo	Guatemala	Malaysia	Poland	Trinidad
Azerbaijan	Cook Islands	Guinea	Malta	Portugal	Tunisia
Bahamas	Costa Rica	Hong Kong	Martinique	Puerto Rico	Turkey
Bahrain	Croatia	Hungary	Mauritania	Qatar	Turkmenistan
Barbados	Cyprus	India	Mexico	Romania	UAE
Belarus	Czech	Indonesia	Midway Islands	Russia	Ukraine
Belgium	Denmark	Iraq	Morocco	Saudi Arabia	United Kingdom
Belize	Dominican Republic	Ireland	Namibia	Senegal	Uruguay
Benin	Ecuador	Israel	Netherlands	Singapore	Uzbekistan
Bolivia	Egypt	Italy	Netherlands Antilles	Slovakia	Venezuela
Brazil	El Salvador	Cote d'Ivoire	New Zealand	South Africa	Vietnam
Brunei	Equatorial Guinea	Jamaica	Nicaragua	Spain	Virgin Islands
Bulgaria	Estonia	Japan	Nigeria	Spratly Islands	Yemen
Burma	Finland	Kazakhstan	Niue	Surinam	
Cameroon	France	Korea	Norway	Swaziland	

Figure 72: Sources of foreign oil for the United States. Highlighted red are sources of more than 2.5% of total. Source: IEA.

Up until 1970, the United States had been one of the world's largest producers[287] of oil, but in 1970, as predicted in 1950 by Shell geophysicist Marion King Hubbert, the U.S. domestic production of oil peaked. Hubbert had presented a paper in 1956 that predicted American oil production to peak between 1965 and 1970. His work was criticized at the time, but he became famous when his theory proved to be correct in 1970 and traditional U.S. oil production actually did decline. These events also provoked major energy strategy changes in the West, including serious research into alternative energy resources and energy conservation.

The embargo on oil deliveries came as a shock to American drivers, who had little knowledge about politics in the Persian Gulf. By early 1974, there were gasoline lines around the country. Even today the event is a major energy milestone in America's, indeed, the West's, history.

American leadership did not believe the Arab exporters of oil, OPEC, could afford to lose revenue. Nixon and Kissinger met with Arab foreign ministers on the morning of October 17, 1973. The Americans were convinced that oil supplies would never be cut, but it was that very day that the Arab oil embargo was arranged. In November, it was discovered that the Saudis were enjoying more oil revenues because of the price rises and it went along with preservation of reserves. The exporting Arabs were not having any financial difficulties because they were more than making up for the lost volume with higher prices. Those prices were up to $12 per barrel by 1974. Some members of the cartel, Iran, Nigeria and Indonesia, increased production modestly, but Saudi Arabia pressed hard for a "Production Limitation Scheme," and OPEC subsequently succeeded in creating a "sellers' market" in oil for the entire period up to late 1980. Over the following decade they increased prices again, but this time more non-OPEC oil was coming out of the North Sea, Alaska and Mexico amid falling Western oil consumption as efficiency measures took hold.

People drove smaller cars with more efficient engines. OPEC no longer had control over the oil supply. There was a "buyers' market" again. In 1999, however, OPEC attempted to take the initiative once more.

At its worst, the U.S. energy trade balance was negative by over $1 billion every day, more expensive than the wars in Iraq or Afghanistan and even more impactful than the trade deficit with the Chinese. Surely, this was enough of an incentive to find domestic supplies of energy and keep the money at home?

The West did not learn in time that the great secret of OPEC was merely to restrict supply and not to rush out with new production. When they did this, prices more than compensated and the depleting reserves operated optimally.

Today, the tables have turned. U.S. oil production is greater than Saudi Arabia's. America's foreign supplies of oil are closer geographically, but about 16% of America's oil still comes from the Middle East.

Demand rising elsewhere on the planet has absorbed America's lower imports. I found the EIA data of barrels of oil by country imported into the USA and had a look at the data since 1993 to May 2015 in figure 73.

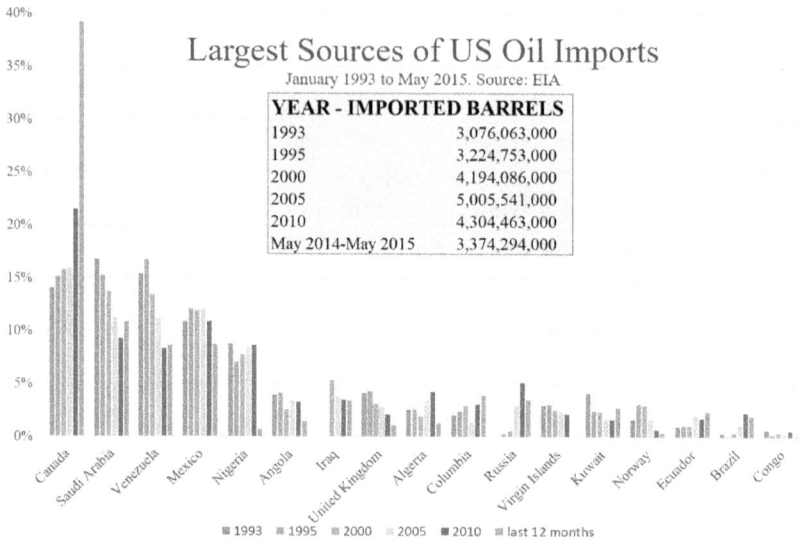

Largest Sources of US Oil Imports

January 1993 to May 2015. Source: EIA

YEAR - IMPORTED BARRELS	
1993	3,076,063,000
1995	3,224,753,000
2000	4,194,086,000
2005	5,005,541,000
2010	4,304,463,000
May 2014-May 2015	3,374,294,000

■ 1993 ■ 1995 ■ 2000 ■ 2005 ■ 2010 ■ last 12 months

Figure 73: Largest of 124 source countries that have sold oil to the United States. Source: Data EIA, chart NEF Advisors, LLC.

Today, due to the success of the fracking technique, the United States is producing more oil than Saudi Arabia and its foreign supplies are now coming mainly from Canada, which has almost doubled its output and is obviously a much better partner than many others even for geography alone, although the Canadian oil has a worse pollution index, like the Maracaibo oil sands in Venezuela. Saudi supplies to the United States have been dropping off slowly and the Nigerian supply has dropped away suddenly.

The results of the embargo were varied, and there was a major renewable energy R&D effort in the United States that resulted in large, commercial wind turbine designs and, of course, silicon solar cells. The geopolitical realities of oil are profound. OPEC has controlled the lion's share of oil production for decades and has been able to shift pricing by altering the supply of oil from their collective wells. There are so many potential trouble spots. OPEC and non-OPEC production changes can have a significant effect on the regular supply of oil. The global economy needs energy as a strategic necessity, but chokepoints galore present potential challenges. OPEC countries that are not friendly to the West are common, but the flow of petrodollars is something they need to protect.

The First Gulf War in 1990: Operation Desert Storm

> *"The most important strategic interest lay in expanding global energy supplies, through foreign investment, in some of the world's largest oil reserves—in particular Iraq. This meshed neatly with the secondary aim of securing contracts for their companies. Note that the strategy documents released here tend to refer to 'British and global energy supplies.' British energy security is to be obtained by there being ample global supplies—it is not about the specific flow."* —Greg Muttitt from his book *Fuel on the Fire,* citing declassified U.S. Foreign Office files from 2003 onward.

Of all wars where energy was a major discussion point, the Gulf Wars were both steeped deeply in oil. At the time of the first Gulf War, there was a major media push by groups angry that the United States was putting soldiers' lives at risk for oil. Demonstrations were common in 1990 with the theme "No blood for oil" and "Blood is thicker than oil" or "Hell no, we won't go. We won't fight for Texaco."

The reason, of course, is that for 60 years since the discovery of Saudi oil and the treaty signed in 1931, the unlikely pair, one a secular republic and the other a conservative Islamic kingdom, were linked by energy and regional stability issues. It is a high likelihood that energy price stability was a higher priority in the minds of the planners of the Gulf War. Persian Gulf oil found its way all around the world, and an increase in price stemming from Middle Eastern chaos was something to prevent if at all possible. A decline in oil production of just 3 million barrels a day corresponded to a decline in U.S. GDP by 2.5%, and 5 million barrels per day resulted in a 3.2% decrease in U.S. GDP and rising inflation and unemployment in both cases. Countries that were holding debt would also be severely impacted by any rise in the oil price. This was true in many respects in Europe and Asia as well. The popular perception was that the war was actually designed to give the U.S. greater control over Middle Eastern oil reserves.

Renewable energy does not have this kind of repercussion. The lack of a feedstock for solar, wind, geothermal and hydro means that the energy can be made without threat of interruption.

Going back in time to as early as June 1961, Iraq was threatening to invade Kuwait. The Emir requested help from the UK, and within a week, 8,000 British troops were deployed as a Brigade Battle Group on the border with Iraq.[288] They were summoned from bases in Kenya, Aden and Bahrain and included the Second Battalion, the Parachute Regiment, coming in from Cyprus. They took up positions on the Matla Ridge, the main approach corridor for Iraqi tanks. A naval task force included two aircraft carriers, HMS *Bulwark* and HMS *Victorious*. Temperatures were in the 120°F to 140°F (49°C–60°C) range, and the real lesson learned was the huge requirement for soldiers to drink water. By July, 2 Para were moved back to Bahrain and were relieved by 3 Para, including Major JJG Cox, my father. The unit remained until May of 1962. The British parachute infantry battalions cycled on duty tours here until April 1967. The Iraqi threat never materialized.

RESERVES - Million Barrels of Oil Equivalent (MBOE)			PRODUCTION - MBOE			YEARS REMAINING (Linear)		
COUNTRY	1991	2015	% Change	1991	2015	% Change	1991	2015	% Change
Iran	205,100	382,200	86%	1,448	2,701	87%	142	141	0%
Iraq	117,754	167,520	42%	105	1,478	1306%	1,120	113	-90%
Kuwait	105,542	113,380	7%	85	1,229	1351%	1,246	92	-93%
Oman	6,148	9,920	61%	279	578	107%	22	17	-22%
Qatar	33,994	187,400	451%	203	1,890	829%	167	99	-41%
Saudi Arabia	294,818	321,380	9%	3,452	5,087	47%	85	63	-26%
Syria	1,700	4,480	164%	172	38	-78%	10	117	1087%
United Arab Emirates	135,324	138,060	2%	1,120	1,793	60%	121	77	-36%
Yemen	5,320	4,980	-6%	72	35	-51%	74	142	92%
Other Middle East	2,410	1,557	-35%	42	133	218%	58	12	-80%

Figure 74: Reserves and annual production of combined oil and gas in Middle Eastern nations in 1991 and 2015. Source: BP Statistical Review 2002 and 2016.

In figure 74, we are looking at both oil and gas reserves and production for the Middle Eastern countries to get a sense about the theater assets that belligerents were fighting over. No other geographical region on Earth has as much oil as the Middle East. Only Venezuela (300,000 million barrels), Canada (172,000) and Russia (102,400) show up in as much scale. On this basis, Kuwait has a similar amount of oil as Russia and so the simple annexing of this tiny country that had been making so much pain for Iraq would in 1991 have essentially doubled the oil reserves of Iraq and made it second in resources only to Saudi Arabia and Venezuela.

The Reagan National Security Council conducted a review[289] of the complex situation that addressed issues such as the likelihood that the Sunni/Shiite conflict could widen and cause the world oil price to rise at a time of vulnerable global economic recovery. The study offered three recommendations: Increase oil stocks, improve the security of friendly Arab nations and embargo further military equipment sales to Iraq or Iran. The plan was implemented by the G7 nations and notably made no mention of developing renewable energy.

Once oil was discovered in Saudi Arabia, it became necessary to find it anywhere it was available in the Middle East. Deposits were very unequally distributed. A group of five oil-producing countries formed OPEC[290] in 1960. One of OPEC's problems as a cartel that controlled the largest volumes of oil sales in the world was that each member state was subject to an agreed-upon quota of oil production. Sell more oil than your quota, and your country received more revenues. Many times, in the history of OPEC, this option overcame the discipline necessary to keep the price high. Iraq accused Kuwait of overproducing oil, and indeed, together with the UAE, they were partly to pay for damages incurred in the Iran/Iraq War. This caused the oil price to decline to about $10 per barrel, resulting in a painful economic loss of about $7 billion for Iraq, exacerbated by their balance of payments deficit of about the same magnitude.

On top of this, Iraq complained that Kuwait was drilling for more oil using slanted drills that crossed its border into the Rumaila Oil field. In 1989, Saudi Arabia and Iraq signed a nonaggression treaty, but the Iraqi debt problem with the Saudis meant that state-backed development projects were unable to go ahead. Iraq demobilized 200,000 (lucky!) soldiers to save money and made efforts to establish an arms export industry. There was a growing enmity against OPEC's control of oil production quotas and expatriate workers from Saudi Arabia and Egypt began to have a tougher time. All this was happening while the rest of the world was watching the fall of the iron curtain that divided the dying communist USSR and Europe, so the spotlight was elsewhere. The United States also received a considerable level of Middle Eastern oil imports at the time.

In July 1990, Iraq presented its list of complaints to the Arab League, but tensions mounted. The U.S. Navy was placed on alert, and the Iraqi army began to mass soldiers on the border with Kuwait. At talks held in Jeddah, Saudi Arabia, Iraq accused the Arab League of siding with American interests and threatened force against Kuwait and the UAE. Egyptian president Hosni Mubarak and April Glaspie, the U.S. ambassador to Iraq, both thought that there was little risk of a war, but the Iraqi position was getting strident. Iraq demanded $10 billion in reparations from Kuwait for breaking quotas and stealing oil from Iraqi reserves. The Kuwaitis actually offered $9 billion, but this was too much provocation for a paranoid Saddam Hussein, who immediately ordered the invasion of Kuwait.

On August 2, 1990, at midnight, the Iraqi army and air force invaded and annexed Kuwait. U.S. president George H. W. Bush accumulated men and materiel in Saudi Arabia and exhorted the community of nations to join in the fight to repel Saddam Hussein's military from Kuwait. The Emir and some of the Kuwaiti military finally managed to get across the Saudi border to the south, and within a day the Iraqis had total control of Kuwait.

On August 7, 1990, President Bush launched Operation Desert Shield, a wholly defensive mission to prevent Iraq from invading Saudi Arabia. The Arab League wanted only Arab nations to work on curtailing the violence and warned about outside interference, but oil was too important a commodity to put the global economy at risk. Only Libya, Sudan and the Palestine Liberation Organization (PLO without a UN vote) sided with Iraq, while Yemen and Jordan led by the Sandhurst-educated King Hussein, normally a very strong Western ally, opposed intervention from non-Arab states in line with the Arab League.

My father was at Sandhurst at the same time as King Hussein, and during training, they became friends, resulting in our family being invited by the king to visit Jordan in 1965. I remember as a nine-year-old the streets of Amman and Jerusalem, the Red Sea and riding a horse down the long gully that leads to the stone-carved, red rock capital of the Nabatean Kingdom, Petra. We were accompanied by one of the king's family and a host of helpers. The king sent his sons to Sandhurst, too, and one day, when I was a Second Lieutenant at Victory College, Sandhurst, in 1980, King Hussein's son, Abdullah, the present Jordanian king, and I met, just once, for tea.

Iraq offered to withdraw from Kuwait if sanctions were lifted and it was permitted full access to some strategic Persian Gulf islands and all of the Rumaila Oil field, which extended underneath part of Kuwait as well. The offer included a willingness to negotiate a favorable oil agreement with the United States that would help the Iraqis emerge from their difficult financial circumstances as well as stabilize the Persian Gulf.

U.S. foreign secretary Jim Baker pulled together the diplomatic consensus necessary to deploy the full force of UN Resolution 678. He talked to King Fahd and the Emir of Kuwait, and the result was that half of the $61.1 billion cost of the war was met by $32 billion paid by both countries. A total of $52 billion was paid by other countries. At the Jeddah, Arab League meeting, Saddam had assured Egypt's president Hosni Mubarak there would be no war, and now Mubarak was determined to participate in his ouster from Kuwait. It didn't hurt that the United States was willing to forgive a $7.1 billion debt Egypt had to the United States. Baker got a $2 billion contribution from the Germans, equipment from the Italians and support from the Turks, Syrians and Russians. Eventually, Germany and Japan contributed another $16.6 billion. The final coalition of 34 countries was 73% American troops, and the total number of soldiers committed was 956,600. Even Argentina, the only Latin American country to participate, contributed a destroyer, a corvette and a supply ship that performed helpful service. They participated in the UN blockade in the Persian Gulf. They made over 700 interceptions and sailed over 25,000 miles in the theater. This significantly helped to overcome the "Malvinas Syndrome," an unfortunate legacy of being the enemy during the Falklands Conflict, which had happened eight years earlier. Argentina was later classified as a major non- NATO ally for its Gulf War contribution.

On January 17, 1991, the Gulf War began. It was given the code name Operation Desert Storm. The world witnessed 100,000 air sorties dropping 88,500 tons of bombs on Iraqi military and civilian infrastructure. The entire coalition push took 100 hours before President Bush declared a ceasefire.

There were many justifications for getting involved in the conflict. Among those that would resurface later were the disapproved Iraqi use of biological and chemical weapons that had been used in the Iran/Iraq war and against Iraq's own Kurdish minority in the Al-Anfal Campaign. There was also supposed to be a nuclear weapons program with a report on it from January 1991 that was declassified in 2001. This is of huge significance, but in retrospect the project must have been rather an early attempt to collect both the intellectual muscle required to develop a bomb along with the necessary metals, enriched U-235 and the means to manufacture it from relatively common U-238 as well as plutonium. The war lasted seven weeks and was also one of the first to involve real-time journalism from members of the press embedded with coalition military units.

On February 27, Saddam ordered a retreat from Kuwait City and President Bush declared it liberated. Retreating Iraqi forces set fire to 737 oil wells and placed land mines around the wells to make it more difficult for the fires to be extinguished. It took some teams of skilled oil-well blowout crews, such as one led by the charismatic Red Adair, in the swan song of his career, to extinguish the Kuwaiti flames at a cost of $1.5 billion, paid by Kuwait. By November 1991, after 10 months, the fires were all out. About 6 million barrels a day of oil production were lost.

An interesting physics fact about very heavy metals is that they make exceptional projectiles and can penetrate much better than lighter metals such as iron. Consequently, the U.S. military had used "depleted" uranium in bullets and armor-piercing rounds fired by tanks that later became a contentious issue due to the supposed health impact of uranium radiation on soldiers and civilians from spent ordnance.

The Second Iraq War: Operation Iraqi Freedom

"Of course it's about oil; we can't really deny that." —General John Abizaid, former head of U.S. Central Command and Military Operations in Iraq, in 2007.

"I am saddened that it is politically inconvenient to acknowledge what everyone knows: The Iraq war is largely about oil." —Former Federal Reserve chairman Alan Greenspan, in his memoir.

"People say we're not fighting for oil. Of course we are." — Senator and defense secretary Chuck Hagel in 2007.

"Iraq possesses huge reserves of oil and gas-reserves I'd love Chevron to have access to." — Kenneth Derr, CEO of Chevron, in 1998.

Al Qaeda became organized after the first Iraq war. One of the first things they did was to come to New York. Here, the bombers sent letters with demands to New York newspapers. The three demands were that the United States stop interfering with Middle Eastern countries' internal affairs and that the United States end diplomatic relations and aid to Israel. The letter also admitted that what was about to happen was terrorism but justified it by saying that Israel exercised terrorist activities all the time with the full support of the United States.

On Friday, February 26, 1993, two Al Qaeda operatives drove a yellow Ryder rental van into the public parking garage below the World Trade Center. They lit the fuse and fled. Twelve minutes later the bomb exploded deep underneath tower number 1 of the World Trade Center. This was the northernmost tower, where my friend Victor was working in the offices of his bank. More than half a ton—1,336 pounds—of urea nitrate exploded, plus bottles of hydrogen placed around the main charge. It was meant to topple tower 1 into tower 2, killing tens of thousands. Had the bomber parked the van closer to the WTC's concrete foundations, the plan might have succeeded. Thankfully, it failed, but it still killed 6 and injured 1,042, who mostly suffered smoke inhalation. The bomber escaped and hours later was on a plane to Pakistan. The personalities involved included one or two who attacked the same towers eight years later. One of the results of the first bombing was a heightened security trained on package delivery and building evacuation. The names of the six victims were included in the memorial gardens built in the decade after what happened in 2001.

There are so many who have stories about the bright, clear, sunny Tuesday, the 11th of September of 2001. My own experience was to go to work at 666 Fifth Avenue, where I was a portfolio manager. The buzz in the trading room and other offices started when the first aircraft flew straight into tower number 1, the northern tower of the twin World Trade Center towers. Later we all saw the remarkable video, captured by a film crew who just aimed their cameras up at the low-flying aircraft in time to catch its impact into the building. We were all dumbfounded by the event and at the time it did not yet feel like terrorism necessarily, although there was no shortage of questions about how a plane could fly into a tall building in such good visibility. Then, shortly afterward, the second plane ploughed into tower number 2, from the south, and while the amazement grew it also sank in that there was little doubt that it was an attack. I went down into Fifth Avenue and joined several others in the street, watching the growing clouds of smoke rise from the buildings five miles south of us. My friend Victor still worked in the north tower on a high floor, and another friend of mine was in an early morning plane waiting for takeoff at Boston airport, one of the Boston planes that had not been hijacked by the suicidal terrorists.

The second tower collapsed before the northernmost tower, but there was only a short time between the two collapses. Victor had left his office almost immediately, ignoring instructions to stay put, and walked the long way down the stairs. He remembers fire fighters who would never be seen alive again running up the stairwells, breaking open vending machines to distribute water and soda to evacuating office workers. Thirty minutes after Victor left the building, it collapsed, still with many people inside. The next evening, we had dinner together and watched the never-ending supply of huge trucks driving down the avenues to start fixing up the attack site.

A report on *Frontline* in 1990 called "The Arming of Iraq" explained that many Western companies had sold them chemical, missile and nuclear technologies. Western governments knew but were indifferent to it at the time, and the United States had been arming Saddam against the Iranians in any case. After the Gulf War, the UN prohibited Iraq from developing or owning weapons of mass destruction (WMD). By the end of 2002, the United States was exploring links between Al Qaeda and the arsenal of chemical, biological and possibly even nuclear weapons that threatened the West. Iraq came under pressure to agree not to pursue WMDs and to submit to inspections. An 11,800-page report on Iraq's capabilities was produced at the end of 2002, saying it had no WMDs, but the International Atomic Energy Agency (IAEA) and the UN said the report was incomplete.

From the energy viewpoint, a report[291] titled "Energy Security" had been written in 2001, commissioned by the U.S. vice president, Dick Cheney. It was published by the Council on Foreign Relations and the James Baker Institute for Public Policy and rang alarm bells about how a global energy crisis could easily disrupt U.S. and global stability and make energy prices volatile. Iraq was a focus of the report.

Just prior to the Gulf War, there was a report that Iraq had purchased 500 tons of uranium yellowcake from Niger and was storing it at the Tuwaitha nuclear complex 12 miles south of Baghdad. After a visit, Ambassador Joseph C. Wilson said that these reports were "unequivocally wrong." In his January 2003 State of the Union address, however, President George W. Bush mentioned that Iraq had sought uranium.[292]

Then there followed an effort by the USA and UK to persuade the international community that there was a real reason to invade Iraq. Secretary of State Colin Powell appeared before the UN to present evidence that Saddam had WMDs, while on the other side, Hans Blix, who headed up the weapons inspection team for the UN, reiterated that no evidence of WMDs had been discovered. The effort to find a coalition was very unlike the first Gulf War. The UN and France opposed any use of the threat of violence to persuade Iraq to comply.

Then, an Iraqi scientist, Rafid Ahmed Alwan al-Janabi, admitted in February 2011 that he had lied to the CIA about biological weapons so that the United States would invade and remove Saddam from power. Antiwar groups across the world sprang into action. About 36 million people formed over 3,000 protests around the world against a war in Iraq. Even Nelson Mandela said, "All that [Mr. Bush] wants is Iraqi oil." By March 2003, Hans Blix had found no evidence of any WMD activities, but the U.S. government said that Iraq had had its chance to cooperate and the United States would proceed with a "coalition of the willing" to get rid of Iraq's alleged WMDs. Kofi Annan said that the war would not be within the bounds of the UN Charter. Lord Bingham in the UK described the war as a serious violation of international law, and Nick Clegg, the deputy prime minister of the UK, declared the invasion of Iraq as illegal. Iraq's foreign minister, Naji Sabri, informed the CIA that Saddam had hidden poison gas in Sunni tribal areas and had ambitions for a nuclear option but that was not active, and no biological weapons were being produced. George Bush was briefed about this new data about the lack of WMD in Iraq. After the war, aging chemical munitions from the Gulf War were discovered including Sarin gas and mustard gas.

The two aircraft that attacked the World Trade Center, American Airlines Flight 77 that hit the Pentagon, and the fourth plane, United Airlines Flight 93, that came down in a field in Shanksville, Pennsylvania, after its crew and passengers fought back against the hijackers were the scope of the audacious attack. Planes were chosen that were full of fuel for a transcontinental flight to the West Coast to create more fire damage after their collisions.

Immediately after the event, the Bush administration was seen to be finding reasons to invade Iraq again. Two days prior to a Senate vote on the Joint Resolution to authorize the use of United States Armed Forces Against Iraq, 75 senators were told in closed session that Iraq was capable of hitting the U.S. Eastern Seaboard with unmanned aerial vehicles (UAVs) armed with biological or chemical weapons. Even Hans Blix was saying that Iraq was not taking heed of the fact that it had to come clean on unaccounted-for stocks of chemical agents, VX nerve agent and large amounts of anthrax.

Secretary of State Colin Powell addressed the UN, saying that Iraq was hiding WMDs. The French government believed there was anthrax and botulism toxin. By March, Hans Blix announced good progress on inspections but found nothing. Bush and Tony Blair of the UK were working together as the "special relationship" led them both down the path of invasion.

The CIA entered Iraq on July 10, 2002, and they prepared for the invasion, attempting to persuade the Iraqi military to surrender and identify all the leadership targets. Bush in his 2003 State of the Union address mentioned mobile biological weapons laboratories. The CIA helped the Kurdish Peshmerga to hold the northern front. On March 20, 2003, the main invasion began under General Tommy Franks, without a declaration of war, and 40 governments, consisting of the "Coalition of the Willing," made some sort of contribution of money, men or materiel. Among the objectives of the invasion was "to secure Iraq's oil fields and resources, which belong to the Iraqi people." This speaks of a more noble approach to the oil. Baghdad fell on April 9. Saddam's 24 years in power were over. There was gratitude toward the coalition invaders, but Pandora's box had been opened and the Shiite and Sunni Baathist politics now had nothing to tamp down their excesses.

On May 1, President Bush landed on the U.S. aircraft carrier *Abraham Lincoln* and gave his "Mission Accomplished" speech, but they still had not captured Saddam. The Iraqi insurgency had begun and was using weapons from caches laid down by the Iraqi military before the war as well as stolen
U.S. weapons. The Coalition Provisional Authority (CPA) under Paul Bremer oversaw the Iraqi government, and Bremer immediately fired 85,000 to 100,000 Baath party administrators, which turned out to be a great mistake. Saddam was eventually found on December 13, 2003.

The U.S. government promised over $20 billion to help reconstruct infrastructure, and it was going to be repaid against future oil sales, but the insurgency volume was turned up by the friction between Shiites and Sunnis, displaced and unemployed soldiers and disenfranchised

Sunni tribespeople from the Baath party. On January 31, 2005, an Iraqi Transitional Government was elected. Fighting continued throughout 2005 and on October 15, a referendum was held to ratify a new Iraqi constitution and elect a new national assembly in December.

By December 2006, there was an average of 960 attacks on U.S. soldiers every week. Coalition forces transferred power to the Iraqi government again. The U.S. role was messed up badly by mismanagement in the CPA, Abu Ghraib prisoner abuses and two cases of murder and rape by U.S. soldiers. Saddam was executed by hanging on December 30, 2006, after being found guilty of crimes against humanity by an Iraqi court.

General Petraeus became the commander of forces left in Iraq and authored the "Surge" tactics, which largely pacified the most violent parts of Iraq. On a Charlie Rose interview, the general made it very clear that the lack of calm was causing further unrest because very little business could be done with the insurgency at such levels. The confidence of the ordinary Iraqi citizen was needed and that required enough soldiers to offset any further negative activity. Eventually, the violence subsided, but Iraqi lawmakers objected to the occupation. On April 9, 2009, tens of thousands of Iraqis thronged the streets of Baghdad in an effort to send a message to the Coalition forces to depart immediately. On April 30, the British formally ended combat operations and Basra was handed over to Iraqi forces. The succession of Obama as U.S. president in 2008 cemented the return of U.S. forces.

In 2009, the Iraqi Oil Ministry awarded contracts to the international oil companies for some of Iraq's oil. A fixed fee of $1.40 per barrel was set to be paid once a threshold volume of oil had been reached. Prior to the invasion, the Iraqi oil industry had been nationalized and closed to Western oil countries. As can be seen from the increase in oil production in Iraq between 1991 and 2015, Iraqi oil production soared over 1,300%, partly due to the efficiency and productivity of the large oil majors, Exxon Mobil, Chevron, BP, Lukoil and Shell, and oil service companies such as Halliburton. The Iraq Hydrocarbons Law, partially drafted by and very friendly to the Western oil companies was not passed, due to enormous public opposition reflected in parliamentary voting. As U.S. elections arrived, the oil companies used the situation to their advantage and managed to have long-term contracts signed that offered them access to the Iraqi oil anyway.

Failure to Share Oil Wealth Creates War

Other OPEC nations share similar problems. After independence from Britain in 1963, the state of Enugu, home of the Igbo tribe, attempted to secede from Nigeria and call themselves the Republic of Biafra. They wanted to keep oil wealth from being shared with the rest of the country. This province contains the oil-rich Niger Delta and the town of Port Harcourt. Two-fifths of the oil was bought by the British, and the ill-distributed cash flows caused significant internal stresses. The Igbo population in southeastern Nigeria were severely mistreated throughout the rest of Nigeria.

The Biafran War was the result of the decision by the Nigerian federal government to secure the oil assets for the greater good of Nigeria. A very sad part of the war was the use of starvation as a deliberate military tactic to gain control which resulted in the horrific famine that killed over 2 million civilians. This resulted in significant reaction within Britain and the first emergence of nongovernmental organizations (NGOs) that acted like focused charities to cope with global challenges like famine. This war highlighted the plight of those who live near bountiful oil, but don't necessarily benefit from its extraction, a phenomenon all too common in the oil age globally.

Strong Oil Export Currency Crowds Out Other Export Industries

In South America, the world's largest oil resource and its eighth-largest natural gas reserve was discovered in Venezuela during the First World War in the Lake Maracaibo and the Orinoco river basin. The discovery led to an economic boom. Political power changed hands several times with civilian and military coups until in the 1960s the first democratic elections were held again, and guerilla groups laid down their arms.

Venezuela was a founding member of OPEC in 1960. During the 1973 oil price shocks that hurt oil importers, Venezuela's economy benefited hugely from the sudden price increase, which went hand in hand with the nationalization of the oil industry, ejecting foreign oil companies from the country. This left Venezuela vulnerable to low oil prices, which slammed the economy in the late 1990s. Venezuela is a major source of western hemisphere oil for the USA. The recovery of the oil price in the early 2000s fueled populist economic policies such as social programs that significantly reduced poverty and inequality.

However, the oil-strong currency caused the collapse of coffee and cocoa industries, leading to a new increase in poverty, with inflation and basic staples shortages, which led to malnutrition among children. The internal price of gasoline is heavily subsidized by the government, making it the cheapest in the world. In 2010, 40% more oil reserves were identified, causing Venezuela to overtake Saudi Arabia with the world's largest crude oil reserves. The current weak oil price is a major reason that the country's export economy is suffering. Venezuela currently suffers the world's highest inflation rate of well over 100%, which resulted in protests and riots in 2014–15.

Instability is really the core characteristic of the global energy regime. As globalization progresses, the global economy has become increasingly interdependent. This means there are concerted efforts by developed and developing economies to avoid shocks to the global system. Unfortunately, a commodity like oil is a perfect foil for political and economic disaster. Oil is a resource that is not widely distributed like wind, sunlight or water. This has resulted in have and have-not countries and energy wars. One of the only ways for such a cycle to end is to bring along renewable resources that are available everywhere in greater or lesser degree. Japan is an economy without much natural energy supply. This is a reason for its early interest in solar cells, nuclear power and importing of oil and refined products from the Middle East.

The United States was an importer of oil for almost all of the period from 1940 to the present day. During this time, a long line of presidents revisited the theme of getting America off foreign oil. This is no small challenge. The world's major oil companies were British, European or American. They were discovering oil around the world and repatriating profits as they did so. Oil money flowing out of the United States, often called petrodollars, built the Middle East and permitted huge architectural and social achievements in oil-producing countries, but it also provided funding for groups that were increasingly focused against the forces of globalization, which have culminated in three Middle Eastern wars, ongoing drone strikes and a sense of complex and accelerating chaos.

Fighting for Precious Water Resources

A world using sustainable resources is likely to experience more stability since energy pricing will be noninflationary and provide almost
infinite reserves with sufficient production and of course without any of
the additional costs of the eight externalities I examine in chapters 3 to
10. An interesting factor is that as temperatures and water supplies get
more distressed, tribes in East Africa have shown increased conflicts
among themselves. A Yale Environment 360 video report[293] shows how
worsening droughts pit groups and nations against each other. Ethiopian
and Kenyan pastoral communities that live on land that also holds the
origins of humanity, have over the last 40 years been suffering from an
increasing drought that affects their cattle and livelihoods. The River
Omo in southern Ethiopia has increasingly been diverted by irrigation
and a dam project, the largest hydroelectric power project in sub-Saha-
ran Africa, which will hold back water and annual flooding on which
hundreds of thousands of tribes in both Ethiopia and Kenya depend.

The Omo River normally drains into Lake Turkana, but a 2°F
(1.1°C) warmer climate since 1960 has accelerated evaporation from
the lake, reducing its level and causing its waterfront to retreat into
Kenya. The Dassanech, Nyangatom and Mursi tribes of Ethiopia and the
Turkana of Kenya now frequently carry out cross-border raids into each
other's territories to kill each other and steal cattle. This story serves
as a clear warning of what a wider resource problem would offer...as
if an example was really necessary. Having been at the sharp end of a
human conflict, I'm convinced that resource struggles are something we
can overcome without a descent into "politics by other means" and by
managing, not the reserves of energy, food and water we are used to but
by sharing the almost infinite alternative resources that vastly outweigh
anything we are using today.

Oil Exploration Incentives in
Argentina and the Falklands

One hundred million to 145 million years ago, in the Lower Cretaceous
period, river estuaries in the continent that was to become South America
drained fresh water from the Andes down to the South Atlantic.

The water was full of organic material and sand and was deposited in the coastal plain as a series of fan-shaped outcroppings. In the intervening time, it was encased in mud and the organic material, crushed by pressure and heat, became oil and gas. It took modern-day 3D seismic technology to tease out a picture of all this accumulation. In 1974, oil reserves were discovered between the Falkland Islands and mainland Argentina. The size of these reserves at the time was not exactly known, but they were strategic. Today we know that there are about 8 billion barrels of oil in the different formations at the site.

While I was at university in Scotland in 1976, Andrew Lloyd Webber and Tim Rice's hit musical *Evita* was first released.

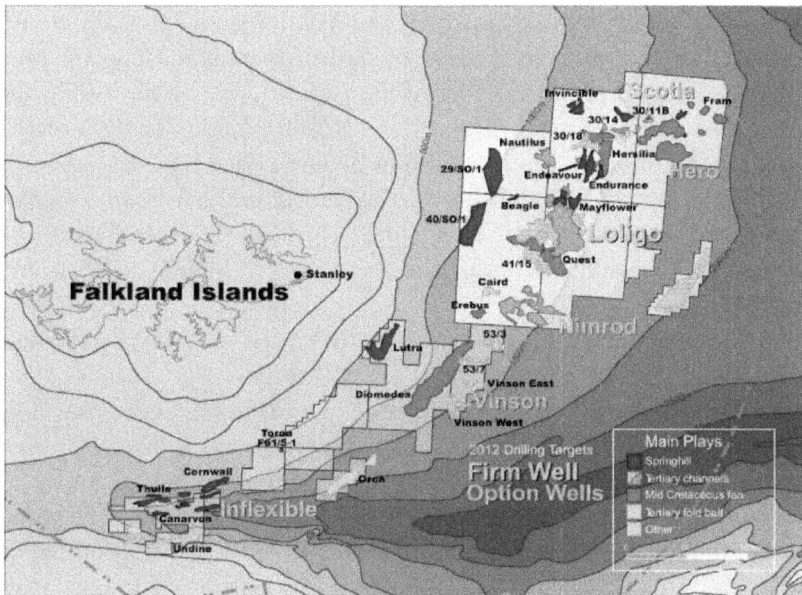

Figure 75: Oil license sites east of the Falkland Islands. Source: Oil Change International, February 13, 2012.

The theme song "Don't Cry for Me Argentina" became a number one hit around the world and remains a familiar tune today. Incredibly, during the Falklands War it was actually banned by the BBC. The plot concerned María Eva Duarte de Perón, the second wife of Argentine president Juan Domingo Perón who served as the first lady of Argentina from 1946 until her death at age 33 from ovarian cancer in 1952. She was affectionately referred to as Evita. The opposition was characterized by aristocratic landowners who felt victimized by a populist personality cult that the Peróns rode until Evita's death and then long after as well. Eva Perón was naturally sympathetic to struggling poor people and labor rights. She became a figurehead to the pro- Peronist trade unions embodied by the Confederación General del Trabajo (CGT).

She ran the ministries of health and labor and founded a charitable foundation that supported women's suffrage. She was also the head of the Female Peronist Party, Argentina's first female political party. She was known for her passionate speeches delivered from the balcony of the Presidential Palace, the Casa Rosada, in support of the poor and jobless. She ran for the position of vice president of Argentina in 1951, supported by the poor, the unions and the Peronists. Opposition from the military and middle class as well as bad health forced her to relinquish this goal. However, just before she died, the Argentinian Congress gave her the title of "Spiritual Leader of the Nation," and on her death, she was given a state funeral. There were 26,000 requests to the Pope to have her beatified and over 2 million people kissed her glass-covered coffin. The period of mourning was extended, and her body was preserved and given a place of honor in the CGT building, where it could be viewed, in a morbid and not dissimilar manner to Stalin's preservation in the USSR.

On June 16, 1955, the tension between the haves and have-nots erupted, and Argentine navy aircraft flew sorties over Buenos Aires, bombing and strafing, killing hundreds of civilians. Within weeks, Juan Perón was forced into exile and the Peronist party was outlawed. General Pedro Eugenio Aramburu Silveti took control of the first of two military dictatorships and two civilian governments that filled the next 15 years. Evita's body was hidden by the military, eventually being buried in Milan, Italy, under a false identity. Perón found exile in Madrid and married for the third time to María Estela Martínez, also known as Isabel. He was called upon to resist the military government in Argentina. General Aramburu was himself eventually kidnapped by Peronist Montoneros, put on trial and executed. His replacement negotiated with Juan Perón and offered to return Evita's body.

The body was exhumed, still in a good state of preservation, which shocked the graveyard workers, and driven in a hearse across northern Italy and into Spain and on to Madrid to rejoin her husband.

The military government backed off. Perón returned to Argentina and won a landslide election for his third presidency on October 12,

1973. Isabel Peron was elected as vice president, and when Juan Perón died at age 78 on July 1, 1974, she became the first female Argentinian president. This presidency was short and chaotic. Her principal advisor was an astrologist and the social and economic problems of Argentina were not addressed. This presidency was notable for a deal between the Montoneros and the government in which Evita's preserved body was finally returned to Buenos Aires, its broken nose and cuts repaired, and placed in a crypt for viewing.

In 1976, a new military dictatorship led by Jorge Videlas replaced the Isabel Perón government. This development was initially greeted with some measure of relief and a confidence that the military could get things moving again. Very soon it was clear that the new regime had an agenda to clear out "left-wing subversives." A period of horrific pressure on the left wing ensued. Between 9,000 and 30,000 university students, journalists, unionists and over 500 pregnant women were made to disappear in an act of cruelty that marks one of humanity's darkest moments and that echoed some of the anti- Semitic, anti-communist attitudes that were extant in the Axis powers during the Second World War. Since there was no longer any requirement for Evita's preserved body, it finally found its resting place in a crypt in a Buenos Aires cemetery and remains a tourist and Peronist pilgrim destination to this day. Congress was suspended, and trade unions, political parties and provincial governments were all banned. This period was called the "Dirty War," and torture and mass executions were the junta's tools of repression.

There was denial of any wrongdoing by the authorities, but mothers and grandmothers assembled in Buenos Aires' Plaza de Mayo to protest the disappearances. Argentine media were silent about the atrocities for fear of reprimand, but the *Buenos Aires Herald*, a newspaper for the cricket-playing and tea-drinking "Anglos," continued to publish photographs of disappeared people at great risk to the editor in chief, British citizen Robert Cox,[294] and Uki Goñi, a fellow journalist. Robert Cox was later arrested and taken to a jail, but with the help of the British Embassy he was released after two days. Goñi reported that Cox saw a Nazi swastika on the wall during his stay. Cox decided for the safety of his family to leave the country in 1979. These men's efforts to help the "mothers of the disappeared" were also resisted by their own staff, fearful of attracting the attention of the authorities. Foreigners at the time basically bought the position, articulated by the military junta, that nothing out of the ordinary was happening and that any complaints were all fabricated.

Eventually, a military officer revealed that he was a party to throwing people out of aircraft over the Atlantic Ocean, something called the *vuelos de la muerte,* or "death flights," an extrajudicial form of killing that had a zero chance of being witnessed. The victims, many after detention and torture, were told they were being transferred to the south, to be released there. This filled many with hope and they were given a "vaccine" but really drugged into a stupor with Pentothal. They were groggy and many needed help to board the plane. Once in the air, a further injection of sedative made them unconscious. They were stripped naked and thrown, one by one, out of the aircraft into the night over the Atlantic Ocean. One navy lieutenant commander, Adolfo Francisco Scilingo,[295] convicted in 2005 in Spain of crimes against humanity, explained there had been 180 to 200 death flights between 1977 and 1978. He had participated in two such flights where 13 and 17 people, respectively, were killed.[296] In later years, it was revealed that thousands of individuals had passed through or died in the Escuola de Mecanica de la Armada (ESMA), the naval engineering school.

In July 2012, Jorge Videlas was sentenced to 50 years in prison for masterminding this period of oppression. One victim, a 16-year-old student organizer, Patricia Isasa, had been arrested and tortured by police and authority figures. For two and a half years, she was held without trial at one of the 585 clandestine detention centers set up during the Dirty War. Today, after a series of long legal battles, she has lived to see six of her nine torturers, who included army officers and judges, sentenced to prison.

Cox's news reports surfaced in the Carter administration, prompting President Carter to send his undersecretary for human rights, Patricia Derian, to Buenos Aires to meet with Cox and demand explanations from the generals, which she did. Cox felt guilty that he was not able to save more lives. It fell to his son to write a biography of his father, *Dirty Secrets, Dirty War*, which was published in 2008 and in Spanish in Argentina, making it possible for Cox to go back to his former home, where he was acclaimed as a hero. In the United States, President Carter exposed the junta and succeeded in isolating it in the world community. It now only needed one more major mistake for the regime to collapse.

In March 1981, Lieutenant General Leopoldo Galtieri visited the United States, where he was warmly received. The Reagan administration liked his stand against communism, and he was described as a "Majestic General" by national security advisor Richard V. Allen. Galtieri had supported the United States in its efforts to shore up the Contras and the Nicaraguan Democratic Forces (NDF). Argentinian advisors were sent there, and many NDF personnel were trained in Argentina. Argentine support became the main source of funding and training for the Contras. On December 22, 1981, Galtieri carried out a coup against General Roberto Viola and assumed the presidency of Argentina. He retained control of the army and placed General Guillermo Suarez Mason as chairman of Yacimientos Petroliferos Fiscales (YPF), the state petroleum company and largest company in Argentina. That general was in tenure at the time of the largest loss of any company in the world up to that point of US$6 billion.

Galtieri introduced a political reform that allowed a measure of expression for dissent, which triggered anti-junta demonstrations. One week before the invasion of the Falklands, on March 30, 1982, at the Plaza de Mayo, which had become the habitual meeting place for the mothers of the disappeared, the military suddenly and violently cleared the square using rubber bullets. On August 20, 2002, the U.S. State Department released over 4,000 documents[297] detailing all the intelligence reports and dispatches from the U.S. embassy in Buenos Aires. They assisted in the prosecution of General Galtieri and other officers for the crimes committed at the time.

In April 1982, I was a lieutenant in the Third Battalion, the Parachute Regiment. We were on our base in Tidworth, Hampshire, UK, freshly returned from a live firing exercise in the Oman, when General Galtieri invaded the Falkland Islands. As a popularity stunt, the invasion was an initial success. Galtieri even appeared on the balcony of the Casa Rosada near where Evita had made her impassioned speeches in support of the people decades before but, in his case, to celebrate hiding the Dirty War behind a veil of populist militarism.

The Falkland Islands are 8,000 miles away from Great Britain and only 400 miles off the coast of Argentina. They are the subject of a continuing sovereignty dispute between the United Kingdom and Argentina, which both claim the islands as their territory. The islands were first sighted by John Davis in 1592.[298] John Strong, an English sea captain, made the first recorded landing in 1690—that is, the first landing recorded by someone who took the record back to the UK. He named the islands after the British treasurer of the navy, Viscount Falkland, before sailing away. In 1764, a group of French settlers from Saint-Malo in Normandy, called the islands the "Malouines" after their parent city.

The following year, John Byron, the grandfather of Lord Byron the poet, established a fort at Port Egmont, on Saunder's Island north of West Falkland. Byron claimed the islands for Britain, unaware of the French presence on East Falkland.

In 1767, France transferred its settlement at Port Louis to Spain, which was the first nation to make a serious effort to establish a settlement, then called "Islas Malvinas." The Spanish made repeated efforts to expel the British and finally succeeded in 1774. In 1775, Britain claimed the entire territory. Britain did not relinquish its claim of sovereignty and in 1811, after 60 years of continuous settlement, the Spanish withdrew their garrison to the mainland to help quell rebellions there, leaving the islands uninhabited. In 1816, the Argentinians asserted their independence from Spain and, in 1820, as a new country, claimed sovereignty over the Falkland Islands and founded a settlement in support of its earlier claim over the islands.

Britain's claim to the islands was illogical from the Argentinian position of geographical proximity, but the strategic importance placed by Britain on the islands at the height of her empire meant that few were going to have the resources to compete once this strategy was set in motion. Britain reasserted its claim to the islands in 1832, then established control and expelled the Argentinian garrison in 1833, resulting in the departure of the remaining Argentinians. Britons then made up a permanent population. Argentina continued to claim sovereignty based on geographic proximity and its settlement in 1820.

Darwin's *Voyage of the Beagle* recounts his stopping in 1834 in the Falklands.[299] The Falkland Islands were notable in that no land mammals, not even small rodents, lived there. The only mammal was the Falkland Island wolf (*Dusicyon australis*), which inspired discussion about how it had ended up on the islands subsequent to the last ice age when there may have been an ice bridge from South America. Three specimens of this only indigenous land mammal on the Falklands were taken back on the *Beagle* to the UK. It was also called the warrah and went extinct in 1876 partly due to being exterminated to protect sheep farming. Darwin's observations about its similarity to other South American mammals, such as the Chilean Darwin's fox, led him to his central conjecture about the role of remote island environments on the evolution of species.

The British invasion of 1833 sparked off a disagreement with Argentina that simmered for decades. The UK claimed sovereignty on the basis of the 1,800 island population's wishes and the history of their earlier settlement.

This last argument appears hypocritical in the face of an event that happened at Diego Garcia, a British colonial possession in the Indian Ocean. Just years before,[300] a slightly larger population of indigenous people from the Chagos Archipelago, which had Diego Garcia as one of its islands, was forced out to allow Britain to lease the island to the United States as an airbase during the Cold War. The British agreed to "administer" the departure of the 1,500–2,000[301] Chagossians in exchange for the forgiveness of a mere $14 million military debt. Those islanders now face the optimistic possibility of being able to return to their island home after the injustice eventually became unsupportable by the United States or UK, but not before much hardship and tragedy befell the Chagossians.

In 1851, the Falkland Islands Company was founded primarily to husband the herds of cattle left by the original French settlers, and sheep farming became a basic part of the Falklands economy. In 1892, the settlement was formally granted the status of a colony and had a population of 2,000. Britain was now a global power and the Falkland Islands offered a major resupply station for shipping going around the cape to the Pacific. The islands met Britain's requirement for a whaling station during the heyday of that industry and continued to operate as a strategic naval base. Once the Panama Canal was opened in 1914, however, there was an immediate cessation of shipping via the Falklands.

In 1964, the issue of the Falklands came before the General Assembly of the United Nations, which in 1965 designated it a "colonial problem," and the disagreement continued to escalate. Argentina said that British control was a relic of colonialism, and the British said that the Britons on the islands risked immediately becoming colonials of Argentina. Britain even considered a transfer of the islands to Argentina in the 1970s…until the potential for natural resources was realized.

In 1982, General Leopoldo Galtieri was running a very unpopular regime. In 1983, it would be 150 years since the original British invasion of the Falkland Islands. The junta calculated that if they could repossess the islands for Argentina, they would win popular support and distract attention from the Dirty War. The Argentinians noticed that a Royal Navy support ship was withdrawn from duty in the region, suggesting that the British were downgrading the status of the islands. The junta finally sent 5,000 troops to "invade" on April 2, 1982. They were ably resisted by a platoon of Royal Marines stationed there, who were eventually overwhelmed and returned to Britain.

Margaret Thatcher immediately imposed a 200-mile exclusion zone around the islands and sent a "task force" of five battalions of soldiers, including the Third Battalion, the Parachute Regiment, my unit, and a naval carrier group to eject the Argentinians. The first British ships arrived in the area by the end of April. The conflict, in which I fought, resulted in 255 British deaths and 655 Argentinian deaths. An unconditional Argentinian surrender was declared on June 14, five weeks later.

Figure 77: The author and combatant, at the Buenos Aires "Malvinas" Falkland War Memorial. Source: The author, November 2009.

In the battle, my platoon was the point platoon for the after-midnight, B Company surprise attack on Mount Longdon, one of five mountains that stood guard over the western approach to Port Stanley, the Falklands capital. We fought all night and, at dawn, came under artillery fire that lasted most of the day in bursts. It was snowing and lightly overcast, and we went through another night looking after the wounded of both sides. I had 36 soldiers in my platoon. During the fighting, Private Gross and Corporal Stewart McLaughlin were killed, while 10 men were wounded. The other two platoons in B Company did much worse than mine, and it was purely out of slightly better luck that we had moved through the least dangerous stretch of the mountain. The following morning of the fourteenth, I remember the word of the surrender being yelled around the mountaintop with no small amount of relief. Berets were tossed in the air and shortly after this we all marched directly into Port Stanley, where I was later able to call my sister in London at 2 a.m. her time using a ship's satellite phone aided by journalist Kim Sabido.[302] Three days later, General Galtieri and his cabinet resigned in disgrace.

On becoming president of Argentina in 1989, Carlos Menem opened peace talks with Britain. In 1990, London and Buenos Aires restored diplomatic relations and hostilities were formally concluded by 1995, after dragging out talks for six more years. This peace treaty addressed the issue of potential oil discoveries to be made in the Falklands area. A compromise on oil was made in 1995 in which the two countries agreed to share, in different proportions, the wealth deriving from any oil discoveries depending on the source location. The British intended to use any wealth to offset the expense of the Falklands Conflict and the maintenance of the garrison, which numbered 1,700 persons to defend 2,200 residents. Licenses for oil exploration were issued in 1996, and exploratory drilling began in 1998.

In 2009, the Argentinians again requested talks about sovereignty of the islands. In December 2014, the Argentinians passed a law claiming sovereignty over many Atlantic islands, including the Sandwich Islands, South Georgia and the Falklands, which was immediately rejected by the British. The arrival in Falklands waters of a British naval frigate was deemed to be a "militarization" of the islands by the Argentinians. In March 2013, the approximately 3,000 islanders voted in a referendum almost unanimously in favor of remaining a British overseas territory. In April 2015, Argentina stated that it ruled out any further conflict but would continue to challenge the British occupation of the islands as "illegitimate."

The geology of the South Atlantic around the Falkland Islands has always been favored as potentially bearing oil. After its North Sea oil reserves began to decline in the 1990s, Britain needed a new source and the Falklands were perfect. The resource amounts to tax income of $180 billion if it's developed. Subsequently, new reserves have been discovered in the North Sea.

Exploration companies continued to explore for oil, and in early 2010 Argentina required all Falklands flag-bearing ships to obtain a permit from Argentina or be blocked from using any South American port facilities that were offered by Mercosur trading bloc countries.

Well Site	mmbbl
Sea Lion	393
Zebedee	281
Isobel Deep	240
Isobel/Elaine Complex	386
Jayne East	85
Chatham	100
Total	1,485

*Figure 78: Falkland Island oil well sites and expected resources in millions of barrels.
Source: Edison Research. Exploration Watch. Drilling Returns to the Falklands. 25
February 2015.*

In 2012, two cruise liners, the *Star Princess* and the *Adonia*, docking in Ushuaia, the southernmost city in the world, were refused port facilities after previously stopping over at Port Stanley in the Falklands.

Three publicly quoted British exploration companies, Premier Oil, Falkland Oil & Gas and Rockhopper (named after a local penguin species), found better oil- and gas-bearing rocks than they expected. In 2010, Rockhopper made a discovery in a well called Sea Lion, a resource amounting to 393 million barrels of oil. Drilling was very active until 2013. In 2015, it started up again, and 33 years after the Falklands Conflict, an oil discovery was made in license block PL004b in wells called Zebedee. A fourth company, Argos, was not able to fund itself despite having access to 52 prospects amounting to 3.1 billion barrels of resources.

This was the first discovery after a nine-month drilling campaign. Falling oil prices at the time caused some explorers to cut back, although an expected reserve of 8 billion barrels, multiples of the oil remaining in Britain's North Sea, makes the site very important. Another company, Borders & Southern, drilled exploration wells south of the islands where tests showed reserves of 4.7 billion barrels, 10 times larger than the Sea Lion wellsite reserves.

The fifth-generation oil rig Eirik Raude drilled six exploration wells before 2016. The cost to drill the site is expected to be $200 million, mostly met by Premier and Rockhopper and FOGL, contributing $25 million and $20 million, respectively. Each well takes about 1 month to drill at a cost of $50 million each. The Eirik Raude rig costs $570,000 per day and is overspecified for the NFB but is on spec for the deeper wells planned in the south and east basin.

During the time that the British exploration companies were incentivized with tax breaks, in 2012 the Argentine government nationalized the assets of YPF, an oil company owned by Spain's Repsol. In 2013, YPF announced a deal with Chevron. In 2011, Argentina had imported oil for the first time since 1984, a further blow to its already famously weak foreign currency reserves. Nationalization and its impact on internal efficiency hit YPF's crude output, which fell in the first quarter of 2013 by 0.7% and natural gas by 3.7%. There was a refinery fire, and energy imports in 2013 amounted to $14 billion, after $9.2 billion in 2012. Repsol discovered huge shale oil reserves in Neuquèn province. The Vaca Meurta field, which will cost $68 to $89 billion to develop, has 16 billion barrels of shale oil and 308 trillion cubic feet of natural gas, which would give Argentina the fourth-largest reserves in the world. The deal with Chevron is designed to overcome the significant barriers represented by Argentina: currency vulnerability, lack of debt funding from the IMF and an unstable government liable to seize assets at a moment's notice.

Argentina is its own worst enemy. Had things gone better, Buenos Aires might have become the de facto base for Falklanders, whose children were beginning to speak Spanish and often went on to study at Argentinian universities. If it had kept its energy assets in private hands and developed them before getting into a borrowing fix, those assets may already have been keeping the peso strong. If the country had also had a healthy incentive program and conditions to encourage investment, other locations on the significantly potential geology of the Atlantic coastal plains might have been better explored. The extreme swings between right- and left-wing politics have taken a deep toll out of an otherwise beautiful, resourceful, highly educated nation.

The Nuclear Geopolitical Externality

In 1968, the Treaty on the Non-Proliferation of Nuclear Weapons (NPT)[303] was signed by the family of countries that were nuclear then. It remains the most widely observed international agreement and depends upon disarmament, nonproliferation and peaceful uses for nuclear power. The treaty agrees that non-nuclear states will not try to become nuclear and will accept the International Atomic Energy Agency (IAEA) safeguards on their weapons capabilities. Nuclear weapons states commit to not transfer nuclear weapons to other states. All states have a right to peaceful use of nuclear energy and ought to assist one another to develop it. There is a conference every five years.

Nuclear power externalities differ from fossil fuel externalities. It clearly takes a lot of fossil fuel to drill, mine, extract, process and distribute uranium for nuclear power. It also takes a lot of fossil fuel to construct a nuclear reactor, so it has to have a measure of pollution with the release of a small amount of CO_2. Just as we say that criticizing renewable energy because it takes a lot of fossil fuel energy to build the machinery or commute the employees is misguided because, in a 100% sustainable economy, only renewable energy will be used to make wind turbines, it turns out that this works for nuclear as well: in a sustainable environment, nuclear power would also have no emissions externality.

In its operation, nuclear power is clean and emissions-free. Nuclear is also the most heavily subsidized power, at least in the USA. The externality that nuclear has that is out of all proportion to the rest is geopolitics. Nuclear has brought us away from war and almost to the brink of war again today. These vignettes on the nuclear geopolitical externality are very real, and they would go away if we switched to thorium molten salt reactors, which have no proliferation risk, 300 years of mild radioactive waste instead of tens of thousands, refueling and separating out of fission products during operation, no pressure dome construction, no fuel rods, frozen salt safety plug, 21,000 hours of production—1,500 of them at full power—16 MW, can use and draw down existing nuclear waste and a fuel that is to all intents and purposes hundreds of times more common and more energetic. The advantages are legion, but instead we have something you can make a bomb out of. One story about this alternative almost caught on but was interrupted by another disaster. Lack of ability to discriminate between different forms of nuclear tarred all nuclear, good and bad, with the same feathers.

Atomic Bombs

"Had I known that the Germans would not succeed in producing an atomic bomb, I would have never lifted a finger." —Albert Einstein, regretting that he had sent President Roosevelt a letter with a recom-

mendation to accelerate research into building an atomic bomb.

While ridding German universities of Jewish intellectuals as early as 1933, the Nazis put a lot of pressure on some individuals. Even Einstein's name was on a list for purging, his books burned and a reward of $5,000 was offered for his assassination. He was listed as an enemy of Germany with the phrase "Not yet hanged." The Nazis effectively destroyed themselves by shooing away the very people that would end the coming Second World War. They acted, thankfully, against their own interest in one of the biggest paradoxes in modern history and one that clearly gave the lie to any value in their brand of fascism.

Lise Meitner was originally a Jewish Austrian who had converted to Christianity at adulthood just after the turn of the century. She was one of the first two female doctoral graduates of the University of Vienna in 1905, despite being a woman. This was possible only because she could be educated privately. She went to Berlin and attended lectures by Max Planck, who had frequently rejected women attendees in the past, but she nevertheless became an assistant to him within a year. Although the Nazis allowed her to keep her professorship for a bit longer and despite her actual Christian faith, things got intolerable by 1938 and she took a very risky trip by train to Holland without her passport. Her laboratory partner, Otto Hahn, had given her a diamond ring that he had inherited from his mother in order for her to attempt to bribe the border guards if necessary. Luckily, she did not need it.

Meitner and Hahn continued their communication and even met once in Copenhagen. It was while she was on a trip to Sweden for a holiday during Christmas of 1938 with her family, which included her physicist nephew, Otto Frisch, that she was the first to correctly analyze Otto Hahn's experimental results from uranium fission. In fact, in 1939, Hahn, Meitner and colleague Fritz Strassman discovered that when they split, or fissioned, a U-235 atom, it yielded 200 million times more energy than was represented by the single neutron that was released at the atom of uranium in the first place.

What a piece of information! The most important historical product of that Berlin laboratory collaboration was first accurately appreciated by a woman who was rejected on the grounds of both gender and ethnicity! Meitner effectively saw that the gas radium or krypton and the barium found after the reaction explained by Hahn were lighter than the original uranium they were bombarding with neutrons. She realized the difference was energy. Meitner understood this while walking in the snow, observing that a large nucleus was like a large drop of water, which will break apart if it is wobbled too much.

She wrote up the results in a paper and with Otto Frisch became the first person to articulate how the atom is split. In 1939, it was impossible for Meitner to be honored within Germany. Although she missed the Nobel Prize awarded in 1945—it went to Otto Hahn, in one of the more glaring errors of omission of the Nobel Prize committee—she nonetheless had the 109th known element, Meitnerium, named in her honor. Otto Frisch confirmed the fission results in separate experiments in January 1939, and they published their work in *Nature* magazine.[304]

The diamond ring was later used by Otto Frisch in his own marriage. Meitner was firmly opposed to the Nazis but also rejected an offer to work on the Manhattan Project to build the first atomic bomb. She and Frisch had also discovered the reason that no stable elements beyond uranium existed in nature. These big nuclei are unable to resist the electrical repulsion of so many protons binding the neutrons and protons together.

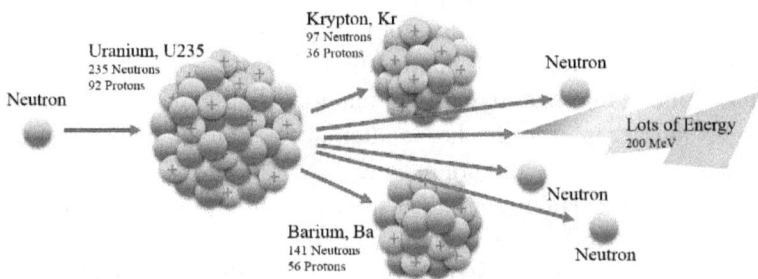

Figure 79: Reaction pathway of uranium-235 bombarded by an initial single neutron (from a neutron source) showing how it fissions or splits into barium and krypton, opening up a chain reaction pathway as each of the three released neutrons hit another U235 atom and repeat the process. The protons carry a positive charge. Source: NEF Advisors diagram.

After luckily avoiding the war, Meitner wrote withering criticism about the passive resistance that Hahn and other scientists had presented to the Nazi regime. Meitner and Hahn nevertheless remained close friends all their lives. If the three excess neutrons collide with three other atoms of U235, then the same event happens three more times, then nine neutrons are released, and so a chain reaction is started. This process exchanges a small amount of matter for a large amount of energy, à la Einstein. Some substances such as the noble gas xenon and carbon absorb neutrons, so they inhibit the chain reaction. This fact is used to control or moderate the intensity of the chain reaction in a nuclear power station, but in a bomb, there is no moderation.

After seeing the German results, Hungarian physicist Leo Szilard realized that the nuclear chain reaction he had posited in previous years had almost been discovered. He moved to America, accelerated his own research activity and conducted experiments on uranium at Columbia University in collaboration with Enrico Fermi. He proposed that a reactor could be built using uranium and carbon. The carbon would absorb neutrons and work like a damping mechanism, permitting the control of any reaction.

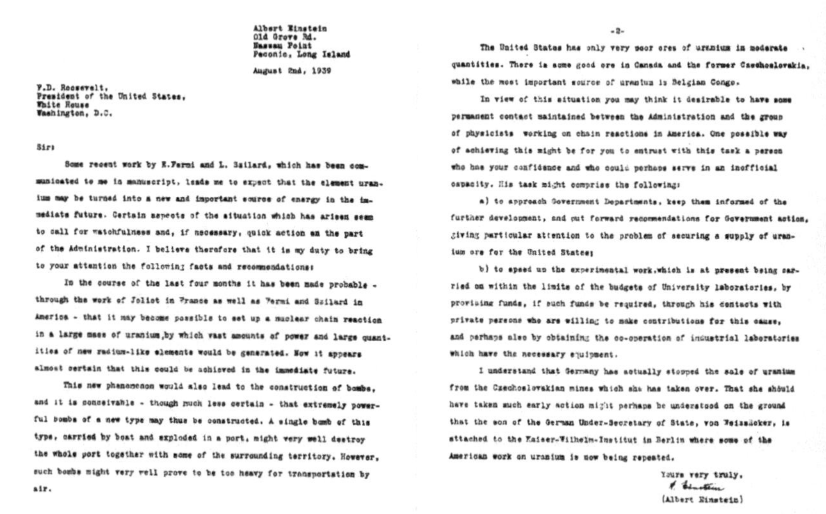

Albert Einstein
Old Grove Rd.
Nassau Point
Peconic, Long Island

August 2nd, 1939

F.D. Roosevelt,
President of the United States,
White House
Washington, D.C.

Sir:

Some recent work by E.Fermi and L. Szilard, which has been communicated to me in manuscript, leads me to expect that the element uranium may be turned into a new and important source of energy in the immediate future. Certain aspects of the situation which has arisen seem to call for watchfulness and, if necessary, quick action on the part of the Administration. I believe therefore that it is my duty to bring to your attention the following facts and recommendations:

In the course of the last four months it has been made probable - through the work of Joliot in France as well as Fermi and Szilard in America - that it may become possible to set up a nuclear chain reaction in a large mass of uranium,by which vast amounts of power and large quantities of new radium-like elements would be generated. Now it appears almost certain that this could be achieved in the immediate future.

This new phenomenon would also lead to the construction of bombs, and it is conceivable - though much less certain - that extremely powerful bombs of a new type may thus be constructed. A single bomb of this type, carried by boat and exploded in a port, might very well destroy the whole port together with some of the surrounding territory. However, such bombs might very well prove to be too heavy for transportation by air.

-2-

The United States has only very poor ores of uranium in moderate quantities. There is some good ore in Canada and the former Czechoslovakia, while the most important source of uranium is Belgian Congo.

In view of this situation you may think it desirable to have some permanent contact maintained between the Administration and the group of physicists working on chain reactions in America. One possible way of achieving this might be for you to entrust with this task a person who has your confidence and who could perhaps serve in an inofficial capacity. His task might comprise the following:

a) to approach Government Departments, keep them informed of the further development, and put forward recommendations for Government action, giving particular attention to the problem of securing a supply of uranium ore for the United States;

b) to speed up the experimental work,which is at present being carried on within the limits of the budgets of University laboratories, by providing funds, if such funds be required, through his contacts with private persons who are willing to make contributions for this cause, and perhaps also by obtaining the co-operation of industrial laboratories which have the necessary equipment.

I understand that Germany has actually stopped the sale of uranium from the Czechoslovakian mines which she has taken over. That she should have taken such early action might perhaps be understood on the ground that the son of the German Under-Secretary of State, von Weizsäcker, is attached to the Kaiser-Wilhelm-Institut in Berlin where some of the American work on uranium is now being repeated.

Yours very truly,

(Albert Einstein)

Figure 80: The letter written by Einstein, prompted by Leo Szilard to President Roosevelt, alerted the U.S. authorities of the potential for a German atomic bomb. Source: The Einstein letter was published in 1945, as part of the Smyth Report, without a copyright notice. It is thus in the public domain in the USA.

Einstein's 1905 formula, $E = mc^2$, had already shown that only a small amount of mass could release a huge amount of energy. The exact mechanism that could demonstrate this was still to be designed and built, but Szilard and Fermi were realizing the first steps by arranging for supplies of uranium and graphite (the carbon that Szilard needed). Eugene Wigner,[305] another alumnus of the Berlin University environment, Hungarian American theoretical physicist and mathematician, and 1963 Nobel Prize winner in Physics, was also party to events. He had met Einstein in Berlin and had emigrated to the States in 1930. He was also a close friend of Szilard and had originally introduced him to Einstein.

Realizing that action had to be taken, Szilard visited Einstein on Long Island. They wrote the letter (figure 80) to President Roosevelt, alerting him of the danger of the Germans being able to develop the atom bomb before the Allies did. German physicists had already excited their warfare strategists with visions of small bombs that could take out a city such as London, and resources were deployed to help make it a reality. The result was that the United States made the development of the U.S. version of the atom bomb an official goal. The energy we are talking about is stupendous.

After early success with the initial Chicago-based reactor in 1941 and after the German invasion of France, other scientists finally acknowledged the need for secrecy and the Manhattan Project was formed. Sir James Chadwick, the discoverer of the neutron, headed up the UK contingent to the Manhattan Project and was present at the first demonstration of the Trinity bomb. He became a central personality in the British team that developed the British bomb after the war.

The power of the atom is dramatically revealed in the huge energy yields of the first atom bomb and then the hydrogen bombs, or H-bombs, that came shortly afterward. The race to develop the first atom bomb was concluded with the first bomb, Trinity, at 21 kilotons, which was successfully tested in Alamogordo, New Mexico, just two months after victory in Europe. The use of the term *kiloton*, literally, "thousands of tons," is a way to scale the size of the nuclear explosion in terms of an equivalent explosion of the more familiar tri-nitro-toluene (TNT). The Trinity event was the equivalent of exploding 21,000 tons of TNT. The plutonium used in the Trinity bomb had cost the United States about a billion dollars to produce, with the inefficient methods available at the time.[306]

The surrender of the Germans in the Second World War removed the original anti-Nazi purpose of the U.S. bomb from the many intellectual Jewish émigrés who had championed, innovated and built it. At least Hitler had deprived himself of the brain trust that would have accelerated his own accession to the nuclear club, and that was a blessing. Suddenly, the driving force behind the development of nuclear weapons appeared to vanish. It was now uniquely the domain of the U.S. military will to win the war in the Pacific theater and choose to use the atom bomb to do so that kept momentum going. It was a true test for the scientists who had produced a weapon that might kill millions of people, but now in a completely different theater of the global war.

Their moral qualms came crashing back to haunt them. After all they had learned, the Manhattan Project scientists knew that the bomb itself was

simple. In fact, the first bomb used in anger on Nagasaki, the Little Boy, did not even need to be tested first. I hasten to add that the use of the word *simple* here is totally relative. To the scientists involved, the technology had become "simple," but it remains intimidating to the rest of us. It can't have been that simple, either, because, as Eric Schlosser in his book *Command and Control*[307] explains, only 1.38% of the uranium in the Little Boy bomb actually fissioned to cause the explosion. This amounted to only 0.7 grams of the material, as the rest was blown away in the explosion. It was, however, enough to kill 80,000 people and destroy two-thirds of the city.

In a fact that now seems trivial but that at the time was life or death, the town of Kokura was saved by clouds twice. Kokura was the backup target for the first atom bomb if Hiroshima had been cloudy. Three days after the bombing of Hiroshima, Charles Sweeney, the pilot, was tasked with bombing Kokura as the main target this time. He flew over the town for an hour above the clouds with the bomb bay doors open, ready to drop Fat Man. Eventually, the poor visibility caused by clouds and drifting smoke from the U.S. firebombing of a neighboring town, Yahata, forced him to fly to the backup target, Nagasaki, where 75,000 people died.[308]

Figure 81: This dramatic image of the Baker Shot undersea nuclear explosion in Bikini Atoll was an atom bomb test similar to the Hiroshima bomb, conducted by the United States on July 25, 1946, to establish the damage parameters on naval vessels and effects on living animals of the power of the atom. The vertical black shape on the right of the blast water column is the US Arkansas, a 171-meter-long battleship and the closest moored ship, caught here suspended 100 feet above the waterline. The plume of water is 6,000 feet high. Source: United States Department of Defense.

After the attacks on Japan, it must have seemed all the more awful to realize that this destruction was what the United States had originally engineered for Germany if that surrender had not already taken place. Leo Szilard and others had been agitating to have a warning offered prior to use of the weapon in Japan, but they were overruled by the military.

In the period from 1945 to 1998, over 2,000 nuclear explosions were conducted by eight countries,[309] each of whom announced their nuclear presence by their initial nuclear detonation. Bombs were detonated in space, underground, in the ocean and in the atmosphere. Only North Korea has tested nuclear devices since that time, culminating with its purported hydrogen bomb test in September 2017.[310] Other countries like South Africa and Israel have nuclear power and have relied on the test data of other countries.

Concern about the cumulative effect on the global environment of radiation and fallout over the latter half of the 20th century caused test ban treaties to concentrate on banning atmospheric testing and eventually any sort of testing. The sudden cutoff of testing in 1998 was a testament to successful global negotiations stemming from the successful 1996 Comprehensive Nuclear Test Ban Treaty (CTBT).

Worldwide nuclear testing, 1945 - 2013

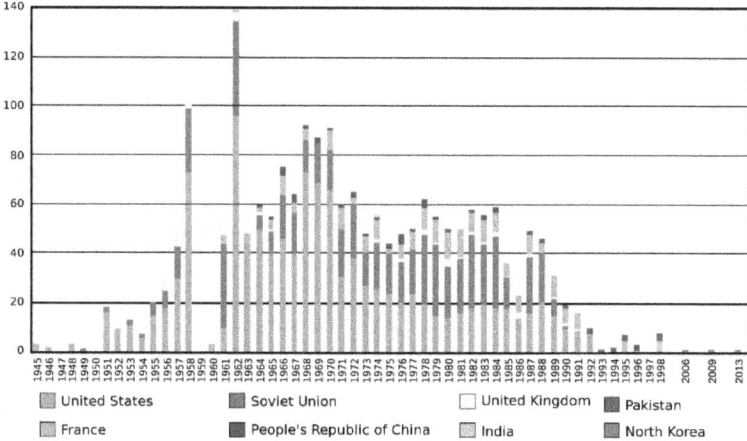

Figure 82: Numbers of nuclear explosions detonated annually by the nuclear "club." Source: "Worldwide nuclear testing". Licensed under CC BY-SA 2.5 via Commons.

North Korea is the latest country to join the club and is the only country to have exploded any devices in the 21st century. The dominance of the United States and USSR reflect the imperatives of the Cold War period, when a strategy of mutually assured destruction (MAD) is credited with having effectively deterred both sides from launching nuclear weapons against each other. In 2014, there were some 30,000 nuclear weapons in the world, with all except 200 controlled by the United States and Russia. However, it only takes one to precipitate a disaster.

You can see that as the Berlin Wall came down, the economically strapped Soviet Union only exploded one bomb in 1990, its last. Interestingly, Russia had also exploded the "mother of all fusion bombs" in October 1961, an enormous 50-megaton fusion monster called the Tsar Bomba, partly designed by famous Soviet physicist Andrei Sakharov. The top of the mushroom cloud rose to a height of 35 miles. Sakharov actually cut its size from 100 megatons because of fears about the effects of fallout.

Figure 83: Yucca Flat, the Eastern Nevada Nuclear Test Site, showing the pockmarks of some of the 739 nuclear explosions carried out in this location. Source: United States Government.

They carefully dropped this 30-ton behemoth (to be differentiated from its 50-megaton explosive yield) on a parachute to give the plane some time to escape. The test was carried out above the Arctic Circle on a deserted peninsula in the eastern Barents Sea. In fact, when the blast wave hit the aircraft, it fell a half mile before recovering controlled flight. The blast yield was 1,400 times as large as the Hiroshima bomb and represented 10% of all the explosives used globally in World War Two.[311]

The nuclear powers tested a lot of atomic and fusion bombs. It turns out that exploding 2,000 atomic bombs at huge cost on friendly territory during the Cold War took its toll on the bomb-making countries anyway. Huge tracts of land remain unusable. Some communities were deeply disrupted.

Bikini Islanders are still not back on their home atoll in Micronesia, and large parts of the Nevada desert have been and will remain off limits for a long time. Large amounts have been paid out in compensation for fallout, radiation poisoning and ruined homelands. Thousands of service personnel have suffered from radiation poisoning, and the effects on childbirth and health have been locally disastrous.

None of this was as significant as the scale of damage to the actual wartime Japanese bomb victims. A recent visit back to the Bikini Islands by Stanford biology professor Stephen Palumbi 60 years after the bomb tests and featured in a documentary called *Big Pacific* revealed that nature had grown back. Corals have recolonized on the edge of the mile-wide hole that "Bravo Crater" made in the bay when the bomb was set off, the "equivalent to 216 Empire State Buildings being blown into the sky." Local organic material still contains high levels of radioactive materials and eating the local fish or coconuts or drinking the water is not a good idea. Today, local animals are unafraid of humans and schools of large fish and sharks populate the reefs abundantly.

One hundred ninety-one countries ratified a global Non-Proliferation Treaty (NPT) in 1970 designed to prevent the spread of nuclear weapons, increase nuclear disarmament and promote peaceful use of nuclear power. All countries have adhered to the treaty except for North Korea, which originally signed in 1985 but withdrew in 2003. The treaty recognizes the UN Permanent Member Security Council, the United States, UK, France, Russia and China as nuclear weapons powers. There is an implicit agreement among members to spread the benefits of nuclear technology among other countries in the world. No non-nuclear power can adopt nuclear weapons. Members originally agreed to meet every 5 years for 25 years, but in 1995 extended the meeting cycle indefinitely.

At the time of ratification, it was expected that there would be 25 to 30 nuclear weapons states within 30 years, but proliferation has proved to be mercifully difficult. Over 40 years later, there are only 4 UN member-states that have never ratified the treaty: India, Israel, Pakistan and South Sudan (which was only founded in 2011). Preventing proliferation has been relatively successful. Only the first four of these are nuclear weapons countries, but little has been done to disarm. The five authorized states still control 22,000 nuclear warheads. The non-nuclear states kept their side of the bargain, but the nuclear states did not. The United States and Russia have indeed reduced their arsenals but still retain more than enough to incinerate most of humanity.

They intend instead to maintain and upgrade their nuclear weapons capabilities. The United States has decided to spend $1 trillion over the next three decades to upgrade its nuclear weapons. Donald Trump has the inclination to commit even more treasure to this purpose. Nuclear states point out that while even one single nuclear power exists, maintaining a nuclear capability is a necessarily defensive act. The United States, however, refuses to rule out "first use" of nuclear weapons against even non-nuclear states. Russia has indicated it will use them against NATO. Pakistan and India threaten to use strategic and tactical nuclear weapons against each other if conventional warfare gets out of hand.

During the Cold War, the threat of a nuclear counter-response kept the use of nuclear weapons in check and arguably prevented Armageddon. It was called MAD, for mutually assured destruction. It became effectively suicidal for any nuclear aggression to occur. After decades, nuclear tensions have eased, economies have grown steadily, and the quality of human life has progressively improved. The close nuclear shaves I have listed have almost driven us to those consequences purely by accident. It's only by the narrowest margins that modern humanity has emerged, relatively unscathed. The nuclear nations, however, have resisted many efforts to change nuclear doctrine to a "no first strike" but have never adopted this position. Dwight D. Eisenhower developed a doctrine allowing first use of nuclear weapons in response to any form of military aggression. Smaller, strategic nuclear weapons that can be fired by artillery made it all the more difficult to adopt a "no first strike" stance. The Russians and Chinese, however, have been "officially" committed to a "no first use" doctrine, and although this is still the position, the West has never actually trusted it to be the case.

In 2013 and 2014, 150 countries met in a series of historic UN-led conferences on the humanitarian consequences of nuclear weapons. A letter[312] written by six Nobel Peace Prize winners evoked the promise of a nuclear-weapons-free world. Even a limited nuclear war using up only half a percent of the existing nuclear weapon stockpiles would cause climate disruption, leading to starvation for 2 billion people. A major conflagration, say, between the United States and Russia, could cause human extinction. The greatest threat to non-nuclear states is the nuclear states. At the end of 2016, the UN General Assembly presented the conclusions on how to pressure the nuclear powers to disarm followed by a formal conference in 2017 to conclude a new "Ban Treaty" against nuclear weapons. Needless to say, there is fierce opposition from the nuclear nations.

Postwar Nuclear Britain and the Special Relationship

After the Second World War, the U.S. government voted to restrict the information gained during the Manhattan Project to the United States. Even though it participated in the project, the UK was not immediately included in the "special relationship" with the United States because it was not yet a nuclear state. As quickly as possible, the UK determined that it had to develop its own bomb capability, and they did so, achieving an operational reactor from which to source essential materials as early as June 1950. They assembled their physics and engineering "boffins" and built a test reactor- based design from experience gained at Hanford in the United States, where plutonium had been produced for the Trinity bomb test. The site chosen in Britain was situated at Windscale in Cumbria in the northwest of England. Since there was little intention of providing electricity for the local grid and the sole purpose of the effort was to acquire nuclear bomb status, the project was rushed.

A lot of clever engineering was done on the fly, including a system that cooled nuclear fuel rods using air, not water. In retrospect, they neglected to think that in the event of a fire, the use of air for cooling was like pouring gasoline onto a fire. The scientists settled on a design that would bombard common U-238 with neutrons, turning it into unstable U-239, which decays to neptunium-239 (Np-239), which has a half-life of only 2.355 days, meaning that within 7 days almost 90% of the Np-239 has become the desired plutonium-239 (Pl-239).

Recurring political demands for plutonium and tritium (fusion and H-bomb material, respectively) for the British atomic and hydrogen bombs led to a near-disaster during the Windscale fire on October 10, 1957. Increasing heat in the piles was often dealt with by using more coolant air. When working the reactor to produce plutonium, it got excessively hot. Becoming conditioned to this extra heat, the physics team developed an understanding of the way the graphite pile was behaving. Eugene Wigner had noticed in the U.S. reactor pile in Hanford, Washington, that strange hot spots in the graphite would occur.

Wigner ingeniously reasoned that there must be a substance produced that was effectively distorting the flow of neutrons. He realized this even before it was later understood that xenon-135, which comes from an isotope of iodine and is the most effective absorber of neutrons known, was produced as a decay product and had the effect of absorbing neutrons, causing the hotspots to occur. Luckily, the xenon-135 would quickly decay, and the problem would go away. It played an almost fatal role in Windscale. The method used by the British scientists at Windscale to deal with the frequent pile overheating was named the "Wigner Release" after Wigner.

One of the design problems with solid fuel rods is that the fission products cannot be easily extracted, except by a formal process of refueling of the reactor, which means closing it down and refueling with fresh fuel rods over a period of days. A liquid reactor, on the other hand, can easily cope with xenon gas buildup or with unwanted fission products by simply removing them from the fissioning molten salt liquid media. Xenon was also a major contributor to the disaster at Chernobyl.

A fire started among the Windscale aluminum fuel canisters, and air was used to cool it down with almost disastrous results, literally fanning the blaze. For three days, the physics and engineering teams struggled to contain the problem, pioneering new ground in nuclear safety and often taking exposure risks as they did so. They tried pushing out the fuel canisters with scaffolding poles, but they got too hot, literally coming out dripping with molten metal. Then they doused the fire with some of a delivery of 25 tons of CO_2 that was nearby. The fire was so hot it actually stripped the oxygen atoms from the carbon dioxide molecule, which resulted in feeding the flames rather than inhibiting them. Eventually, they succeeded in removing many of the canisters and poured water on the reactor, finally extinguishing the fire.[313]

Since the events happened prior to the United States accepting the UK back into the special relationship, Prime Minister Harold McMillan covered up the almost-disaster. Once the UK was successfully back in the nuclear club, however, the cover-up story was modified, but in a subsequent enquiry, the Windscale management team were still "blamed" for "errors in judgment."

In actuality, the Windscale crew deserved hero status for producing the plutonium and tritium required for the British bombs despite pioneering new technology and enabling the special relationship with the United States to continue, dealing with the excessive political pressure and narrowly averting the first potential nuclear disaster.

Much radioactivity had been released, however, and the event was the first in a series of nuclear reactor events that have dogged the progress of the nuclear industry and reactor design, including, among the big incidents, Three Mile Island in the United States, Chernobyl in the USSR and, more recently, the Fukushima Daiichi event subsequent to the Tōhoku Japanese earthquake and tidal wave on March 11, 2011. These and thousands of other "incidents" have resulted in a deep negative opinion of nuclear power that has caused it to be limited to approximately only 4.41% of total global energy use in 2017,[314] even though, in reality, it is one of the least harmful of all energy sources.

Peaceful Nuclear Power

> *"...to find the way by which the miraculous inventiveness of man shall not be dedicated to his death but consecrated to his life."* — President Eisenhower, Atoms for Peace, 1953.

At the United Nations in a speech on the future of the atom, Eisenhower declared his "Atoms for Peace" remarks, heralding a less weapons-filled future for uranium.

Figure 84: Enrico Fermi and Leo Szilard with the Chicago "pile" CP-1, the first-time humanity created a nuclear event. It went critical on December 2, 1941.

In the 1950s, various members of the Manhattan Project who were behind the intellectual progress continued research into the peaceful usefulness of nuclear power. In this case, *peaceful use* described the use of neutron chain reactions in fissile materials such as plutonium and uranium to obtain heat and thereby electricity.

Various labs were set up around the United States. Enrico Fermi and Leo Szilard had built the original Chicago-based stack and made the first-ever demonstration of a controlled chain reaction, or "going critical" in a real "pile" of graphite with uranium fuel rods. Eugene Wigner was also present at this demonstration. No use was made of the heat in this circumstance, however, and it was more of a confirmation demonstration so that the Manhattan Project could go ahead and make the bomb. Scientists returned to this original design to find a way to provide controllable nuclear thermal energy that could be used to generate steam and consequently electricity.

The Department of Energy quickly capitalized on the existing centers of excellence, establishing research laboratories. Fermilab, near Chicago, was named for Enrico Fermi, who directed it for some years. Oak Ridge National Laboratory (ORNL) in Tennessee was also set up and is the laboratory that developed a nuclear configuration, the MSRE, which is a focus of my main nuclear theme: thorium nuclear power.

In 1964, the very first nuclear electricity generation public grid connection was made by a 5-megawatt Obninsk reactor in Russia. Then the British followed their air-cooled plutonium-producing reactor (ACR) at Windscale with a water-cooled design. In the United States, they had used water-cooled reactors, or WCRs. These "piles" of reaction-moderating graphite with holes in which the fuel rods were inserted were the basis of many new designs. In 1954, the "Pressurized Pile Producing Power and Plutonium," or PIPPA (yes, really), a project launched by the UK government to provide the larger amount of plutonium required for its full nuclear deterrent, involved building out eight new reactors. Subsequently, Calder Hall, very close to Windscale, was the first nuclear reactor deliberately designed to generate peaceful, commercial electricity and had a capacity of 200 megawatts. It was connected to the grid in August 1956 and officially commissioned by Queen Elizabeth II in October 1956. The Shippingport reactor in Pennsylvania was the first American nuclear power station intended to produce electricity for the grid, to which it was connected in 1957. Just over a decade after the

Second World War closed with a nuclear bang, peaceful nuclear power switched on.

Nuclear Swords to Nuclear Ploughshares

> *"There are some innovations in nuclear, there is modular, there is liquid, and innovation really stopped in this industry quite some time ago, so the idea that there are some ideas laying around is not all that surprising."* —Bill Gates.

In June 2013, I went to see Gerald Feldhamer[315] in New York. Feldhamer had received an appointment to West Point. His brother received an appointment to Annapolis but due to poor eyesight was unable to become a pilot. He served in the Air Force as ground crew and flew as a "guest" in bombers over Europe in the Second World War and survived. At the start of the war, Gerald Feldhamer was already in the National Guard, so he was not called to wartime duty. In 1978, after being a partner at H. L. Federman & Co., a member of the New York Stock Exchange, Gerald Feldhamer founded his own investment company, Feldhamer Capital Corporation.

The very next year, in 1979, the United States experienced the pressurized water reactor (PWR) accident at Three Mile Island in Pennsylvania. A coolant problem led to overheating and a partial melt-down. Fission products xenon and krypton, both gases, were released, but because these elements are not absorbed by animals or humans, there was no further danger. No casualties followed, but the event resulted in a distaste for nuclear power that initiated resistance that eventually went global and that has stymied the industry ever since. This story illustrates how ignorance of a technology creates a monolithic fear of it, despite the facts, which in this case are that there are very few actual health problems and very little actual overall mortality from nuclear power—even less than from wind or solar.

Feldhamer Capital had done very well by introducing the Subaru motor car brand from Japan to the United States. Through connections with Israel, Feldhamer came upon a project led by Dr. Alvin Radkowsky that appeared to have all the hallmarks of a successful venture: the revisitation of the use of thorium in a nuclear reactor designed with peaceful power generation in mind. Conventional nuclear reactors generate plutonium-239 because of uranium fission. Plutonium-239 is also fissile and can be used as fuel as well. The problem is that it can also be used to make a bomb and presents a significant nuclear proliferation risk.

The thorium cycle produces plutonium-238, which can't be easily made into a bomb and which has many other peaceful uses. Radkowsky said in 1984:

> *A thorium reactor's plutonium production rate would be less than 2 percent of that of a standard reactor, and the plutonium's isotopic content would make it unsuitable for a nuclear detonation.*

Dr. Alvin Radkowsky was an American-born physicist who worked closely with Admiral Hyman Rickover on U.S. Navy nuclear submarine and ship engines as chief scientist of the U.S. Navy Nuclear Propulsion Program. He also was the designer of the earlier-mentioned first commercial-scale civilian nuclear power station in the United States at Shippingport, Pennsylvania. He was one of the founding scientists of the U.S. civil nuclear sector. He was nominated for the Nobel Prize and developed the system, used by nuclear submarines, to remove used uranium fuel rods from the boat's reactor. Edward Teller had been his advisor when he was a physics student at George Washington University, and later, Teller encouraged Radkowsky's thorium approach because it couldn't make a bomb.[316]

Radkowsky lived for some years in Israel and taught physics at Tel Aviv University. While there, Radkowsky developed a method of using thorium, an abundant and safe actinide metal, as a source of energy in a light water nuclear reactor. The design was nonproliferative, which meant that it had an important advantage over uranium-fueled reactors in that it could be established worldwide without any geopolitical danger caused by the stockpiling of atomic weapons materials such as enriched uranium or plutonium.

Dr. Radkowsky started a company called New Power Technology (NPT) in Israel (the initials also deliberately spell out Non-Proliferation Treaty). New Power Technology's strategy was to support and develop the use of thorium as an alternative nuclear fuel with all its safety and built-in advantages.

Yitzhak Moda'i, the Israeli energy minister, explained:

> *The Israeli Government is very pleased with the designation of New Power Technology to assist us in the development of Dr. Radkowsky's progressive work in the field of nuclear energy. This outstanding project would not only benefit Israel but all mankind.*[317]

Feldhamer Capital was appointed by the Israeli Energy Ministry to commercialize New Power Technology following an agreement between the Energy Ministry and the Universities of Tel Aviv and Ben Gurion, where the research was conducted. The advisory committee and board of directors for this small company with a big product were an extraordinary collection of the world's most famous Manhattan Project and Nobel Prize–winning physicists, all willing to lend their names to this humane venture. They were signing on to a project that had as its goal the expansion of the use of nuclear power as a peaceful means to improve humanity by providing plentiful power and in a way that did not entail the nuclear baggage of bombs, the geopolitics of scarce uranium fuel or the threats of destruction and long-term dangerous waste.

The anxiety born from the recent accident at Three Mile Island had diminished some by then, and the idea of using thorium was ready for the mainstream. Figure 85 shows on the top row Mr. Steven Katz, Gerald A. Feldhamer and Bernard Levine, and on the bottom row from the left, Professor Edward Teller, father of the hydrogen bomb and founder of the Lawrence Livermore Laboratory, Nobel Prizewinner Professor Eugene Wigner of Princeton University, Dr. Alvin Radkowsky, Nobel Prizewinner Professor Hans Bethe of Cornell University and Professor Herbert Goldstein of Columbia University.

This extraordinary group had come together because they had been at the heart of the development of America's atom bomb which ended the Second World War. Dr. Radkowsky had persuaded them all, relatively easily, to back the NPT strategy to develop a new thorium reactor fuel rod because of a deep commitment to peaceful, cheaper, safer nuclear power using thorium fuel.

Figure 86: Photographs of the board of directors of New Power Technology. With the permission of Gerald E. Feldhamer, 2016.

Uranium and plutonium are energy elements that share the ability to become either swords or ploughshares. These men didn't want the swords anymore. Thorium had all the advantages that uranium lacked and more, and you couldn't easily make a bomb out of it. Thorium could only be a ploughshare.

They all came from their homes for a meeting together in New York. Professor Hans Bethe flew in from Ithaca. Feldhamer Capital covered their expenses. On Friday, February 3, 1984, the *Jewish Times* printed an article on page 41 covering the gathering with a photograph of the Israeli energy minister Yitzhak Moda'i, Gerald Feldhamer and Dr. Radkowsky celebrating the coming together of this extraordinary board. Feldhamer recalls that the manager of the Regent Hotel in Manhattan exclaimed that there was more brainpower in one room than there was in the rest of the world! It was impossible to buy this remarkable support, and with the benefit of hindsight, all members of the board were operating on a high level of intellectual honesty and at a key time in history. New Power Technology was capitalized with half a million dollars, and Dr. Radkowsky held the majority of shares while the members of the board were each given a piece. Gerald Feldhamer controlled the management of the company.

They marketed Dr. Radkowsky's thorium reactor as a form of nuclear power which did not pose the sort of danger of proliferation and nuclear weapons that the Cold War had made commonplace. The policy of mutually assured destruction (MAD) and the missile standoff with the Russians in Cuba helped sensitize the public about the weapons danger of nuclear power and this was an angle they could play on. Dr. Alvin Radkowsky was very available and attentive to the business and to Gerry and they developed a strong friendship. Feldhamer also had a good friendship with Edward Teller and they would often dine together in New York. Mr. Feldhamer initiated discussions with large companies such as GE and Hewlett Packard and many others and there was much interest in the strategy.

Gerald Feldhamer remembers that Dr. Radkowsky used to stand on his head for exercise. He held that Professor Wigner was an intellectual giant, even among this exalted group, for all for his contributions to physics. The Radkowsky design was an attempt to replace uranium within existing LWR (Russian) reactors with a thorium fuel cycle, which offered greater proliferation resistance. Enriched uranium drives the nuclear chain reaction and provides neutrons for the thorium fuel to convert to U-233. In these circumstances the production of plutonium, already difficult, is reduced to only one-third of that of a comparable uranium fuel cycle and potentially much less on a per megawatt basis since more of the thorium is actually burned up, reducing the plutonium per megawatt. In addition, the Radkowsky design produces plutonium-238, the wrong isotope for making bombs but very useful in other ways. On top of this, thorium is a much better stand-in than the U-238, without which there would be no plutonium anyway.

The project appeared highly worthwhile, and they thought there was a strong probability that they would shortly be able to secure a contract to build a thorium reactor almost anywhere in the world because of the lack of proliferation danger. They expected such contracts would move the company steadily forward, scaling up to hundreds of installations. It was not to be.

In 1970, a small Ukrainian town called Pripyat, with a population of 50,000 and built prior to the collapse of the USSR over a port village on the Pripyat River in the Ukraine, was close to a Russian nuclear power station called Chernobyl. The World Nuclear Power Association tells us that this power station was one of 11 Soviet-era reactors built using a design known as Reaktor Bolshoy Moshchnosti Kanalnyy (RBMK), or "High Power Channel-Type Reactor." It was the ninth "nuclear" town in Russia. While the town and neighboring cities were closing down for the night on April 26, 1986, when electrical demand would be low, the managers of the nuclear power station had arranged for a test of the water pumps. The junior personnel on duty in the control rooms were about to face the greatest challenge of their lives. During the test, loss of cooling water to the nuclear stack led to a meltdown after midnight and an explosion that lifted the 1,000- ton containment building off its foundations. A sudden increase in power output caused a reactor vessel to rupture in a series of steam explosions, exposing the graphite moderator material to the air, which caused it to ignite. Plumes of radioactive particles leaked into the atmosphere, and the fallout affected an extended geographical area over Pripyat, Belarus, the Ukraine...and Western Europe.

The fallout was detected next in Sweden, when the security door for another nuclear power station paradoxically would not open to staff because it sensed a radioactivity leak. The workers had been walking on ground where radioactive fallout from Chernobyl had collected on their shoes, triggering the alarm. This confirmed that radioactivity had escaped. The source was soon discovered, and the secretive USSR had to confess to its worst nuclear accident.

This event set off alarm bells against nuclear power everywhere. The event itself caused the deaths of 31 fire fighters and plant staff who were overexposed to radiation.

A release of iodine into the atmosphere was deposited back on the ground by rainfall, and then cows eating grass became sources of dangerously radioactive milk, which caused a spike of about 6,000 thyroid cancers in children, an otherwise rare disease; a group of 15 had died by 2015.[318]

The Chernobyl accident is considered to be the worst nuclear accident in history. During the lifetimes of all the people associated with the event, a predicted total of about 4,050 deaths among those experiencing increased irradiation are expected. Today the World Health Organization tallies the expected death count from Chernobyl at a total of 9,000, while, antinuclear Greenpeace say they expect 90,000 to die once all is said and done. The USSR spent over $200 billion cleaning up the damage. Arguably the worst of the effect was the evacuation and resettlement of over 350,400 people in terms of cost, loss and depression. Its effect on the nuclear power industry was as consequential to the fate of the global nuclear power industry in the late 1980s as was the 2011 Japanese Fukushima accident.

The Chernobyl disaster stopped New Power Technology in its tracks. It did this even though the thorium reactor design could never have created a similar disaster scenario. After the Fukushima accident in 2011, the nuclear industry had to cope with an extra security and safety cost, and the new technology of hydrofracturing, or fracking, was introducing abundant new sources of oil and natural gas, making fossil fuel energy cheaper than nuclear.

Dr. Radkowsky did not give up. In 1992, the Radkowsky Thorium Power Corporation (RTPC) was formed to once more address the huge potential for thorium to replace uranium in solid fuel rods and further develop his successful designs. By 1994, a collaboration was started with the Kurchatov Institute in Moscow, where extensive R&D evaluation was carried out. In 2001, RTPC changed its name to Thorium Power, Inc. In 2005, the company obtained U.S. government funding with the goal of disposing of plutonium from Russian light water reactors with good results. A year later, Novastar Resources, interested in commercial use of thorium fuel, completed a reverse merger with Thorium Power, whose shareholders were now the majority shareholders of Novastar. This was followed by a name change to Thorium Power, Ltd. In 2008, the company provided consulting services with a team of experts to the United Arab Emirates, which planned a nuclear program. This service broadened to other global clients. In 2009, the company's shares were listed on the NASDAQ stock exchange (LTBR), and the name was again changed to Lightbridge Corporation. Thorium Power Inc. is still a wholly owned subsidiary of Lightbridge and controls the intellectual property rights. Lightbridge continues to advise the UAE on its nuclear program and has expanded its client base to include South Korea.

Dr. Radkowsky died on February 17, 2002. However, his zeal to reduce the proliferation of nuclear weaponry never died. Radkowsky thorium fuel, or RTF, remains his legacy and is being championed today by the Indian government, among others.

Nuclear Accidents

> *"We came very close [to nuclear war], closer than we knew at the time"* —Robert McNamara, U.S. Secretary of Defense, in 2002, of the 1962 Cuban Missile Crisis.

The history of both the weapons and commercial public use of nuclear power is littered with accidents. The stories of some of the accidents are among the most gripping accounts of close calls with disaster that exist. Interestingly, the incidents concern missiles, reactors and movement of nuclear weapons by aircraft almost exclusively. While the Russian nuclear navy has plenty of accidents to report, the U.S. nuclear navy has persisted, since Admiral Rickover's initiation of safety measures and technology standards, without any accidents at all.

Admiral Hyman G. Rickover was an unusual character who rubbed many of the people he met the wrong way, but his gruff manner was more than offset by his appetite for work and "doing his duty." He pushed himself forward to work on submarines and just after the war was assigned to Oak Ridge National Laboratory as a military colleague for Alvin Weinberg.

In 1945–46, after the Manhattan Project resulted in two atomic bombs being dropped on Japan, two incidents underlined the severity of nuclear radiation. Originally, a third atom bomb was due to be dropped on Japan by August 17, 1945. The Japanese surrender came on August 15. The core of that bomb, named the Demon Core, was first to cause two peacetime nuclear deaths.

On August 21, 1945, Haroutune "Harry" Krikor Daghlian Jr., a Manhattan Project physicist, was working with the small, 3.5-inch sub-critical mass of plutonium that formed the spherical core of the bomb. He was building a wall of tungsten carbide bricks around the core to reflect back neutrons to see whether they could reach criticality with a smaller quantity of plutonium. Daghlian noticed that the neutron flux was increasing and was close to criticality. He decided not to place one more brick to avoid going too far by invoking a sustainable chain reaction within the fissile plutonium and releasing millions of neutrons. While moving the brick away, he accidentally dropped it, setting off the sustained criticality. Daghlian was fatally exposed. Not feeling anything initially, he was able to recover the brick and take apart the remaining bricks from the wall to terminate any further reaction. Rapid medical care was unable to prevent him later going into a coma and dying on September 30, 1945.[319]

Another similar fatal accident occurred on May 21, 1946, despite new safety measures, when Canadian scientist Louis Slotin was demonstrating the core to an assembled group of scientists. He accidentally dropped a screwdriver on the core, also setting off a criticality, a sustainable chain reaction of fissioning plutonium, releasing neutrons, which irradiated Slotin, who was standing next to the core. Slotin died on May 30, and seven observers developed associated problems decades later because they were within a few feet of the event and received relatively high levels of radiation.[320] There have been over 60 known criticality events, 22 of them outside nuclear reactors, which have caused 13 deaths since 1945. (This must be compared to the annual death rate from coal!)

With the weapons, the risk manifested from much higher bomb yields than expected, as in the case of the Castle Bravo test at Namu Island on Bikini Atoll in the Pacific. It was the first test of a lithium deuteride hydrogen bomb and the first example of a deliverable hydrogen bomb. It was detonated on March 1, 1954, at 6:45 a.m. The lithium part of the bomb was made up of two isotopes of lithium, 6 and 7. The latter lithium material was understood to be nonreactive, but in the event, it did react. The resulting explosive yield of 15 megatons was three times the 5-megaton target. This is a thousand times the size of the bombs used on Japan. This was the famous case where radioactive fallout fell on a Japanese fishing ship, the *Daigo Fukuryu Maru,* with one fatality.[321]

This caused an uproar in nuclear-bomb-sensitive Japan, and the United States paid $15.3 million as compensation. The sailors were paid $5,550 each and given Hibakusha status, the same government support given to Hiroshima and Nagasaki survivors. The fallout went east and also fell on the Rongelap, Rongerik and Utirik atolls, which were still populated with 20,000 islanders. The day before the test, the wind was to the north, but on the day of the test the wind sheared to the east of Bikini, showering those atolls with radioactive fallout. There were no plans to evacuate the islanders if the explosion had been as expected, but with 15 megatons, an emergency evacuation plan took three days. This was too long. The air was full of fallout, a snow-like radiation that profoundly affected the health of all concerned. The Cold War had started, and everything that could be learned about atomic bombs was a high priority. A schedule designed to beat the Russians was more important than ensuring the safety of the unfortunate islanders. The nature of the radiation danger inspired Neville Shute to write *On the Beach* in 1957.

During the Cold War, there were cases where bombs almost went off or where nuclear missiles were almost launched in response to atmospheric or celestial events or flocks of birds. In one case in the United States, an armored truck was parked over the doors of an intercontinental ballistic missile silo to prevent it from leaving the launch pad, even though all the other protocols had been observed. On another occasion, Boris Yeltsin, then Russian president, was awoken with less than 30 minutes to decide whether to launch a counterattack against what was being interpreted as an attack by the Americans. He decided not to counterattack, possibly against much resistance, and literally saved the world.

In another story, released as a movie called *The Man Who Saved the World*, a Soviet satellite confused sunlight reflected off high-altitude clouds with a U.S. ballistic missile launch. The Soviets were using a combination of radar and satellite data to confirm whether missiles had indeed been launched, but tensions were extremely high on this occasion, with a real expectation of a surprise U.S. attack. This was also just after the shooting down of Korean Airlines Flight 007, a Boeing 747, over Sakhalin Island on September 1, 1983. The flight was on its way from JFK in New York City via Anchorage to Seoul, South Korea. One passenger, Larry McDonald, was a U.S. House of Representatives member from Georgia, one of 269 passengers and crew to die. The Russians mistakenly thought the plane was on a spy mission and only released the flight recorders in 1993, well after the Berlin Wall had come down.

On September 26, 1983, Lieutenant Colonel Petrov[322] had just come on duty with his team at the secret Serpukhov-15 control center near the Kremlin when an alarm broke the silence after midnight. Their command post lit up with the word "Launch." His team looked at him in a high state of anxiety to see what his decision would be. Within minutes, a second siren joined the first, indicating another U.S. missile was on its way. Soon, five missiles were causing alarms to blare.

Petrov needed to know whether the computer's information was reliable. He knew his decision would spark a retaliatory response from the Soviet missile command, and he remembered the extreme stress of having to have the correct response. The world relied on the judgment of one man, and he had 15 minutes to decide whether to alert the Soviet nuclear response command, who would order a counterattack that would end the world.

It turned out that radar only picked up missiles well after they were launched, and Petrov knew the radar was more reliable than the satellite. Petrov reported that there was no radar confirmation and that a U.S. launch could not be confirmed, but he and his team had to wait 20 endless minutes to learn whether they were correct. Later it was shown to have been the

> *"I am not a hero. I was just at the right place at the right time"*
> - Stanislav Petrov

satellite registering a launch signal by mistake from the reflection of sunlight on high clouds. None of this emerged until the fall of the Soviet

Union, and the events transpired in such secrecy that even Petrov's wife may not have known of her husband's heroism. Lt. Colonel Stanislav Petrov died, unheralded, of pneumonia in Moscow in May 2017 at the age of 77, a fact only realized when a German journalist called him to wish him a happy birthday in September and was told of his passing by his son.

This happened on the American side as well. Robert M. Gates wrote a book[323] about the Cold War in which he relates the incident where Zbigniew Brzezinski, the national security advisor to President Carter, told him how he was awoken one night with a terrible burden of office:

> *As he recounted it to me, Brzezinski was awakened at three in the morning by [military assistant William] Odom, who told him that some 250 Soviet missiles had been launched against the United States. Brzezinski knew that the President's decision time to order retaliation was from three to seven minutes. Thus, he told Odom, he would stand by for a further call to confirm Soviet launch and the intended targets before calling the President. Brzezinski was convinced we had to hit back and told Odom to confirm that the Strategic Air Command was launching its planes. When Odom called back, he reported that 2,200 missiles had been launched. It was an all-out attack. One minute before Brzezinski intended to call the President, Odom called a third time to say that other warning systems were not reporting Soviet launches. Sitting alone in the middle of the night, Brzezinski had not awakened his wife, reckoning that everyone would be dead in half an hour. It had been a false alarm. Someone had mistakenly put military exercise tapes into the computer system.*

On October 6, 1973, in Nixon's last year in office, the Arab armies, backed by Russian equipment and training, struck at Israel, which was able to reverse the pressure. Before hostilities ceased, there was a point when President Nixon was unable to respond coherently to a nuclear crisis.[324] Watergate had preoccupied him at home, and his lack of focus had been exploited by the Soviets. Gerald Ford had yet to be confirmed by the Senate as vice president, so his role was absent. On the evening of October 24, Secretary of State Kissinger was convinced the Russians were about to send forces in to help the Arabs. He could not get through to the president to talk about this crisis. Nixon was upstairs in the Whitehouse, tired and slurring his words and barely aware of it. The Russians impressed on Kissinger that if America could not work with them, they would act unilaterally, and a response was demanded. There were intelligence reports of Soviet air and naval movements.

Kissinger attempted to get a response from the president again but was told not to wake him. Kissinger then called and warned the Russians. He summoned the intelligence and military chiefs to the White House Situation Room. Kissinger had to do everything. He put the military on DEFCON III, one up from the DEFCON IV, where DEFCON 1 is war. Bombed-up B-52s prepared to take off, and ballistic missile silos along with aircraft carriers and nuclear submarines all took up prepared positions, able to launch their missiles into the USSR. Amphibious ships based in Crete started to move; B-52s based in the Pacific were recalled. The 82nd Airborne was placed on alert and a stern reply to the letter from Brezhnev was sent that said that unilateral Soviet action in the Middle East would be against the Agreement on Prevention of Nuclear War. The letter was sent out in Nixon's name. Brezhnev responded that same day and the threat was over. Nixon was elated and went off to Camp David. The world had yet again been on the brink of a nuclear crisis

After the Soviet Union broke up in 1991, more details of Russian nuclear accidents emerged from interviews with the Russian newspaper *Pravda*. One story that came out, and it's not chronological because of the release of the story, concerned the first Russian nuclear submarine, the K-19,[325] with 139 crewmembers. It was already so unlucky it was nicknamed "Hiroshima" for its cursed bad luck, involving the deaths of many of its construction workers. The Russians were fighting to catch up with U.S. nuclear submarine technology and making shortcuts in the design of these highly complex machines. In 1961, on its initial voyage a reactor coolant leak revealed that a backup cooling system had never been installed. A shadowing U.S. destroyer offered to help the sub, stranded on the surface, but the commander of the K-19, Nikolai Vladimirovitch Zateyev, refused the American help. This echoed the later refusal by Russia to use Scandinavian sub rescue help when the much more modern and nuclear Kirsk submarine was sunk in shallow water. It was a dreadful story that ended after more than a week with the needless deaths of the survivors of the crew after a submerged explosion.

In K-19, Zateyev was forced to install, successfully, in real time, a backup coolant system, in the process exposing 22 sailors, including the entire engineering crew, to a fatal dose of radiation. Interestingly, the captain, fearing a mutiny due to the high radiation suffered by all on the ship, had thrown all the small arms overboard except for a few weapons for trusted senior officers. Eventually, the crew were recognized for their success and the dead were honored for their courage in solving the real-time nuclear crisis, although there was sticky resistance from the unsympathetic Russian naval command to awarding honors to a crew that only dealt with an "accident." A 1972 fire caused by the same submarine's nuclear reactor caused the deaths of 28 more crew members. The K-19 was finally decommissioned in 1991 and scrapped in 2002.[326]

Another amazing incident,[327] and arguably the most dangerous of all, happened on October 27, 1962, just a year later as the Cold War played out at highest intensity. An American U2 spy plane was shot down over Cuba and another was missing over the USSR. Tensions were very high. The executive officer on the K-19, Vasili Arkhipov (effectively played by Liam Neeson in *The Widowmaker*, directed by Kathryn Bigelow), was now executive officer of a diesel-powered B-59 Soviet Foxtrot-class submarine armed with nuclear torpedoes, off Cuba. Eleven U.S. Navy destroyers and the aircraft carrier USS *Randolph* were under orders to enforce a naval quarantine off Cuba and were unaware of the nuclear torpedoes.

The Americans located the sub and the American destroyers, led by the USS *Beale*, dropped practice depth charges, which had only the explosive power of a hand grenade, but which presented a real bombardment of some scale to an exhausted submarine crew. This process was entered into only after warnings were issued and was intended to persuade the sub to surface for identification. Exhausted, lacking contact with Moscow and hiding at greater depth from the American ships meant that all radio contact with the sub was lost. The three most senior officers needed to agree on any use of nuclear weapons in this particular situation, although normally it would only require the captain and the political officer for an effective quorum. Arkiphov, as executive officer, was second in command but of equal rank because he was in command of the larger submarine flotilla and very influential after his courageous participation in the K-19 events of 1961. The B-59s batteries were very low, and its air-conditioning had failed. The crew of the B-59 believed that a war had already begun, and Captain Valentin Grigorievitch Savitsky decided he wanted to launch one of the submarine's 10-kiloton nuclear torpedoes at the USS *Randolph*. He was reported by the Soviet intelligence report to have said, "Were going to blast them now! We will die, but we will sink them all. We will not disgrace our navy." Ivan Semonovich Maslennikov, the political officer, sided with the captain. Arkiphov was brought into the decision-making process to obtain his approval. In the fear-filled, sweltering control room of the submarine, hours of argument went one way and then the other.

Finally, only Arkiphov prevailed against the launch, eventually persuading Captain Savitsky to surface and await orders from Moscow. His reputation from saving the K-19 by personally exposing himself to dangerous levels of radiation that eventually contributed to his death as an admiral in 1998 saved the day. The launch would likely have resulted in the mutually assured destruction of the United States and Russia as nuclear missile arsenals in their thousands would have been launched. Arkhipov's decision evaded an almost certain nuclear war that would have ensued if that torpedo was fired. Later it was determined that the Soviet crew had never received the American warning. Thomas Blanton, director of the U.S. National Security Archive at the time, said, "A guy called Vasili Arkhipov saved the world."[328]

In 2013, a document called "Goldsboro Revisited,"[329] written in 1969 by Parker F. Jones, a supervisor of nuclear safety at Sandia National Laboratories, finally became declassified and available to the public via the Freedom of Information Act. In it, we discover that on January 24, 1961, a B-52 bomber was on a flight with two live Mark 39 multi-megaton thermonuclear hydrogen bombs on board, each of which was 260 times more powerful than the Hiroshima bomb. Another way to put it was

Figure 88: Sign in Goldsboro, North Carolina, indicating the almost-disaster that happened there when two live hydrogen bombs were dropped accidentally and almost armed themselves.

they contained more explosive yield than all the munitions ever detonated by humankind.

The bomber was flying over Goldsboro, North Carolina, when it broke apart, killing three out of the eight crewmembers. Both bombs were in safety mode, but the mechanical stresses in the aircraft breakup pulled lanyards, which in turn yanked safety pins and initiated the weapons' arming sequences. In the process, one of the bombs behaved almost exactly as you would have wanted it to in wartime: its parachute opened, and the trigger mechanism engaged. In the second bomb, a detonation signal was initiated toward the core of the bomb, but because the parachute did not open, it was not armed. Three of the four safety mechanisms failed to operate and all that saved North Carolina and countless potential casualties from fallout, and much anguish, was a final switch. One bomb fell into a field and the other into a nearby meadow. The report was unearthed by investigative journalist Eric Schlosser[330] and was described in his book *Command and Control*. This event led to an antinuclear movement in Britain, where the bomber had been headed.

Another accident is famously known as the Damascus Incident[331] and took place in September 1980 in the Little Rock Air Force Base in Van Buren County, Arkansas. During the evening of September 18, a technician working in the silo at launch complex 374-7, which housed a USAF Titan II antiballistic missile, accidentally dropped a socket from a socket wrench, which fell 80 feet down the side of the missile, eventually hitting the skin on the missile's first-stage fuel tank. It began to leak, causing the area to be immediately evacuated. By 3:00 a.m. the following morning, an explosion shot the warhead several hundred feet to near the base entry gate. All the safety features operated correctly, and no radioactivity was released. One Air Force airman was killed, and others seriously injured while the launch complex itself was destroyed. It was later decommissioned all for a cost of about $225 million and the land sold.

Also, there is the Chernobyl accident, which I describe in detail in the next section because it was instrumental in influencing the public and myriads of private nuclear concerns. No new nuclear power stations have been constructed in the United States since Three Mile Island. Chernobyl put the cap on that, and only now with climate change demanding a non-CO_2- emitting power source is nuclear coming back into its own.

It's instructive to look at the impact all this bad news had on ongoing efforts to use nuclear power. Many accidents are dangerous at a distance because they release radioactive gases such as xenon. In the Fukushima reactor situation, where Japanese planners had underestimated the danger from tsunamis, the reactors overheated, caused meltdowns and released large quantities of radioactive gases, including cesium and iodine, which can both be incorporated into the human body.

Once Bitten Twice Shy

All the global accidents and incidents had unintended consequences. It was the furor that surrounded the construction and closing down of the Shoreham nuclear plant on the north side of Long Island. The Shoreham nuclear plant was designed to be a 540-MW plant and was later upgraded to 820 MW. Initial cost expectations were that it would be completed for about $75 million. Little did Long Island Lighting Company (LILCO) president John Touhy realize that he was signing his company's death warrant. In 1973, during the plant's construction, there were cost overruns on design changes ordered by the Nuclear Regulatory Commission (NRC) as well as deep union complicity in prolonging work. The plant was seen as a perfect boondoggle and union labor was pretty open about not wanting the project to finish.[332] A Racketeer Influenced and Corrupt Organizations (RICO) action was successfully brought by Suffolk County against LILCO to try to get things moving. In one case, a witness moved to Alaska and testified only by videotape to try to protect himself.

In March 1978, after the partial core meltdown at Three Mile Island in Pennsylvania, more design changes from the NRC followed and a requirement for detailed emergency response planning was added. The release in the same month of the film *The China Syndrome*, about an emergency nuclear plant core shutdown, only heated up negative public opinion.

In June 1979, the largest public gathering in Long Island history occurred, with 15,000 attendees demonstrating against the Shoreham nuclear reactor. These protests continued for years. In 1981, the NRC declared the plant safe for operation, but the Suffolk County legislature voted 15–1 that, if necessary, the county could not be safely evacuated, and Governor Cuomo ordered state officials not to approve any evacuation plan.[333] Despite this, LILCO got permission to run 5% power tests on the new nuclear equipment. This was the only time this capital-intensive power plant would ever run.

Hurricane Gloria in 1985 (a foreshadowing of Hurricane Sandy in 2012) caused Long Islanders to go without power for weeks. LILCO could not get it right and even the chairman, William Catecosinos, contributed to the public outcry by remaining on an Italian holiday with his family rather than returning to address the problems. In April 1986, Chernobyl's nuclear power plant explosion sent a radioactive plume of gases around the globe and resulted in the permanent evacuation of over 300,000, an entire city, to a new location. The hurricane, Three Mile Island and then Chernobyl created a very unstable situation for LILCO. It was the proposed evacuation plan, addressing the logistics of a mass departure from the eastern end of Long Island, that was the last straw. All this accumulated in a crescendo of public opinion that cemented the demise of the plant.

It's amazing to think that Long Island had a power-producing asset that would have reliably generated 820 megawatts with a high capacity factor over a long period of time, but that irrational fear shut it down. There was no public confidence that the superior safety record of nuclear in the United States, despite Three Mile Island, would guarantee against a need for a mass evacuation.

The wrong reactor design, complete with fuel rods, containment structure and coolant system, were also part of this, but I have no doubt that if a molten salt and much safer and cheaper reactor had been constructed at the time that the same antinuclear forces would have closed it down too. However, if the molten salt reactor had been used as a power source from the beginning, there would not have been the same (I can't really say "any") nuclear accidents, and there would have been a qualitatively different response from the public.

In the midst of all of this concern over nuclear power in the middle of Long Island, it was overlooked that the Brookhaven National Lab (BNL) had three reactors of its own that were operational at the time. Cuomo had no legal authority to close the Shoreham plant, but he could make a deal and did so on June 6, 1989. He persuaded LILCO to sell the plant—for just one dollar—to the Long Island Power Authority (LIPA), a state-owned agency formed precisely for this purpose. Long Island ratepayers were given the task of paying off the $6 billion cost of the plant with a 3% surcharge on their electricity bills over 30 years. In 2006, the NY state comptroller authorized a further increase of 17% to energy bills, saddling Long Island residents with the fourth-highest utility bills in the entire USA. Shortly after the "deal" with LIPA was made and the debt removed from LILCO, the LILCO share price shot back up to $25 from the $5 per share of when it was discounting the Shoreham expense. Shoreham's equipment was auctioned off and it became the first commercial American nuclear plant to be decommissioned. The NRC finally terminated the license of the Shoreham plant in 1994. Part of the reason that Hurricane Sandy had such a deep effect on Long Islanders in 2012 was that LIPA is still addressing the Shoreham debt with the majority of the cash flow from their operations and was not capable of responding quickly to the damage done by the hurricane.

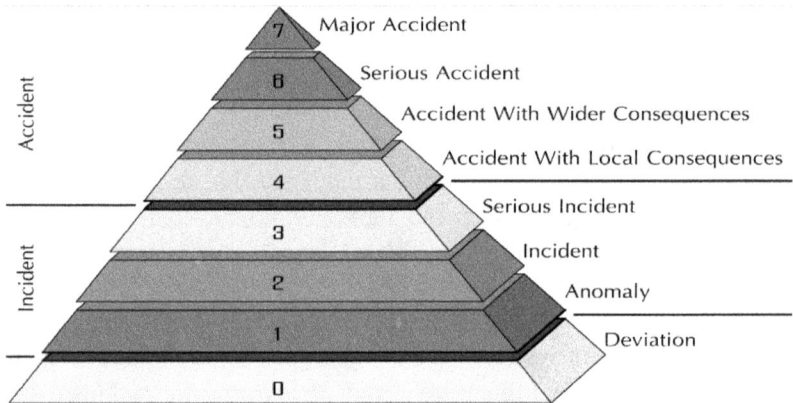

Figure 89: The International Atomic Energy Agency (IAEA) International and Radiological Event Scale (INES). Source: IAEA.

Shoreham had a twin sister, the 660-MW Millstone 1 nuclear reactor that was approved in 1966 and built in Waterford, Connecticut. This reactor was an entirely different story and in fact ended up as the top-performing boiling water reactor out of several hundred in the world for several years (1983, 1988 and 1993) but was closed anyway in 1995.[334]

In 1990, the International Atomic Energy Agency (IAEA)[335] introduced the International Nuclear and Radiological Event Scale (INES) as a public relations exercise, not as an objective scientific tool. Like the earthquake scale, it is logarithmic, where each level is 10 times the severity of the last.

Deciding on which level of severity is appropriate is less of a pure quantitative process than with earthquakes. This means that the severity on the INES is assigned long after the event. Fukushima and Chernobyl were both level 7 events, while Three Mile Island and the Windscale fire in the UK were level 5 events.

The accidents at Chernobyl and Fukushima resulted in Austria, Germany, Spain, Sweden and Switzerland enacting laws to cease construction on new nuclear plants, with many others debating whether to join them. Germany has 17 operational reactors of which almost half have been permanently decommissioned with a goal of completion by 2022. Italy voted to stay non- nuclear. Japan has a plan to restart some of its 54 reactors and continue to construct those already begun.

LEVEL	EVENT	YEAR	COUNTRY	DEAD	INJ	NOTE
Level 7	Chernobyl	1986	Russia	56	4,000	65,000 evacuees and 30 k exclusion zone
	Fukushima Daiichi	2011	Japan	3	0	earthquake and tsunami overcame cooling. Exclusion zone
Level 6	Kashtym	1957	Russia			100 tons of TNT explosion, release of 80 tons of radioactive material
Level 5	Windscale Fire	1957	UK	240	0	Air cooling didn't work releasing radioactive dust
	Three Mile Island	1979	USA	0	0	loss of coolant and partial meltdown releasing radioactive gases
	First Chalk River	1952	Canada	0	0	Reactor core damaged
	Lucens	1969	Switzerland	0	0	Partial core meltdown
	Goiania	1987	Brazil	4	28	Theft of medical ceasium source
Level 4	Sellafield	1955	UK	0	0	5 incidents from 1955 - 1979
	Idaho Falls SL-1	1961	USA	3	0	Reactor reached fast criticality killing three operators
	RA-2	1983	Argentina	1	2	Buenos Aires Experimental reactor criticality
	Jaslovske Bohunice	1977	Czechoslova	2	0	Contamination of reactor building
	Tokaimura	1999	Japan	2	207	Criticality accident at reprocessing facility
	Saint Laurent des Ei	1980	France	0	0	Frances worst accident. Cooling system failure
Level 3	Thorp Plant	2005	Sellafield	0	0	
	Paks	2003	Hungary	0	0	
	Vandellos	1989	Spain	0	0	Fire destroyed control systems
	Davis-Besse	2002	USA	0	0	High pressure coolant release
Level 2	Blayais	1999	France	0	0	Flood
	Asco	2008	Spain	0	0	Backup generator failed
	Gundremmingen	1977	Germany	0	0	Weather caused short circuit of high tension power lines
	Shika	1999	Japan	0	0	Dropped fuel rods. Incident covered up until 2007
Level 1	Penly	2012	France	0	0	Primary circuit leak after reactor fire
	Gravelines	2009	France	0	0	Incident during fuel bundle exchange on reactor 1
	TNPC (Drome)	2008	France	0	0	Leak of 4,800 US gallons of water containing 165lbs of uranium
	Fukui Prefecture	2009	Japan	4	7	Steam explosion

Figure 90: Nuclear accidents and incidents. Below and including level 5 mentioned incidents are just examples from a much longer list. Source: IAEA.

By 2013, Australia, Austria, Denmark, Greece, Ireland, Israel, Italy, Latvia, Liechtenstein, Luxembourg, Malaysia, Malta, New Zealand, Norway and Portugal remain firmly opposed to nuclear power. At least a hundred old and smaller nuclear reactors will be closed by 2026. Closure of nuclear power stations forces a closer look at renewable energy. China, Brazil, Finland, France, India, Japan, Russia, South

Korea, the UK and the United States all have new reactors being built, 27 in China alone. In the United States, nuclear is unable to compete with cheap natural gas and has difficulty finding investors in nuclear reactors.

Up until the 1970s, nuclear power stations were about the same cost as coal-fired power stations, but the cost increased fourfold by safety measures adopted after disasters. Post-Fukushima, we are experiencing the same freeze in nuclear development that occurred after the Chernobyl event.

Adoption of the Wrong Kind of Nuclear Power

> *"Seems that somewhere in the 60's & 70's, lots of scientists and politicians stuck their heads in the sand and blindly forged ahead with the WRONG nuclear energy technology......leaving the world "driving around" in the product equivalent of the AMC Gremlin or Reliant Robin while the Ford Mustang / Ferrari Daytona was left parked on the drawing board!"* —Unattributed quote, http://forums.bistudio.com

After the success of Edward Teller's hydrogen bomb and other atomic bombs, the metal required, plutonium, was only one type available of three nuclear pathways, and thorium was not one of them. The United States chose to maintain a civil nuclear power technology that was also capable of supporting the military requirements for defense. Research funding for thorium ceased in 1972 when the experimental molten salt reactor at the Oak Ridge National Laboratory was finally closed down after operating successfully for six years. They got another year of performance out of it, however. The current reactor technologies were all designed before computers hit the scene, so there were no 3D simulations of reactor designs, and before we were able to do a lot more with our current knowledge of particle physics. At the end of 2011, 32% of the world's nuclear reactors were over 30 years old. Very few significant innovations in conventional reactor design have occurred since the very first years of active design in the late 1950s.

As climate change has become an increasingly important issue and more has been learned about the physics of nuclear power, more, younger workers have moved into the industry, and there are now several new ventures proposing safer, more efficient and more economic nuclear power, some of them without any proliferation risk of nuclear weapons and some without nearly as much or any radiation risk.

The Pollution Solution Was Dilution

Nuclear waste has for decades been dumped in the ocean. A hundred thousand tons has disappeared into the oceans near Europe since the 1960s. Greenpeace highlighted this in the 1970s, and there was an outcry. Environmentalists were initially criminalized and attacked. The military dumped a lot of nuclear waste as well. Waste was loaded onto ships to be transported to a "safe" location where it would be dumped overboard and drop to the sea bed. This meant that the sailors on the ships were being exposed to radioactivity, so captains had to reach the dumpsites within a certain time to minimize exposure. If the clock ran, out the radioactive waste needed to be dumped then and there, wherever the ship was located. There is no certainty that waste was dumped in the planned locations.

Much of the atomic waste from Sellafield went into the Atlantic, the English Channel and the Irish Sea. The practice was then officially banned. Minister of State Michael Meecher, MP, was against the dumping. Protests were later successful. Greenpeace earned its spurs. Countries everywhere banned the practice.

They had assumed that if you dilute radioactive waste, it's safe to dispose of. This approach was quickly proven wrong. The first evidence that it doesn't dilute was discovered after the Germans examined the Atlantic dumpsites. They found nine barrels of plutonium had leaked and increased concentration on the seabed, in fish and in the water. In the English Channel, many nuclear waste barrels corroded and are only in 100 meters of water. Many were dumped very close to Alderny, an island in the Channel Islands group with only 2,400 inhabitants. There was a case of childhood leukemia—one case is already too many—but information on the island about cancer was kept from the public. There turned out to be several cases, which was statistically important.

Authorities say the radiation is far too low to cause harm, but the radio-activity finds its way back to the land.

Dumping in the ocean was banned in 1993. Then pipes were built where the radioactive effluent was distributed into the ocean. In the Hague, Greenpeace activists demonstrated underwater. The Hague processes nuclear waste from all over Europe. Predictably, Areva's numbers for the detection of radioactivity are much lower than the activists' numbers. Plutonium and cesium are found in ocean mammals like gray seals and porpoises.

One pipe leading directly into the ocean from the UK is in the north of England. In 1997, Greenpeace highlighted it. Nuclear waste is still dumped into the sea and called "land-based disposal." Nuclear waste from Sellafield can now be found on the Norwegian coast. The dangers are greater than the people in charge care to admit. The usual cases of cancer were shown to be statistically significant. The different parties find the evidence that they want to find to back up their positions. Radiation risks exist in everyday things like aircraft and masonry as well as background radiation, and there are more important risks by far than these.

Mutually Assured Destruction (MAD)

Having exhausted itself militarily and financially to win the Second World War, which culminated with the dropping of the rapidly designed and constructed atom bombs on Hiroshima and Nagasaki, the United States was in no mood to play off different nuclear technologies and chose the pressurized water reactor as its base nuclear technology. Admiral Hyman G. Rickover adapted it to the U.S. Navy, then ally Great Britain started to demonstrate nuclear detonations and a hydrogen bomb of its own in Australia using scientists who had participated in the Manhattan Project. Then the Russians got hold of the nuclear bomb, and the clash of ideologies appeared to come right to the precipice. Would communism use the bomb against democracy and capitalism? The situation became really tense and as you will read in chapter 5, disaster sometimes came so close that we are all forever lucky just to draw our next breath.

In the end, what saved us was the truly awful picture of destruction visited on almost every city in both the West and the East that demonstrably nobody wanted. Every decision about security was taken in the light of the assurance that if a single armed nuclear missile was to come either way, the response would be total. This precarious stalemate came to be called mutually assured destruction (MAD).

Instead of blowing each other up, there were arms limitation talks and a "getting to know you" phase that softened East-West relations. By the mid- 1980s, the Russian centrally planned economy could no longer compete with the West and we got Perestroika, the restructuring of the Communist Party and Mikhail Gorbachev's *glasnost*, or open-ness. The democratization of the Soviet Communist Party resulted in the Soviet satellite countries suddenly coming unbound. Old anxieties resurfaced and played themselves out. These included the Serbian strug-gle against Islamic minorities in Yugoslavia when President Tito died and the Ceausescu presidency in Romania. History has shown there are those for whom certain destruction is no deterrent, however.

Iran's Nuclear Ambitions

"Petroleum is a noble material, much too valuable to burn. We envision producing, as soon as possi-ble, 23,000 megawatts of electricity using nuclear plants." —Shah of Iran, March 1974.

On March 5, 1957, an agreement between the United States and Iran was announced. The United States was promulgating Eisenhower's "Atoms for Peace" program, and together with European governments it made Iran a beneficiary country. The Shah of Iran announced plans for 23 nuclear plants by the year 2000. Iran signed treaties in 1968 saying it'd never use nuclear, biological or chemical weapons, making any nuclear activity subject to IAEA verification. This was perfectly fine until the 1979 Iranian Revolution, but even after the revolution, a categorical reli-gious decree in the form of a fatwa was issued by the Supreme Leader of Iran, the Ayatollah, Ruholla Khomeini, against the use, stockpiling or development of nuclear weapons. He considered nuclear weapons as a suspicious Western innovation. Many of Iran's best scientists fled the country in any case. Later versions of this fatwa[336] only forbid the "use" of nuclear weapons but said nothing about their development or production. After Ruholla Khomeini's death in 1989, the new Ayatollah,

Ali Khamenei, and then Iranian president Ali Akbar Hashemi Rafsanjani sought to return to nuclear development, but with the Russians as suppliers of choice.

In the late 1980s, the Iranians were complicit in obtaining nuclear fuel cycle information from nuclear power Pakistan via a top Pakistani scientist, Abdul Qadeer Khan, who is also suspected of having shared information with the North Koreans, but in his own defense always pointed to his senior politicians as requesting his help in doing so.

Iran has always had a nuclear program located at several research sites. They have two uranium mines, a U.S.-supplied 5-megawatt research reactor fueled by highly enriched uranium and three uranium enrichment plants where fertile U-238 is separated from the sought-after fissile U-235. The Russian nuclear company Rosatom helped the Iranians build two reactors, the Bushehr 1 reactor, commissioned on September 12, 2011, and a 360- megawatt reactor being built at Darkhovin with the help of the Russian nuclear company Atomenergoprom. Although U.S. intelligence said in 2007 that Iranian efforts to make a nuclear weapon had ceased in 2003, by 2012 they and others were worried that new research could help the Iranians make bombs, but they were not actively doing so yet. Iran has established a goal of pursuing development of its peaceful nuclear power program and developing more uranium mines. In November 2011, the International Atomic Energy Agency, which had never found cause in earlier years, criticized Iran for making plans for detonators for nuclear weapons prior to 2003. Iran's apparent noncompliance led to international sanctions that are only now being relieved by an international agreement led by the United States to forestall Iranian military nuclear ambitions.

The risk of terrorism damaging energy supplies is ever present. Running risk scenarios is the constant activity of security forces. Current circumstances remain dynamic with the Arab Spring in 2014, which has aged and soured with the horror of the Syrian civil war, the rise of ISIS, the cycle of elections in Egypt and the persistence of the Palestinian-Israeli issues. Iran has been a challenge over the years, on the one hand claiming a requirement for peaceful nuclear power, but on the other being mistrusted by the rest of the world for wanting its own nuclear weapon. I have always thought that a thorium molten salt nuclear reactor, which cannot make bombs, would have been a perfect play for Iran. If they rejected it, it would indicate with transparency their true inclinations. They have a new and moderate president and have negotiated a deal with the United States on nuclear power even though the Iranian mullah's anti-Westernism appears to be entrenched.

The Threat of the Suitcase Nuclear Bomb

William C. Potter, director of the Center for Non-Proliferation Studies at the Monterey Institute of International Studies, said in a report that he'd spoken to a senior advisor to President Boris Yeltsin who in turn stated that in the late 1970s and early 1980s the Soviet KGB acquired an unspecified number of small nuclear weapons weighing less than 75 pounds that were never included in any inventory.

In a May 1997 private briefing to a delegation of U.S. members of Congress, former Russian security advisor Aleksandr Lebed said that he believed eighty-four 1-kiloton suitcase bombs, measuring 24 by 16 by 8 inches, were missing from the Russian nuclear arsenal. On September 7, 1997, the U.S. TV show *Sixty Minutes* offered an interview with Lebed, and he claimed that more than 100 suitcase-sized 1-kiloton bombs—enough energy to kill 100,000 people or take out Capitol Hill or the Kremlin or the Houses of Parliament—were missing. You can't destroy Washington, Moscow or London with them, but they can do a lot of damage. Lebed claimed that a total of 250 of them were manufactured. In later interviews, he used different numbers but with the same message. U.S. authorities quickly checked with the Russians, who claimed that there was no real issue and all nuclear weapons were accounted for.

It also turned out that the producers of the *Sixty Minutes* story were husband and wife team Andrew and Leslie Cockburn, who had just written a book on nuclear terrorism called *One Point Safe* and were also producers of an exciting George Clooney, Nicole Kidman movie called *The Peacemaker*. The commentator on *Sixty Minutes* was also a former National Security Council staffer, Jessica Stern, who was also a paid consultant on *The Peacemaker* and the model for the role of Nicole Kidman. Although this did not prove that the suitcase bombs did not exist, it really helped the story fly.

On October 3, 1997, however, Lebed's claims were bolstered by Soviet scientist Alexei Yablokov,[337] an environmental advisor to President Yeltsin who served on the Russian Security Council. He explained that being manufactured for the KGB meant that they were not accounted for as if they were in the Russian Defense Ministry control. Yablokov explained that he was earnestly trying to publicize the issue to prevent a disaster. He testified to the Research & Development Committee of the House National Security Committee, chaired by Representative Curt Weldon.[338] Chechnyan and Palestinian leaders announced that they had possession of the weapons, but nothing was verified.

The thing about nuclear is that a little material gets turned into a lot of energy, and in a bomb, even a small one, that can be plenty enough to destabilize a country. During the Cold War, the United States and other nations developed "tactical" nuclear weapons that were smaller strategic nuclear bombs that could fit on a 115-mm artillery shell as opposed to the larger ones that fit in the belly of a large bomber aircraft. Artillery tactical nuclear weapons were also installed on nuclear ballistic missiles. Such was the power available that miniaturizing the components still meant that a smaller weapon could pack a significant punch. Only the United States and Russia are known to have produced nuclear weapons small enough to fit in a backpack during the Cold War. Israel is suspected of having produced a weapon small enough for a suitcase. The Nagasaki and Hiroshima bombs were equivalent to 16 and 21 kilotons of TNT, respectively. The Mk-54 SADM backpack nuclear weapon was an 11 by 16-inch cylinder that weighed 51 pounds.

Members of the U.S. Special Forces have practiced infiltration exercises with these bombs with a mission to detonate the weapons. Each weapon had 6 kilotons of power. For ballistic missiles, including those fired from submarines, there is intense control. Codes and commands are required from multiple persons before such large weapons can be detonated. With the suitcase bomb, no such controls exist. However, from one point of view, we can all relax a bit. Every few years the active components of nuclear weapons have to be replaced. Yablokov explained that he believes this to be a very low likelihood and that, therefore, there is a high possibility that any bombs from this source will no longer operate.

There was a ready requirement for a small tactical nuclear weapon that would be able to handicap or take out a harbor, factory, airport or key base. The smallest U.S. warhead to be fielded was the W-54Y2 Mod.1, with a yield of 0.01 kilotons, or just 10 tons of TNT.

In the USSR, a special nuclear bullet based on the element Californium was developed. Bullets were made in 14.3 mm, 12.7 mm and 7.62 mm. This last is the same as bullets for the AK-47 Kalashnikov assault rifle, but it was not made for this weapon but rather the machine gun. In nuclear weapons, the explosion requires a critical mass of fissionable material. For U-235 or Pu-239, that mass is about 2.2 pounds. This is too large for normal bullets but okay for artillery rounds.

When the transuranic compound Californium-252 was discovered, with a critical mass of only 1.8 grams, the bullet possibility came to life. Californium also decays with 5 to 8 neutrons at once versus only 2 to 3 neutrons for U-235 and Pu-239. It was very effective. Californium was produced as a by-product isotope in a nuclear reactor. It was shaped into a 5- to 6-gram dumbbell-shaped piece of metal that was crushed into a ball shape by an explosive charge. Unfortunately, Californium continuously decays, releasing heat that poses the risk of an unintended discharge. Storage of such bullets required a power-consuming (200-W) cryogenic facility to keep them at safe temperatures. A special, cumbersome (242 pounds) cryogenic device was required to keep the bullets cold enough to use safely.

Quite a lot of planning was required to get the weapon into the field and properly operated. The prospect of bullets unfreezing over a 30-minute deployment and causing home casualties and the radiation risk very likely meant these weapons could only be fired at maximum range. Only three such bullets could be fired consecutively. One bullet could destroy a tank by melting its armor. Hitting a brick wall evaporates about a cubic meter of wall material, and three bullets could demolish a building. It was also noticed that if the bullet hit a tank of water, this would prevent any explosion since the water would absorb the neutrons.

India and Pakistan's Nuclear Détente

> "As long as the world is constituted as it is, every country will have to devise and use the latest devices for its protection. I have no doubt India will develop her scientific researches and I hope Indian scientists will use the atomic force for constructive purposes. But if India is threatened, she will inevitably try to defend herself by all means at her disposal." —Jawaharlal Nehru, June 26, 1946, soon to be India's first postcolonial prime minister and just prior to Pakistan and India separating the following year.

"If India builds the bomb, we will eat grass and leaves for a thousand years, even go hungry, but we will get one of our own. The Christians have the bomb, the Jews have the bomb and now the Hindus have the bomb. Why not the Muslims too have the bomb?" —Zulfikar Ali Bhutto, in 1965.

This story is not about fossil fuels but is still about CONG and concerns the geopolitics of the nuclear externality of proliferation and the conditions that led Pakistan to become a nuclear power.

The year 1947 saw the termination of British occupation and administration of India, which had endured since 1757, for 190 years. Up until 1858, India was governed by the British East India Company after which it was taken over by the British Crown in the person of Queen Victoria. It was called the British Raj. At Partition, India was composed of regions controlled by the British and the self-governed, princely states, of which there were 565. The end of British rule was also the moment that two religious communities on the Indian subcontinent, the Muslims and Hindus, separated, creating India and East and West Pakistan.

Pakistan was where the Muslims migrated to, and Hindus generally chose to move east to India. In the process, over 12.5 million people were displaced and over a million people died from difficulties experienced during the migration, from hunger and thirst to being attacked by local inhospitable communities. Additionally, territorial disputes sprang up from Muslim leaders acceding territory to Pakistan where there was a significant Hindu population. Despite their common languages, cultures and history, Pakistan and India have endured a long distrust stemming from events during Partition. India emerged as a secular nation with a Hindu majority, while Pakistan became an Islamic Republic, whose constitution nonetheless guaranteed freedom of religion.

I had the privilege of living in Pakistan as a boy, first in Quetta, a dry, elevated town in Baluchistan in 1964, and then again from 1969 to 1973 in Rawalpindi and the nearby capital, Islamabad. I remember the cover of the *Rawalpindi Times* showing a black-and-white photograph of Neil Armstrong setting foot on the moon in 1969, and in 1971, I remember the British- manufactured Indian Air Force Hawker Hunter aircraft flying low overhead. Blackout conditions were in order at the time, a rule forbidding any light from the city of Rawalpindi from betraying the location of the city below to Indian pilots relying on visual cues to find infrastructure to destroy. This was something I was previously aware of only as a precaution taken during the Blitz in London during the Second World War.

My father, a British Army career officer, was working in the British High Commission as the assistant military attaché. The fighting was part of the 13-day-long Indo-Pakistani War of 1971, which took place during the time of the Bangladesh Liberation War.

The war took place on both the western and eastern borders of India, where West Pakistan and East Pakistan were situated. This was a particularly horrible war. Instead of oil and energy as an underlying cause, it was instead the wealth created by the jute industry that supported much of the economy of both Pakistans. Despite this, the Bengali wealth creators were largely sidelined, even racially discriminated against, and this naturally led to a developing Bengali nationalist fervor.

On December 8, 1953, the U.S. Atoms for Peace initiative arrived in Pakistan, and the Pakistan Atomic Energy Commission was inaugurated. In 1955, the United States and Pakistan agreed to the peaceful and industrial use of nuclear power. Pakistan maintained its Atoms for Peace trajectory despite several voices for weapons development during the 1960s. They discovered a domestic resource of uranium and started mining it. In 1965, the foreign minister Zulfikar Ali Bhutto met in Vienna with nuclear engineer Munir Ahmad Khan to persuade him to bring his Western education in nuclear power back to the homeland. Khan went back to Pakistan, and within years, by 1965, a 5-MW reactor installed by the United States, like the one in Tehran, was commissioned and upgraded to 10 megawatts in Islamabad and a second small 30-kW training reactor was commissioned in 1989, funded by the IAEA. Canada built the first of Pakistan's 137-MW civilian power- producing reactors in Karachi, completed in 1972. It was inaugurated by President Bhutto. Plans for a further 400-MW reactor have been frozen since 2009.

In 1966, a coalition of Bengali nationalist groups in East Pakistan, led by Sheikh Mujibur Rahman, wanted to realize six demands of West Pakistan to end its perceived exploitation of East Pakistan. In the Pakistani election of 1970, the East Pakistani Awami League won 167 of 169 East Pakistani seats out of 313 seats in the Pakistan Parliament, the Majlis-e-Shoora, and won a majority in the lower house. Sheikh Mujibur Rahman presented the six demands to Western Pakistan and claimed leadership of the government of Pakistan. In Western Pakistan, the Pakistan People's Party was led by Zulfikar Ali Bhutto, who had won only 80 seats in the parliamentary election. Bhutto refused to allow Sheikh Mujibur Rahman the Pakistani leadership and pressured President Yahya Khan to do something.

The Pakistani president, Yahya Khan, was a general who had served with distinction as a British Indian army officer in the Second World War. He also had a close relationship with President Nixon, and his contacts with the Chinese led to relationships that resulted in the eventual opening up of China to Nixon. Now, however, President Yahya Khan called in the military, mainly composed of West Pakistanis, to suppress dissent in East Pakistan. This led to a rapid polarization. After all-party talks in Dhaka in March 1971 unsurprisingly failed to arrive at an understanding of West Pakistan's position, a military crackdown was ordered. Sheikh Mujibur Rahman was arrested and flown to West Pakistan and the Awami League was banished. This ushered in the Bangladesh Liberation War, a horrible genocide of the Bengali people by West Pakistan, in which many sources are conflicted about the numbers of dead, at between 300,000 and 3 million. West Pakistani soldiers fought as if in a war but against unarmed civilians. Ten million refugees escaped into India and rape camps were set up by the Pakistani military. The Mukti Bahini, a force of regular troops and guerilla soldiers, fell in with Bengali forces, along with many soldiers and police.

My father was in the British embassy in Dhaka during part of the event and remembers counting bodies of students at Dhaka University, the result, they said, of an effort to rid the Bengalis of their intelligentsia in what was called Operation Searchlight. Few remember that at the time there was also a cyclone storm raging in the Ganges delta causing surge flooding, isolation, lack of energy, food shortages and many refugees. My father remembers that it was the Mukti Bahini who were largely responsible for using tactics that brought down the level of discourse for the entire conflict.

There is a very interesting video[339] of Prime Minister Indira Gandhi speaking to a journalist in which she is supremely confident of the situation just prior to the Indo-Pakistani War, a state of mind that turned out to be justified. Many Bengalis were Hindu and naturally fled to Indian refugee camps in the neighboring Indian states, placing a strain on the Indian economy. The lessons of the Israeli Six-Day War were in the heads of the Pakistani planners, who felt the threat of Indian intervention.

The Pakistani air force launched preemptive attacks the evening of Sunday, December 3, 1971. Eleven Indian air force bases were the initial targets, but unlike the Israelis who used hundreds of planes, the Pakistanis only used 50 aircraft. Immediately, India declared war and fought back with airstrikes. India's navy sailed right into the port of Karachi and destroyed ships and commercial vessels, significantly disrupting Pakistan's ability to respond by sea. East Pakistan was blockaded, and a Pakistani submarine sank an Indian navy frigate in the Arabian Sea. A Pakistani submarine was also sunk. India gradually assumed control of East Pakistan. West Pakistan received support from Saudi Arabia, Jordan and Libya, among others. The war ended the afternoon of December 17, 1971, after the East Pakistan capital, Dakha (Dacca), fell into Indian hands. The independence of Bangladesh was confirmed.

India had flown 1,978 missions in the east and about 4,000 in the west, losing 65 aircraft against the Pakistan air force's 2,840 missions, in which 72 aircraft were lost. It was a major defeat for Pakistan. The jingoistic Pakistani public were dealt a blow by India that left them seething with hostility that remains to this day.

On December 16, 1973, 93,000 Pakistani armed forces surrendered to the Indian army. The incident encouraged Soviet support for India, while the United States and China supported the Pakistani side. India had penetrated some way into Western Pakistan and stood to make significant territorial gains. Pakistan had lost half its population with the loss of its eastern portion and a considerable economic edge, and now they also held a lower-profile role in Southeast Asian affairs. The Pakistani POWs, who India treated according to the Geneva Convention, were released by India only after the Simla Peace agreement was signed, which meant that Pakistan had to agree to recognize independent Bangladesh and accept a protocol for other conflicts in the remaining western province of Kashmir. India returned most of the territory it had occupied. A demoralized General Yahya Khan surrendered power on December 20, 1971, to Bhutto, who became the first Pakistani civilian president since Partition.

As a result of the war, Bhutto called a meeting on January 20 in Multan, central Pakistan. He authorized the nuclear weapons program in order to never allow another invasion from India. Many Pakistani scientists working in the United States or Europe were recalled for this top-secret project. Bhutto wanted a bomb ready by 1976. The Pakistan Atomic Energy Commission (PAEC) had 20 laboratories, and its leader, nuclear engineer Munir Ahmad Khan, who was recalled from his job as director of the nuclear power and reactor division of the International Atomic Energy Commission in Vienna, Austria, quickly fell behind schedule in generating enough fissile nuclear material. Bhutto added Abdul Qadeer Khan, an expert in centrifugal enrichment of U-235, to the team after summoning him from his European job in 1974 and placed him in command of the Kahuta project.

India's first atomic test in 1974 lent even more urgency to the project. By 1978, they had a moderate quantity of fissile material. Finally, after years of delays in obtaining the fissile material and trying different methods, on May 28, 1998, Pakistan detonated five atomic devices called Chagai I in Baluchistan in the west of Pakistan, becoming the seventh country in the world to do so. Abdul Qadeer Khan claimed that they had generated enough fissile material to explode a bomb by 1984.[340] Today Pakistan is said to have between 100 and 120 nuclear missiles and is currently bidding for three nuclear power stations from China to meet a goal of providing 8 gigawatts by 2030.

India's involvement with nuclear energy started almost as soon as it could have, in March 1944, with the formation of the Institute of Fundamental Research, led by Dr. Homi Bhabha. A short war with China in 1962 refocused efforts to develop a nuclear weapon. "Smiling Buddha" was detonated in 1974 using plutonium generated in the Canadian-supplied CIRUS reactor. "Operation Shakti" followed in 1998, and sanctions were imposed on India by Japan and the United States, which have since been lifted. The Cold War encouraged India to use a strategy based on "credible minimal deterrence" called the policy of "no first use." During the escalation of tension with Pakistan, the Indian "no first strike" policy has always remained the same. In 2010, India's policy shifted to a "no first use against non-nuclear states," and in 2013 the National Security Advisory Board said that whatever the scale of a nuclear attack against India, there would be a massive, unacceptably large retaliation. India interestingly did not sign the Treaty on Non-Proliferation of Nuclear Weapons, because it said it creates an unfair distinction between the five states permitted by the treaty to possess nuclear weapons while requiring all other states not to.[341]

In 2008, India signed an agreement with the International Atomic Energy Agency that allowed it to return to a trading role with other countries for nuclear cooperation and exportation of uranium. In June 2014, after a five- year delay, India finally signed an IAEA protocol agreement, and in April, Canada signed a uranium supply agreement with India. Today India has about 90 to 100 nuclear missiles along with the capacity to deliver them by air and by submarine.[342]

The Democratic People's Republic of North Korea

North Korea is a unique situation but illustrates the nuclear geopolitical externality beautifully. It withdrew from the Treaty on the Non-Proliferation of Nuclear Weapons (NPT) in 2003 and is the only nuclear country that is not a party to the Comprehensive Nuclear Test Ban Treaty (CTBT). It detonated bombs in 2006, 2009, 2013 and, supposedly, hydrogen bombs in 2017, long after the rest of the world agreed to stop. It has also been able to enrich uranium and generate stockpiles of plutonium. It launched a long- range missile successfully in 2012 and 2017, showing it could deliver the package if necessary.

North Korea has claimed it has 38.5 kilograms of weapons-grade plutonium that it extracted from spent nuclear fuel rods in 2008. South Korea is not a nuclear weapons power and talks to denuclearize the Korean Peninsula were held in 2003 between six nations eager to see a solution: North Korea, the United States, China, Russia, Japan and South Korea—the six-party talks. These talks were suspended in April 2009, but in February 2012, Pyongyang agreed to suspend nuclear tests, enrichment and long-range missile tests in exchange for food aid from the United States.

Right away, North Korea entered into a dispute about a rocket with the United States and held a nuclear test detonation at its Punggye-ri nuclear test site in February 2013. It also started up all its civilian reactors, including one in Yongbyon, where a 5-MW reactor can produce 6 kilograms of plutonium annually. This was confirmed by satellite imagery in August 2013. North Korea remains an unpredictable element within a global nuclear weapons security context. The country is a bleak example of a centralized government attempting to keep control of everything and succeeding to the great and tragic loss of the population. Hopes for an eventual thawing such as that experienced by Germany after the Berlin Wall fell are not high. It remains astonishing how so few can do so much harm to their people, and because they are behind a nuclear weapons curtain, MAD prevails for now.

With the Trump administration being very unpredictable, however, there have been signs of increasing interest by Korea in having its people discover what the United States is really thinking in order to make the best decisions. It almost feels like formerly hostile countries are trying to protect the United States in its moment of weakness, although the threat is overt and will not go away without regime change or some other force of change.

Conclusion

The thing that's most amazing about the nuclear industry is in the history of its development. The incredible steps taken by humankind to get to the next level of understanding. However, the nuclear industry began with promise and is foundering on mediocrity. Its promise was the almost endless source of cheap, reliable energy. The technologies chosen were compromises between convenience and cost. The thorium option was the most promising option. Almost all if not all the problems experienced in nuclear power are absent in thorium molten salt reactors.

Most efforts to harness thorium fuel were drop-in replacements for the uranium fuel rods, along with all the original design disadvantages. This was also the wrong way to use thorium and did not take advantage of its beneficial characteristics. From the facts on the table—unless there is something I do not know that disqualified thorium—it was the most promising source of long-term clean energy and remains that way today. You can say that if our use of fossil fuels has condemned us to a future where the climate will effectively deteriorate, then it all comes down to a fateful decision made in the 1950s to use the wrong nuclear option. The nuclear we use today is only a fraction of what we may have been using if we'd gone a different direction. The current version of uranium nuclear power possesses every one of the external costs of fossil fuels but with a very different emphasis on waste, subsidies, health and geopolitics.

The push for fusion power is a worthwhile goal. A fraction of the government money spent on fusion could have revivified thorium molten salt reactor technology to the benefit of all. Nuclear can operate at high capacity, without a need to clean up the plant. Waste is an incredibly big externality, and while there have been minimal fatalities with nuclear power, the actual failure of the promise to be the mainstay power source, the failure to get rid of a need for coal, an original hope, is all due to the initial military focus on weapons. The fusion goals will likely be achieved before 2040 but plans for a buildout of the official ITER technologies are so long term as to be generational in scope. Any hope for a fusion solution soon must come from a smaller breakthrough, and there are many promising contenders.

It's not too late to simply replace coal-fired power stations with thorium plants, which promise to be much smaller and more compact per kilowatt- hour of production. All that would be needed is a project to get two small ones back up and running on the same prototype plans that already exist. Two plants, because sabotage might discredit the project. It would not take a decade and billions, and we still have the expertise, as it were, "in house."

Chapter 6: Health

Health

The first thing that came to my mind when I considered negative health effects was considerable skepticism. I doubted that there really were any such effects because I am healthy and most of my friends are, too, and we live at the height of the CONG usage empire. It just didn't seem likely that there was any real way that you could consider this subject and find any suggestions that CONG is a health nightmare, but I was very wrong. It is arguably the worst of the externalities, and this chapter represents a highlight of my research.

One of the best indexes of the impact of an energy source on people's health is how many people get killed by it. As you can imagine, the death rates are very different for the different sources of energy. Many groups, including the World Health Organization (WHO), Centers for Disease Control and Prevention and National Academy of Science, have studied this issue. The number of deaths from an energy source is sometimes referred to as an energy "deathprint," to borrow from the footprint metaphor.

ENERGY SOURCE	DEATHS/TWH	GLOBAL ELECTRICITY PRODUCTION
Coal - China	280,000	75% of China's electricity
Coal - Global Average	170,000	50%
Oil	36,000	36%, 8% of electricity
Biofuel/Biomass	24,000	21%
Coal - US	15,000	40% US Electricity
Natural Gas	4,000	20%
Hydro - Global Average	1,400	15%
Solar - Rooftop	440	<1%
Wind	150	~1%
Nuclear - Global Average	90	17%

Figure 91: Deaths per terawatt-hour for different energy sources. These numbers are a combination of actual direct deaths as well as the estimated epidemiological deaths taken over time per terawatt-hour of produced energy. Source: Brian Wang, www. NextBigFuture.com.

There is no question in all of this research that the result of burning things, even carbon-neutral things, as much of the biomass combustion that happens in the developing world is, has significant impacts on human health.

Figure 91 represents deaths as a combination of actual deaths along with epidemiological estimates, and the numbers are rounded. Coal, oil and biomass release carbon particulates into the atmosphere when they are burned and collectively release carbonaceous particulates. Our lungs are a delicate organ that keep us oxygenated and also release CO_2, one of the products of our metabolisms, for a lifetime. They are soft, wet gas exchange mediums that are very sensitive to chemicals and particulates.

In 2005, the *Lancet* journal published an article showing that annual coal pollution deaths in China exceeded 400,000.[343] Actual deaths from coal in China in 2012 were 670,000[344] and in 2014 were at least 350,000 because they have accelerated their coal-burning programs to comply with the 13th Chinese "Five-Year Plan" to develop their economy. China burns half the coal in the world, and almost all its pollution comes from coal. An article in the *Telegraph*[345] quoted Chen Zhu, the Chinese minister of health and a professor of medicine, who said that studies by the World Bank, the World Health Organization and the Chinese Academy for Environmental Planning pin the number of pollution deaths at 350,000–500,000 per year. The quote is sourced from a December 2014 issue of the *Lancet*. Most Chinese coal plants do not have scrubbers or systems to clean the worst of the emissions from the burned coal. It's not just deaths, though, it's also the nonlethal health effects and lost days of work. The death rates from coal are much lower in the United States than in China due to a combination of safety regulations and the Clean Air Act.

In the United States, the Clean Air Act has saved more lives than any other piece of legislation in history. Coal use still kills about 10,000 people in the United States annually, while 1,000 die from the impact of natural gas. Workers regularly fall off windmills and solar rooftops, but because the impact of those sources is still small, the total deaths per terawatt-hour are also small. The nuclear deathprint is small because so much electricity is produced per death, even if you include the impact of Chernobyl and Fukushima projections and uranium mining deaths. It is the slow realization of this fact about nuclear power along with its relatively low emissions profile that is changing the antipathy of groups such as Greenpeace and the Sierra Club toward nuclear.

World Consumption of Coal, Oil and Natural Gas in 2017 Million Tons of Oil Equivalent (MTOE)		
Coal	3,731.5	27.62%
Oil	4,621.9	34.21%
Natural Gas	596.4	4.41%
Total CONG	8,949.8	66.24%
Total World Energy Consumption	13,511.2	100.00%

Figure 92: Quantity of fossil fuel energy that still makes up our annual consumption. In this case, nuclear power is part of the difference and is almost carbon free, so it works. Source: BP Statistical Review 2018.

In the United States, about 12 deaths came from the nuclear weapons complex during this time period. New nuclear building regulations require passive redundant safety systems and the ability to resist a worst-case scenario no matter how unlikely. The impact of an energy system comes from the various parts of its life cycle. There is the prospecting and exploration stage for fossil fuels, followed by the extraction phase, or mining. Coal mining is famous for imposing a significant and horrifying annual death rate. Then there is the production phase. In coal, again, it's all about the products of the burned coal, the gases and impacts on the environment. It ought to be said here that biomass, a carbon-neutral source of fuel, also results in particulate matter when it is combusted.

Coal was implicated from the start! King Edward I of England banned the use of soft coal in London in 1306 after noble visitors complained of the air quality. Barbara Freese in her book *Coal: A Human History*[346] made it very clear that coal has been a love-hate relationship for humanity. It's interesting that coal is associated with the poor and oil with the rich, but there was one time, in 1952 when the Great Smog of London occurred, relating Britain's capital with the current problem in China.

An anticyclone, high-pressure system had settled over a windless, freezing London in November and December of 1952. This meant that warm chimney gases from burning coal for warmth would not dissipate and remained close to the ground. Large coal-fired power stations belched thousands of tons of particulates, vehicle exhausts, sulfur dioxide, fluorine compounds and hydrochloric acid into the trapped, still air. Although familiar with smog, London was now blanketed in a thick, dense smog unlike any previous event in its long history. Pedestrians shuffled to feel for invisible curbs in 1-meter visibility and even theaters and film halls were affected by visibility. A performance of *La Traviata* at Sadler's Wells was canceled after the first act because the audience could hear, but not see the players. Underground trains and traffic were stopped, and ambulances were no longer able to collect the afflicted. The only game postponed in the history of Wembley Stadium happened during the Great Smog. Four thousand people are said to have been killed and a further 25,000 London inhabitants affected by hypoxia, pneumonia, toxic lung injury and acute bronchitis, which often suffocated the affected person. This event led to Britain's passing the Clean Air Act in 1956. The event caused a major rethinking of air pollution and its impact on the economy.

In the summer of 2017, Delhi in India can be added to large cities like Beijing, China, that have been suffering from a smog blight, added to the 40°C heat waves that India has been experiencing.

A Taiwanese study linked particulate pollution between 2001 and 2014 with sperm quality in Taiwanese men aged 15 to 49 years, showing what you might expect, that high levels of air pollution are strongly linked to poor sperm quality. Researchers from the Chinese University of Hong Kong had a good sample size of 6,500 men and found a strong, robust association between fine particulate PM 2.5 air pollution and "abnormal sperm shape." Although these results are relatively small, even underwhelming, add them to the results of pollution and the possibility of a relationship between pollution and the 50% drop in fertility over the last 40 years arises. Many of the components of particulate matter, such as heavy metals and polycyclic aromatic hydrocarbons, have been associated with sperm damage in previous studies.

A report by the University of British Columbia showed that 5.5 million people a year die from atmospheric pollution, of which 300,000 deaths per year in China are due to coal pollution.[347]

EXPOSURE (μSv)	RESULT
0.09	Living within 50 miles of a nuclear power plant
0.09	Eating a banana
0.10	Airport scanner
5	X Ray
40	A Flight from LA to New York
250	Yearly release limit from a nuclear power plant
2,000	Cranial CT Scan
3,100	Annual exposure from natural surroundings
13,000	Smoking 1.5 packets of cigarrettes a day
50,000	Maximum permitted dose for nuclear workers in the US
400,000	Dose causing symptoms if exposed in a short timespan
1,000,000	Induces nausea, vomiting if exposed in a short timespan
4,000,000	Lethal dose if received in a short timespan

Figure 93: Exposure to radiation measured in milliseiverts (μSv), showing that eating a banana is as dangerous as living within a 50-mile radius of a nuclear power plant. Source: Business Insider.

In countries where coal is a large part of the generation capacity, healthcare costs can be a large externality. Cohen and Pope in 2002[348] and Cohen in 2005 showed that these additional costs rival the total U.S. energy costs since healthcare expense is some $2.6 trillion while electricity costs are in the range of $400 billion. This data is so easy to read and find replicated elsewhere that it boggles the mind why the Trump administration can get away with such an unenlightened and deeply ignorant path of restoring coal power, unless they are aided and abetted by the fossil fuel industry.

The other way to illustrate this starkly is to say that it costs 2,000 lives to keep the electricity burning in Beijing, but 200 lives in New York. One article by a Greenpeace antinuclear campaigner unsurprisingly seeks to increase the effective death rate of nuclear by downgrading the deaths from wind or solar. It's probably worth mentioning that new nuclear builds would be safer and that the Fukushima designs were 50 years old.

The Greenpeace activist also wants to impose 33,000 deaths from Chernobyl on the basis of thyroid cancer sufferers facing certain death, but in fact longevity there has been good and only about 15 of these patients have actually succumbed. Most importantly, it's not clear whether the Chernobyl background rate of thyroid cancer is greater than normal since there is a similar ratio of sufferers in many other nonexposed locations. Alarm at radiation exposure has been much greater than evidence justifies. The extremely low death rates from nuclear are best exemplified by the table in figure 91, from a video prepared by the Business Insider website. Reports also leave out the fact that wind and solar equipment involve mining at early stages in their cycles and that many deaths are left out. The world uses the three deadliest energy sources, coal, oil and natural gas, which combined, account for 86% of the world's total 2016 energy usage.

The benign and abundant renewable energy systems we have discovered that don't have any significant negative effects are ready and waiting to be deployed. We are squandering the time we have left and eroding the old carbon equilibrium, which brought us so much historical climate calm. We know all about it but are paralyzed by incumbent and profitable CONG technologies and the willingness of their people to misinform about this extra carbon in the system. Instead, we need to inform people about the facts that we, non-ideologue, rational humanity, have fought long and hard to discover. Releasing fossil carbon is like opening Pandora's box. We are releasing ^{12}C from millions of years of natural and difficult-to-replicate sequestration. We are putting it into the oceans and atmosphere, and most of us already know the ills we face because of this action.

Most of our energy today still comes from burning something. Solid and liquid fuels combust and release gases and dangerous particulates into the atmosphere. Those gases and solids, it turns out, are a cornucopia of dangerous substances. Our electricity supply, our cars, most of the trucks, ships and planes and boats are all releasing very nasty things into the same atmosphere that we then breathe into our lungs. Perhaps I should say we think nothing of it when we walk down the high street with our children. Large vehicles driving past belch diesel exhaust with all its awful, active components into the air we are breathing. Even diesel exhaust that has been cleaned up over the years represents lungsful of harm. Today you can occasionally still see clouds of black smoke billowing from a vehicle as it accelerates the engine to start moving. It will seem amazing in the future that we took this for granted and simply breathed this all in or ran the risk of doing so. Sometimes the traffic is snarled, and we even hold our breath as we walk past a particularly smoky, cranky bus or truck engine, perhaps a crane engine or earthmover or a truck close to the sidewalk.

What does this really say? In some locations around the world, the air is trapped in a locality and forms layers due to topographical features such as hills, mountains and river valleys, which make a sort of bowl, with the town in the bottom. These sites accumulate exhaust gases like ozone (O_3), filling the space between hills and having a more pronounced and irritating effect. Los Angeles is a case in point, with winds out of the Pacific that run up against the mountains east of the city, keeping lower levels of air static.

In 2016, coal was responsible for just under 30% of the world's energy consumption, and in the United States, it has fallen from above 50% to about mid-40%. It is an arch suspect for human health problems and is responsible for particulate and gas emissions that are trapped in city air.

Let's have a look at coal first. A Harvard doctor, Dr. Paul Epstein, and his team wrote a report in 2011 called "Full Cost Accounting for the Life Cycle of Coal" for the New York Academy of Sciences.[349] It was an examination of the costs of each stage of handling coal, from its mining and transportation, processing and combustion to the waste stream it generates and the effects it has on the environment and people. The price you pay for a ton of coal is not influenced by these external costs, but they are borne by us anyway in the long term, and many people pay the ultimate price today. The life cycle analysis calculated the link between all negative aspects of coal production and combustion with its impact and a cost for that impact based on reasonable assumptions. Coal releases many more substances than just CO_2. It also releases mercury, lead, cadmium, arsenic, manganese, beryllium, chromium and the gases nitrogen oxide and sulfur dioxide. Coal crushing, processing, washing and combustion releases tons of particulate matter and chemicals that contaminate water and ecological systems. Coal accidents and direct miner illnesses as well as production practices like mountaintop removal (MTR) in Kentucky and Virginia, which remove forests and fill valleys with rubble, bring significant destruction.

The team estimated that the impact was fully a third to a half-trillion dollars each year we continue to use this fuel, $330 billion to $500 billion annually. This adds 17–27 cents per kilowatt-hour generated. Many of the externalities are also cumulative. If the price of coal were to be integrated with these costs, according to the researchers, it would double to triple the price of coal electricity, easily making solar, wind, efficiency and other non- fossil electricity generation cost competitive.

Coal utilities earn significant profits even on low prices for electricity. They are known to be cash cow industries but could not survive if they had to internalize the external damage they do. If the cost of coal electricity was to reflect these externalities, renewable solutions would pay for themselves in a year, and if the incentives for renewables were taken away, they would still be competitive if only because they don't damage health. When the utilities and manufacturing sectors cry out about the subsidies available for renewable energy, the truth is that we would happily forgo all of them if they started covering the cost of the fossil externalities. That would be the level playing field! It's getting better. Renewable energy gets cheaper all the time and is an economic choice today.

Coal is the major bad character in the lineup of fossil fuel suspects. BP's statistical review 2018 showing figures for 2017 explained that in 2017 coal consumption was responsible for burning 3,731.5 million tons of oil equivalent; it released 30 trillion tons of CO_2 in 2017.[350] There is a paradox at work here, though. Since the arrival of fracking as a method to release natural gas and oil trapped in otherwise difficult-to-reach geology, the large quantities of natural gas discovered and brought online in the United States have meant that the electric utilities have replaced much of their coal burning with cheaper-per-kilowatt-hour natural gas to increase margins. This has pushed coal down from over 50% of the generating capacity in 2009 to 33% in 2016, resulting in less carbon release for a kilowatt of electricity than the Europeans have achieved in all these years of legislating the carbon policies they have adopted. However, coal burning is still a global staple of reliable energy production and still delivers the knockout punch and highest percentage of carbon emissions. When coal was 50% of U.S. electricity, it also amounted to 81% of CO2 emissions.[351]

It is clear that fossil fuels and nuclear power have a huge disadvantage in respect to health, not only of human beings but of the plants and animals susceptible to the diverse chemical and radiological impacts they have increasingly been forced to endure by the industrial activity of the last 300 years. Coal mining and burning are not producing just CO_2 and methane.

It's important to remember that mining, pulverizing, transporting, washing and burning coal, then disposing of fly ash residue, result in unsafe chemicals being released, including CO_2 and methane along with carbon monoxide (CO), nitrous oxides (NOx), sulfur dioxide (SOx), sulfur, particulate matter of different sizes (PM10 and PM2.5), mercury, lead, cadmium, beryllium, chromium, arsenic, antimony and selenium, much of which is designated toxic or carcinogenic.

Figure 94: Top four lines of images resulting when putting the words "Fish Consumption Warning" into Google images. Source: Google Images.

Coal washing results in a slurry that in turn results in hundreds of billions of gallons of toxic water held behind walls and dams. Sometimes these lakes are lined but often not, and spills are common. A 309-million-gallon spill in Martin County, Kentucky, occurred in 2000. Nineteen chemicals used in the wash are known to be carcinogenic, and 24 are linked to lung and heart damage, with much testing still to be done. Fly ash is the product of coal combustion and contains toxic chemicals and heavy metals known to cause birth defects, reproductive disorders and nerve damage leading to learning disabilities, diabetes and kidney disease.

Burning coal leaves evidence across the countryside. A federal study[352] produced by the U.S. Geological Survey in early 2009 showed the results of mercury testing in 300 streams across the United States by testing over 1,000 fish, one from at least every stream, between 1998 and 2005. There was not a fish without traces of mercury, but only 25% had levels considered unsafe for human consumption. It prompted the interior secretary Ken Salazar to say, "This science sends a clear message that our country must continue to confront pollution, restore our nation's waterways, and protect the public from potential health dangers." The report observed that levels of mercury have increased by three to five times in the last 150 critical years.

Once in the waterways, mercury turns into methyl mercury and then works its way into the food chain helped by bacteria. The highest levels were detected in streams along the coasts of the Carolinas, Georgia, Florida and Louisiana. Largemouth bass in South Carolina saw a high concentration, but the worst contamination was in Nevada near a group of gold mines. All states bar Alaska have issued fish consumption warnings. This research demonstrated how widespread and urgent was the problem for human health and on the environment from mines of all sorts.

There are some 1,300-coal combustion waste (CCW) ponds in the United States that are mostly poorly constructed and are a source of leaching chemicals into groundwater supplies or bodies of water. The EPA monitors the production of CCW and expects that, as coal is exported rather than burned locally, the production of toxins will continue to grow and accumulate in the environment. High levels of NOx and SO_2 cause wheezing, exacerbation of asthma, shortness of breath, nasal congestion and pulmonary inflammation, hearth arrhythmias, low birth weight and increased risk of infant death. Nitrogen-containing emissions lead to health damage via several routes. When combined with volatile organics, they can form particulates and contribute to ground-level ozone. NOx can reduce resistance to infectious diseases. Ozone and particulate matter increase susceptibility to respiratory illnesses like pneumonia and bronchitis, and ozone is also corrosive to lung tissues as well as being a heat-trapping GHG. Particulate matter goes deep into the lungs and can contribute to heart attack or stroke as well as increase the risk of premature death in infants, young children and older people. Mercury is well documented to be deposited in waterways and accumulated in seafood. It brings adverse neurological and reproductive effects.

The U.S. Mine Safety and Health Administration (MOSHA) and the National Institute for Occupational Safety and Health (NIOSH) track all health impacts on coal miners. Major accidents involving gas leaks, explosions, flooded tunnels or subsiding tunnels occur on a regular basis, with over 100,000 deaths in the United States alone since 1900. Seventeen miners died in Appalachian mines in January 2006, and 29 miners died in April 2010 in the Upper Big Branch Mine in West Virginia. The horror is really bad outside America, where China puts up with a regular death rate of 3,800 to 6,000 annually from mine accidents but also has to cope with 10,000 cases of black lung disease each year. Diseases kill many more miners than accidents do and are led by black lung disease (or coal miner's pneumoconiosis or progressive massive fibrosis pneumoconiosis), which leads to chronic obstructive pulmonary disease. In the 1990s, over 10,000 former U.S. miners died from this cause and the incidence of this disease has increased. Since 1900, pneumoconiosis has killed more than 200,000. Long- term support of these ill and dying miners and their inability to contribute in other ways lead to significant economic hits to federal and state funds outside the direct costs of the coal industry.

Coal mining methods also bear scrutiny. Commodities that are on the surface cost far less to produce and so are good high-margin businesses. Seams of coal are available like this in Appalachia, which includes Kentucky, Virginia, West Virginia and Tennessee. It leads to a mining technique called mountaintop removal (MTR), and it has completely transformed, and ruined, thousands of square miles of countryside, mountains and valleys. The effects on wildlife and water supplies and resulting deforestation, pollution and contamination with heavy metals are also associated with significant health effects on local populations, with cancer clusters reported.

A study[353] by Hendryx and Ahern found that the mortality of residents from all causes was highest in heavy coal mining areas in Appalachia and lowest in non–coal mining locales outside Appalachia. Low-birth-weight (LBW) babies occurred 16% more often in heavy coal mining areas. Low birth weight reflects neurological damage that can lead to chronic disease such as diabetes and hypertension. All this translates to almost 11,000 excess deaths in coal mining areas versus non–coal mining localities annually. All these coal evils were totted up and compared with the benefits of mining coal for the Appalachian area in 2005 U.S. adjusted dollars. The results, not including ecological damages, were $74.6 billion negative ($4.36/kWh) versus $8.08 billion positive. Transportation of the coal, mostly by train, was deemed to have killed 246 people, mostly members of the public, and using the valuation system adopted, this added a further $1.8 billion or $0.09/kWh to the externality bill of coal in the United States. As mechanization improved, the need for employees dropped, resulting in local unemployment in the Appalachian coal mining industry increasing by 56% between 1985 and 2005. Mining only represents 0.8% of jobs in West Virginia.

One interesting angle on health impact of fossil fuels is the effect of climate on populations. The heat waves literally kill thousands, and rising temperatures result in an almost formulaic expectation of heat wave mortality, especially where hospital infrastructure or even cooled public facilities are insufficient. Other extreme weather events such as hurricanes and tornadoes and floods have a significant effect on human and animal populations. Disease vectors such as mosquitoes carry the viruses and parasites for a suite of diseases such as malaria and dengue fever, which annually kill millions.

Sobering data on farmer suicide[354] in the United States and Australia is yet another angle on the pressure of CONG on populations. Coal brings climate change which brings drought which stresses farmers to the point of suicide. The impact of drought on farming economics leaves hundreds of thousands in financial turmoil. Defaulted loans, homelessness and inability to look after their families turn many to suicide. The *American Journal of Preventive Medicine* tells us that suicide rates are second highest of any occupation between 2003 and 2010, with 5.1 suicides for every million workers, just behind police and fire fighters with 5.3. Often firearms are easily available to these communities. Between 1981 and 1986 in the United States, 60,000 farmers were left homeless due to foreclosures. In India, farming has become one of the most dangerous professions, where the rate of suicide is almost as high as in the United States. India has a hot, tropical climate, but current heat waves and unexpected torrential rains can ruin a crop on which a farmer is gambling everything. In the 20 years since the Indian government started keeping records, 300,000 farmers have taken their lives, leaving wives, children and debt. Eighty percent of India's farmland relies on flooding during the monsoon season. Suicide for farmers is caused by drought. In 2013, Germanwatch,[355] a group that maintains a global climate risk index, ranked India as one of the top three countries most affected by extreme weather events, which have increased by 50% in the last 50 years. In 2009, 17,000 farmers killed themselves, a six-year high, after 18,241 in 2004, coinciding with the worst drought the country had had since 1972. This of course brings to mind the droughts in the Middle East, where they arguably have accelerated the breakdown in social structures.

Health Effects of Air Pollution

In 2016, the world burned 5.3 billion tons of coal. China burned 50.6% of it, India, 11%, and the United States, 9.6%. Tens of thousands of people go to emergency rooms every year due to coal pollutants, but it's very difficult to pin down the exact numbers. Lung cancer, heart attacks and other nonfatal complaints result in hospitalizations and lost work or school days.

One of the culprits is particulate matter (PM). Microscopic solids and liquid droplets small enough to be suspended in air are composed of nitric and sulfuric acids, organic chemicals, metals, soils or dust particles and allergens such as pollen or mold spores. $PM_{2.5}$ is a designation that describes the actual size of a particle as 2.5 microns in diameter or less (2,500 nanometers). A micron is a micrometer (μm), which is a thousandth of a millimeter, or a millionth part of a meter.

The fine particles can travel deeper into the lungs than larger ones, which are filtered out by the nose and larger airways. The ocean of cilia, small hairs on the inner side of the trachea, cause many larger particles to be driven toward the throat, where they are either coughed up and ejected or swallowed to the stomach, but cilia stop working when particle sizes drop.

Negative Effects of Coal Combustion Emissions and Associated Activities	Monetized estimates from literature in US$2008 and c/kWh of electricity					
	Low		Best		High	
	US$2008	c/kWh	US$2008	c/kWh	US$2008	c/kWh
Land Disturbance	54,311,510	0.00	162,934,529	0.01	3,349,209,766	0.17
Methane from Mines	684,084,928	0.03	2,052,254,783	0.08	6,840,849,276	0.34
Carcinogens (water waste)						
Public Health burden of	74,612,823,575	4.36	74,612,823,575	4.36	74,612,823,575	4.36
Appalachian Communities						
Fatalities of the public due to coal transport	1,807,500,000	0.09	1,807,500,000	0.09	1,807,500,000	0.09
Air Pollutant Emissions from combustion	65,094,911,734	3.23	187,473,345,794	9.31	187,473,345,794	9.31
Lost productivity from mercury emissions	125,000,000	0.01	1,625,000,000	0.1	8,125,000,000	0.48
Excess mental retardation	43,750,000	0.00	361,250,000	0.02	3,250,000,000	0.19
from mercury emissions						
Excess cardiovascular disease	246,000,000	0.01	3,536,250,000	0.21	17,937,500,000	1.05
from mercury emissions						
Climate damage from CO2 and	20,559,709,242	1.02	61,679,127,726	3.06	205,597,092,420	10.20
N2O combustion emissions						
Climate damage from black carbon emissions	12,346,127	0.00	45,186,823	0	161,381,512	0.01
IEA 2007	3,177,964,157	0.16	3,177,964,157	0.16		
AMLs	8,775,282,692	0.44	8,775,282,692	0.44	8,775,282,692	0.44
Climate Total	21,310,451,807	1.05	63,939,503,861	3.15	215,948,532,974	10.72
Total	175,193,683,965	9.35	345,308,920,079	17.84	517,929,985,035	26.64

Figure 95: The complete costs of coal in 2008 dollars and cents per kilowatt-hour. Source: "Full Cost Accounting for the Life Cycle of Coal" study, Paul R. Epstein, for the New York Academy of Sciences, 2008.

Exposures to $PM_{2.5}$ is linked to an entire collection of ills, making one think of a preindustrial time when air quality was a premium, although longevity was still poor. Some of the issues include premature mortality and cardiovascular and cardiopulmonary mortality, as well as the range of respiratory ailments and lung function symptoms. Diabetes symptoms are enhanced by $PM_{2.5}$, and even sudden infant death syndrome (SIDS) has been associated. $PM_{2.5}$ has no minimum level beneath which it is safe, so prevention is about elimination of $PM_{2.5}$.

A study[356] completed in 2016 titled "Magnetite Pollution Nanoparticles in the Human Brain," published in the *Proceedings of the National Academy of Sciences*, showed the impact of even smaller particulate matter derived from combustion engines in inner cities. Traffic fumes alone can impact not just your heart and lungs but also your brain. Urban environments are full of particulate matter, which includes tiny particles of magnetite, an oxide of iron. These particles were first found in the human brain in 1992. Particles smaller than 200 nanometers in diameter can get into the brain via the nose and olfactory nerve. They have been found in abundance in inner-city residents' brain tissue. Normally, in nature, particles of magnetite are angular, but when fuel is burned at high temperatures it forms rounded particles or nanospheres, like those discovered in the brain. These particles affect tissues due to their magnetic fields, oxygen-reducing activity (redox) and toxicity and are perhaps implicated in neurodegenerative diseases such as Alzheimer's.

It turns out that an IPCC calculation for the externalities of coal very much coincided with the "best" figure for climate impacts, bringing the entire Epstein study into a fine focus for its likely accuracy.

Negative Effects of Coal Combustion Emissions and Associated Activities	IPCC 2007, US Hard Coal	
	US$2008	c/kWh
Methane from Mines	2,188,192,405	0.11
Climate damage from CO2 and N2O combustion emissions	70,442,466,509	3.56
Climate damage from black carbon emissions	3,739,876,478	0.19
Total	76,370,535,392	3.86

Figure 96: IPCC 2007 report on the external climate cost of burning anthracite coal. The IPCC number is just a little higher for climate than the number offered by Dr. Paul Epstein. Source: IPCC 2007.

This best number showed that all the damages of coal, including health damages, were over $345 billion, and this added over 17 cents per kilowatt- hour to the cost of the coal electricity if this damage were to be internalized. It has to be said, though, that the complete picture most likely still leaves this number looking conservative since it misses much of the full cost of the toxic chemicals on ecological systems, chemicals and heavy metals on plants and animals. It also doesn't capture the whole morbidity picture of health effects of released gases or climate forcing caused by increased tropospheric ozone concentrations. What this clearly and conclusively shows is that coal electricity prices ought to be at least three times higher than they are today if external costs were included. This report looked at the holistic, climate, health and other negative effects but with a health emphasis, and I have not separated these out in the book.

This study also showed that if emissions from coal are solved using carbon capture and sequestration (CCS) methods, that the huge CO_2 release into the environment, albeit initially into empty mines, drill holes and other destinations, nonetheless has the result of doubling the price of coal electricity anyway as well as having huge health side effects, as the CO_2 has its chemical way with whatever rock formation it finds itself stored in. What this study immediately points out is that complaining that wind and solar are too expensive is no longer an acceptable behavior and instead we need to exhort our fellow human beings to accelerate our progress toward a sustainable world.

The European Environment Agency (EEA) completed a study,[357] "Air Quality in Europe—2016," focusing on EU exposure to PM2.5 and found 85% of urban dwellers particularly at risk of aggravated heart disease, asthma or lung cancer. In 2013, more than 430,000 people died prematurely and 17,000 died of ground-level ozone exposure. No surprise, it's nations like Poland, Bulgaria and the Czech Republic, heavily dependent on coal energy, where the worst data comes from. The report estimated that healthcare costs due to the coal burning were $4.8 billion annually. In the UK, air pollution costs the economy more than $27 billion. It's true to say that between 2000 and 2014, levels of PM10 and PM2.5 fell but PM still causes too great a damage to human health in Europe. The Royal Colleges of Physicians and of Paediatrics and Child Health in the UK says that about 40,000 early deaths are caused annually by outdoor pollution. Children living in cities can have up to 10% less lung capacity than normal children, and the damage is permanent. The World Health Organization says that 9 out of 10 people on the planet live in polluted air. Outside developed nations, it's indoor air that presents the main killer since it's full of fumes from heating and cooking fuels.

It turns out that coal can be labeled as humanity's most dangerous energy source. Even if the direct coal death rate was only 13,000 annually, its current level, which is very much down from its high, there would have been 3.4 times as many deaths from global coal mining alone as there were from nuclear energy, even including the impact of the United States dropping the first two atomic bombs on Hiroshima and Nagasaki, and that is just since 1945! The 300,000 indirect Chinese deaths from coal-related problems in 2014 alone is a number greater than all the nuclear deaths since the inception of the nuclear industry.

If we go beyond the IPCC report and look elsewhere for evidence, it turns up quickly. Mining is a dangerous activity, but we have become complacent about the subject. The deaths at U.S. mines have dropped significantly over the years and safety standards have improved, but that's in the regulated United States. Early-20th-century numbers were 1,000 deaths per year in the United States, which fell to 451 in the 1950s, 141 in the 1970s, and now to around 35 per year. Much of this can be explained by mechanization leading to fewer miners per ton of coal extracted. Globally, today, 12,000 miners are killed every year.[358] The International Labor Organization (ILO) tells us that mining employs 1% of the global labor force but also gives us 8% of total fatal accidents.

Safety standards inevitably differ among countries. Most safety records are improving, but in China and Russia this is not so much the case. China mines 3 billion tons of coal a year and runs the world's largest mining industry, producing 46.9% of global output, but is responsible for 80% of the global mining deaths. Life, it appears, is cheaper there than it is in developed nations, and the operators are concerned about money. Job insecurity means there is nowhere to complain. The industry is also one in which many accidents are covered up or simply not reported.

After mining accidents, we can look at other impacts of coal on human health. Coal combustion releases a lot of chemistry into the atmosphere that is not good for us or animals to be in contact with. In the United States, we can look at the coal mining country in Appalachia, in parts of Virginia, West Virginia, Pennsylvania and Ohio, which has recently seen an unprecedented bounce in the deadliest mining disease, black lung disease. Forty thousand miners have lost jobs since 2010, and 600 mines have been closed down. From 2010 to 2015, the National Institute for Occupational Safety and Health (NIOSH) reported[359] just 99 cases of black lung disease.

Black lung clinics in the area, however, have reported 962 cases so far this decade, and because only 17% of working miners are tested, the real number is probably much higher.

COAL POLLUTANTS	COAL ILLNESSES
Aldehydes	Anemia
Antimony	Asthma
Arsenic	Brain Abscess
Benzo-a-anthracene	Breathing Passages
Benzo-aapyrene	Bronchitis
Beryllium	Cancer
Cadmium	Coalworker's pneumoconiosis
Chromium	Delayed response to visual stimulus
Chrysene	Dermal irritation
Cobalt	Eye Irritation
Dibenzo-a-anthracene	Heart Attacks
Dioxin	Immune System Compromised
Ethylbenzene	Impaired Devopment
Flouranthene	Kidney Disease
Formaldehyde	Learning Impairment
Furans (tetrachlorodioxin)	Liver Disease
Hydrogen Chloride	Low Birthweight Baby's
Hydrogen Cynanide	Lung Function
Hydrogen Flouride	Lymph System
Lead	Memory Impairment
Manganese	Nasopharyngeal Cancer
Mercury/Methyl Mercury	Neurological Damage
Nickel	Nose Irritation
Nitrogen Oxides	Particulate Matter
Particle Phase Organics	Premature Death
Particulate Matter	Reproductive Impairment
Polycyclic Aromatic Hydrocarbons	Siliconiosis
Radium	Strokes
Selenium	Throat Irritation
Sulfur Dioxide	
Toluene	
Uranium	
Xylene	

Figure 97: List of coal pollutants and illnesses they cause. Source: American Lung Association, March 2011.

When your job depends on good health, the tests are not popular, especially when the employer doesn't want its workforce analyzed for the disease. It is officially against the law to fire miners for getting an X-ray, but from the miner's point of view, giving any pretext to the employer to fire you is sufficient disincentive. Now, however, out-of-work miners flock to the clinics to apply for benefits. Many require a lung transplant. The disease takes a decade or more of exposure to develop and it causes its victims to struggle for each breath. Tiny coal dust particles are breathed in as a cloud, and many of them are so small they can reach the individual alveoli, the structures that allow oxygen into the blood and CO_2 from blood into the exhaled breath. The body is good at fighting pathogens that would be absorbed, but a lump of carbon is not absorbed and accumulates, all the while exciting the inflammatory response and clogging up the path to reasonable oxygenation. It is incurable and fatal. Severe cases require oxygen all the time to avoid suffocation.

The disease appears with another disease called silicosis that is specific to silicon. Black lung PMF is caused by inhalation of large amounts of coal and rock dust over time. Coal is often embedded in thinner seams in rock such as quartz, and the dust from cutting operations is particularly toxic to lung tissue. Respirators, water sprays and ventilation are supplied, but filters get clogged rapidly and the dust conquers all attempts to control it. Coal mining in masks creates discomfort and extra difficulty not dissimilar to the impairment experienced by soldiers operating in nuclear, biological, chemical suits (NBC).

Mining companies fought to resist new regulations that came through after a decade of work and have hugely reduced the amount of airborne coal and silica rock dust, which have dropped by 50% since 2009. Treatment for resulting lung diseases costs from half a million to a million dollars over time, so there is a lot of money about to be spent correcting employers' lapses. A fund called the Black Lung Disability Trust Fund, supported by a coal excise tax, is due to be cut in 2017. The fund is currently $6 billion in debt and struggling to cope with more numerous claims than expected. Needless to say, the Affordable Care Act (ACA), otherwise known as Obamacare, had provisions for miners in it and is now facing dissolution itself.

A report called "Coal's Assault on Human Health," written in 2009 by Alan Lockwood, Kristen Welker-Hood, Molly Rauch and Barbara Gottlieb from a group called Physicians for Social Responsibility, said that coal pollutants impact all major body organ systems and contribute to four of five causes of mortality in the United States: heart disease, cancer, stroke, and respiratory disease. The researchers squarely blamed every step in the coal life cycle for its impact on respiratory, cardiovascular and neurological systems. The report also looked at impacts on human health from global warming. Even though the mining accidents have been declining, the American Lung Association states that the death rate from coal pollution still rests at about 13,000 individuals a year in the United States alone as we try to manage 386,000 tons of 84 different hazardous air pollutants coming from 400 plants in 46 states threatening the health of those who live close or far.

Dutch scientists from the Erasmus University Medical Center have discovered a link between brain abnormalities in babies and particulate matter pollution that is below the "safe" thresholds described by the European Union. The study[360] shows exposure to these fine particles is linked to impacts on the growing fetus and brain cortex, the part of the brain that controls impulsive behavior. This can result in cognitive impairment that has long-term consequences such as mental health disorders and low academic achievement. While even supposedly acceptable air pollution levels have always been found to correlate with impacts on the lungs, heart and other organs, this is the first time it has been associated with slower fetal growth and brain development.

When we consider that the U.S. pollution health situation is almost a paradise compared with the situation in China, we can only hope that similar controls are established in China, although we also know that the Chinese embrace of renewable energy is one measure to combat the intense pollution they experience.

There is less information about mortality from emissions globally. A study in the *Proceedings of the National Academy of Sciences* (*PNAS*) in the United States found that coal usage in northern China shortens the average life span by 5.5 years compared to that of an individual living in the south. This concept builds on the idea that there are "immediate" fatalities from accidents in energy production and there are "latent" deaths, caused by the specific consequences of a particular type of energy production. Some of the largest cases of immediate death have been the hydroelectric dam failures discussed previously that have mainly occurred in China, such as the Banqiao and Shimantan Dams, which killed 171,000 people and displaced 11 million from their homes in 1975. They were among 62 dams that failed catastrophically at the time.

In the case of coal, the latent deaths are much higher, with 600,000 Chinese miners suffering from pneumoconiosis or black lung by 2004, a figure that increases by 70,000 annually. In the United States, the regulatory environment quickly cut the death rates, but pollution-related latent deaths still number between 10,000 and 30,000 per annum due to sulfur dioxide and nitrogen dioxides. The World Health Organization, in 2011, said that the urban outdoor air pollution from burning fossil fuels, and notably biomass as well, causes 1.3 million deaths per year while indoor pollution from burning biomass and family cooking results in an estimated 2 million premature deaths.[361] In 2013, this pollution killed 3.7 million people around the world.

The World Resources Institute (WRI)[362] said that 1,199 coal plants were to be built by 59 countries, 76% of them in China by Huaneng and Guodian and also in India. State-owned companies are the dominant coal power producers in China, Turkey, Indonesia, Vietnam, South Africa, the Czech Republic and many other countries, leading one to hope that governments that commit to doing something about it can lead directly to solutions. Australia is a big producer for the Pacific market and is increasing its production to 900 million tons per annum. So much pressure is now facing the coal industry that many of these plants will never be built.

A recent Greenpeace report[363] summarizing water consumption by the global fleet of coal plants showed that the world's 8,359 installed coal plants consume enough water to meet the basic needs of a billion people, or 19 billion cubic meters (m^3) per year. This amount is 18.3 m^3 per person per year, or 50 liters per day. A 500-megawatt coal-fired plant, if using the water to run through its cooling system once only, can swallow an Olympic-sized swimming pool every three minutes.

In 1990, the U.S. Congress passed the Clean Air Act requiring the EPA to clean up toxic substances from the air. It took two decades for action to result, but finally the electric utilities were forced to spend money to install the "maximum achievable control technology" to reduce emissions on plants generating 25 megawatts or more.

In the United States, cheap natural gas has been chipping away at coal's dominance in the electricity generation market for over a decade now. The extra costs of clean air rules and compliance made coal less economically attractive and the life cycle of the fuel certainly did not match up. Consumers are party to this effort as they become aware of the issues, and groups like the Sierra Club have made a huge impact.

Today there are two employees in the solar industry for every employee in the coal industry. An article in *Fortune* magazine on January 16, 2015, explained that there was a total of 173,807 solar jobs in the United States but only 93,185 workers in the mining industry.[364] The transition so far has been without any increase in electricity rates. Some critics have said it's useless to be a unilateral actor in this coal phase-down while China and India and many other countries continue to grow their consumption. This, of course, is like saying, "We'll stop killing our people if you stop killing yours!" Coal states like Wyoming, Montana, Colorado, West Virginia and New Mexico depend on coal revenues, and like Colin Marshall, chairman of Cloud Peak Energy, a major U.S. coal producer, called Obama's anti-coal stance "pander[ing] to special interest groups."[365] Yes, that's true and in this case the special interest group is everyone not interested in coal, a very large group! Freezing access to coal leasing on federal lands naturally confuses those in the Powder River Basin, which has a trillion tons of coal running through the ground in 100-foot-high seams. We do need fossil fuels, though, to complete an orderly and seamless transition to sustainability.

Patrick McGinley wrote an article in the July 8, 2016, issue of *The Conversation*, where he observed that over 50 U.S. coal companies have filed for bankruptcy since 2012. Competition and more stringent environmental regulations impacted demand for coal, and then prices dropped. Not aware this might be a one-way trip, speculators spent billions to finance acquisitions. Those investors have now asked for $30 billion in bankruptcy protection in order to restructure. As an indicator of how serious this is, Peabody Coal's market capitalization dropped from $20 billion in 2011 to only $38 million in July 2016. An obvious consequence of these bankruptcies is that the coal companies are likely to be able to avoid paying for their huge liabilities as embodied by the Surface Mining Control and Reclamation Act (SMCRA), which was passed by Congress in 1977 for land reclamation. Thousands of acres of land and millions of gallons of polluted mine water would cost billions to remediate. The SMCRA was designed to oblige coal companies to integrate the externality costs of reclamation by using bonds or other financial guarantees. The coal companies' promise to reclaim despoiled land is little more than an IOU backed by company assets.

Nuclear Energy and Health

Nuclear power accidents have been caused by the adoption of the wrong nuclear technology. It became expensive and was prone to accidents that caused the public to want to avoid using it. If you collect the top 20 nuclear and radiation accidents that involved fatalities,[366] including nuclear submarines and even radiotherapy, you get a good sense of the relative danger people risk in the service of providing power in the nuclear industry.

Figure 98 includes the top accidents. It's interesting to note that 16,000 deaths followed the Japanese tsunami, but only 6 due to nuclear power, and those were mainly from circulatory problems. It's also key to establish that with nuclear power, the health effect is almost always only a relatively tiny number of deaths, whereas for coal there are a lot of lingering sick people for an extended period of time. As long as there are no radiation exposures, the nuclear industry has very little impact on overall human health. The brave fire fighters at Chernobyl lost their lives to put things right and many exposed victims live with the intense burn injuries they suffered during radiation events.

In 1976, Peter Beckmann collected studies on health problems of various fossil fuel power sources and published these in his book *Health Hazards of NOT Going Nuclear*.[367] He made the point that if the Shoreham plant, covered in chapter 5, had been operational for 20 years, then some 1,500 people may have been saved during the period and were effectively killed therefore by the antinuclear presence, supported by a media that was presenting a much scarier message about nuclear than it actually made of their exposure to the real but unperceived negatives of coal-fired power.

NUCLEAR DEATHS AS A RESULT OF ACCIDENTS	
Hiroshima	166,000
Nagasaki	80,000
Kashtym	8,015
Chernobyl	4,073
Radiological	121
Windscale	33
US Nuclear Complex	12
Fukushima	6
Total	258,260

Figure 98: Deaths from the various global nuclear accidents. Source: Benjamin Sovacool.

The Three Mile Island incident was studied by the U.S. Nuclear Regulatory Commission, the Environmental Protection Agency, the Department of Health and Human Services, the Department of Energy and the State of Pennsylvania, along with other independent studies—all of which showed that the dose of radiation to the 2 million individuals living in the area of the meltdown was only 1 millirem. Passengers on a coast-to-coast jet flight will be exposed to a total of 6 millirems from cosmic radiation.[368] Nuclear power is less than one-third as dangerous as coal in the developed world and hundreds of times safer than coal in China. Additionally, it is highly likely that if the molten salt thorium reactor had been adopted, with no need for a pressurized containment, continuous refueling, drastically lower radiation issues and processing of fission products, there would have been a significantly better safety record, along with other perks such as a lower nuclear waste radiation risk and of course a much more economic cost. Point made!

Inflammability

Any energy source is dangerous precisely because it packs a punch and needs to be handled carefully. Oil contains 138,000 BTUs per gallon and is flammable. Its refined products are often even more flammable. Fires from accidents in transportation or handling of oil-related fuels are a cost that a move to electricity largely avoids. I make the point here in the context of the transportation industry, where liquid fuels are mostly used. It's definitely the case that biofuels, which are carbon-neutral fuels currently made from food but soon from algae and sources of cellulose, are often just as flammable and prone to accidents.

As the domestic production of oil has increased, shipping oil by train has become a standard practice but has resulted in frequent accidents from heat applied to the tanker cars by accidents such as derailing. The resulting explosions are some of the most dramatic images of direct CONG externality, immediately releasing CO_2 into the atmosphere in large quantities and often occurring in or near small towns. In the United States, the conversation is about building pipelines that would obviate the use of trains and getting oil produced from Canadian tar sands and from "frackin' the Bakken" to the coast for refining and export to foreign markets.

In 2015 alone, crude trains, christened "bomb trains," have crashed in Ontario twice and in Virginia and Illinois, forcing evacuations of local residents and creating spectacular burnouts. In 2010, 29,605 oil tank cars were used, and the number rose to 493,126 in 2014. The industry claims that background accident rate for oil trains is no greater than for any other customer. A derailment of a grain train, however, doesn't result in a destructive fireball. The *Los Angeles Times*[369] found 13 such accidents in the United States and Canada since July 2013 when a runaway oil train horrifically crashed in the center Lac-Mégantic in Quebec, killing 47 residents. These accidents occur on rivers and bridges and in communities, but there has been no loss of life since that horrible event. Some cities have rail hubs where these trains pass regularly. Chicago is such a town, and it sees these hazardous cargoes daily, knowing it's only a matter of time before a 100-car train jumps the tracks and burns right in the middle of a residential neighborhood. Over 1 million barrels of oil from the Bakken deposits in North Dakota ride the rails every day in the United States. If the bombs had been set by terrorists, reactions would be stunningly different from the complacency with which these "banal" but spectacular and increasingly common accidents occur.

The National Fire Protection Association[370] 2007 report discusses accidents to do with gasoline at the 117,000 service stations in the United States. The bulk of accidents involve family automobiles. Between 2004 and 2008, U.S. fire departments responded to an average of 5,020 fires on service station properties per year, with an annual death toll of 2, 48 injuries and $20 million in property damage. Most fires were vehicle fires and were started by combustible liquids. The most common reason was smoking materials.

USGS Forecast for Damage from Natural and Induced Earthquakes in 2016

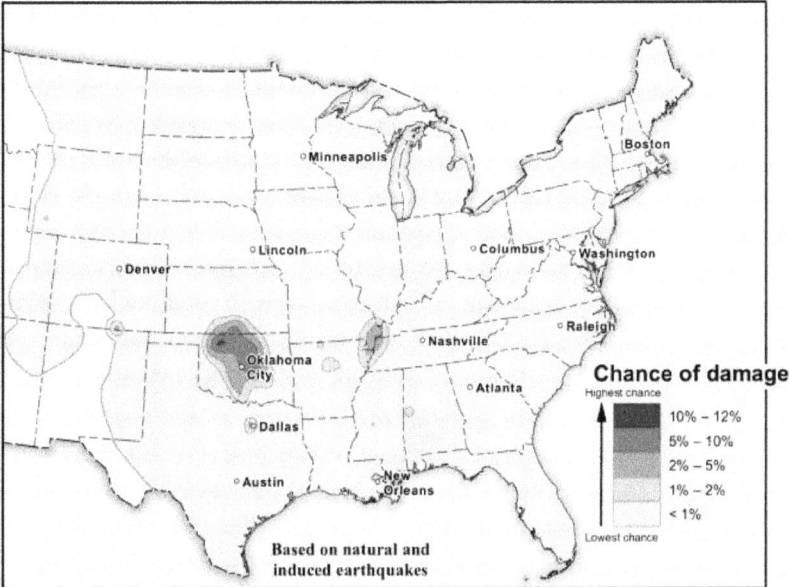

Figure 99: USGS map showing the potential for earthquake damage in human-induced earthquakes in 2016. Oklahoma City is in the heart of oil country and experiences human-caused quakes almost daily. Source: USGS.

There is also the little matter of a different kind of damage done by fracking and drilling, and it concerns the effect of depleting a geological resource and leaving a space, where the only thing that can happen is a collapse of some sort. Whereas it's common knowledge that mines can collapse, it's less well known that depleting oil or other resources changes the nature of the mass density of the geological formations that results in earthquakes. The U.S. Geological Survey indicates that in states such as Idaho, earthquakes in the 1960s might have featured at a rate of about 63 a year with an average Richter score of 3.0. In 2015, there were over 600 earthquakes, with a top earthquake weighing in at 5.8 whose shaking persisted for an entire minute, knocking food off supermarket shelves and causing 100-year-old stone facades to collapse.

As if further evidence was needed, another study[371] on the effects of climate change on human health was published in the *Lancet* in October 2017. It summarized the following observations: since 2000 there has been a 46% increase in weather-related disasters, diseases have spread, hundreds of millions of people are newly exposed to heat waves each year and rising temperatures have led to an accelerating drop in labor productivity, which cost almost a million jobs in 2016, exposed 11.2% more people to dangerous levels of pollution and led to an increase in allergic reactions, especially in the United States due to ragweed pollen. The report states that the delay in dealing with the problem over the last quarter century has jeopardized human life and livelihoods and made climate change not a future threat but a threat that, with storms, floods, droughts, wildfires and heat waves, causes disastrous costs and impacts humanity with physical and mental health consequences. Preventing illness and injury is far cheaper than dealing with the avoidable consequences when people become sick. Modifying climate for the better will improve human health and cut economic costs significantly.

Chapter 7: Price Volatility

Price Volatility

> *"Nothing has been a more reliable indicator for an upcoming recession as the price of Oil. Every major bear market, every major economic decline has been preceded by a large spike in oil prices. The 73-74 recession, [the] recession of beginning 80's and the recession of 2000. Oil prices jumped 80% between 1999 and 2000. Oil prices have been the most important indicator of major economic disasters. Whenever Oil prices rise about 80% from year ago levels, a fair chance does exist that a recession/ bear market will follow."* —Stephen Leeb.

A basic pillar of economics is that a commodity's price will climb as it becomes scarce in supply. We have already seen the effects of price increases on CONG, as demand from developing countries translated into more production capacity at great cost.

Today, though, fossil fuels are in a commodity bear market. All the low- hanging fruit has already been picked, and now oil reserves are only to be found deep in the ocean, in freezing environments or in new rock strata.

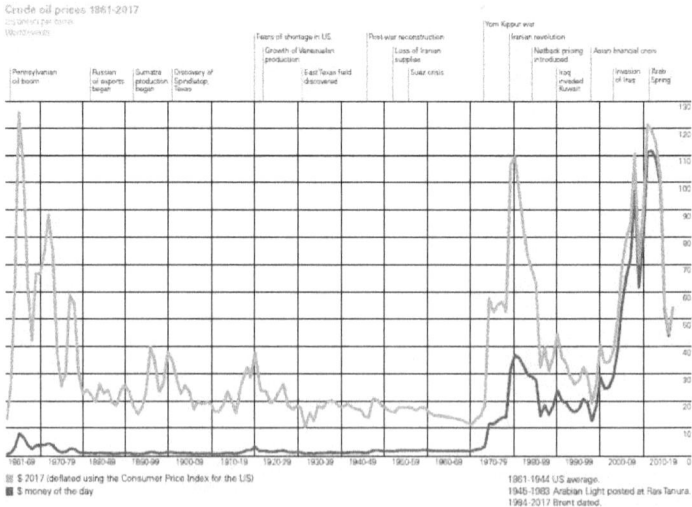

Figure 100: Historical price chart showing the impact of inflation-adjusted and nominal pricing of oil since the start of the industrial revolution. Source: BP Statistical Review of World Energy 2018.

The BP oil catastrophe in the Gulf of Mexico was a testament to the search for new reserves in locations which are anything but easy.

This BP Statistical Review chart, in figure 100, describes the price increase experienced by oil over the last century in the dark green line for nominal and in the light-green line for inflation-adjusted pricing, which shows that oil has been very expensive right at the start of the industry and again after the oil shock that followed the Yom Kippur War and the Iranian Revolution and political disturbances there. Much of this is because the main source of oil for many years and many customers has been the Arabian Peninsula. The price of oil has been through a thorough stress test with the global economic collapse of 2008, when it climbed to $147 per barrel and then, at the depth of the crisis, was seen as low as $35 per barrel. Everybody needs to use energy at about the same minimal rate, even in a weak economy. Children need to be driven to school, commutes need to happen, planes need to fly, and the world needs business as usual, providing a price floor.

The sensitivity of the price of oil, though, has caused a lot of concern to OPEC, which has a significant vulnerability if prices are not kept at a certain level. Additionally, the effort to find new reserves has resulted in higher per barrel costs, meaning that if such reserves are in fact developed, cheaper oil, caused by an economic slowdown, will cause a steep drop in production.

If the growth of alternative energy were to be rapid enough, for example, with the arrival of successful electric vehicles using a non-CONG energy source such as hydrogen from water and sunlight, or simply solar PV electricity from your roof and garden (which is now happening frequently), a situation I believe could easily manifest, then the price of oil might drop, not because the economy is weak but because demand for oil is falling as competitive energy sources penetrate the transportation market. Currently, the oil market is low due to a global commodity bear market combined with overproduction of oil.

Of course, it's not just energy that makes up part of the basket of goods that the inflation rate monitors. Food is in there too. One surprising impact from the evolution of our energy choices was the increase in the price of corn, which started to climb when American ethanol subsidies made it clear to farmers that it was well worth their while to plant more corn but use almost as much for ethanol as for food. This was covered in the renewable energy externalities chapter (chapter 2).

Part of the picture is that when oil prices climb, the economy takes a hit, as though it was suffering greater taxes. This direct relationship between economic growth and the oil price was best illustrated by authors Stephen Leeb and his wife, Donna Leeb, who demonstrated that every American recession was highly correlated to a high oil price. The point here is that as the depleting supplies of CONG are tapped, they become scarcer and are likely to continue to climb in price to reflect more competition to acquire the energy. In actual fact, along with this price increase, renewable energy also becomes more competitive and starts to replace CONG BTU for BTU. Oil prices weakened considerably in 2015, resulting in a potential boon to the United States and other oil-importing economies.

The world market for oil has a new source of supply: North America. Between the Canadian oil sands and the U.S. shale production, the impact has been to reduce the oil price. Additionally, the Saudis have been pumping extra oil, making them the other group pumping over 10 million barrels per day. It's not sustainable. Observers have said that a lower oil price would curb fracking and U.S. supply. Saudi Arabia is a stable Sunni powerhouse in the Middle East and surrounded by hostility. Iran, Iraq, ISIS, Yemen and the Syrian crisis all conspire against it and it needs the oil funds as an insurance. In general, it needs a higher oil price, but a low oil price has impacts it would also welcome. It handicaps its oil-producing antipathetic neighbors and it provides a reason in the West for those against renewable energy to explain to us that renewable energy is not economic.

A low oil price acts as a super stimulus to the economies of the West, like a fall in taxes. As growth returns or maintains, it supports the oil price too. Now the United States has a gas pedal, too, and will produce into strength in pricing and the economy. However, many of the fracking companies that have pushed the United States into the oil exporter role are in the red and cannot go on forever in this stage, and the cost per barrel of fracked oil is high. The moving of sand and water to fracking sites is expensive and fraught with new regulatory friction to prevent earthquakes and drinking water contamination. The price of renewable energy continues to fall. Once it is mature, at lower levels than now, it is likely to sustain at lower levels, with marginal externalities, and with very predictable long-term stability, providing a boon to the economy.

Chapter 8: Pollution

Pollution

"We're not going to be able to burn it all." —
President Barack Obama about fossil fuel reserves
identified in the ground containing carbon which
would enter the atmosphere and transform our
planet.

"To suddenly label CO_2 as a 'pollutant' is a disser-
vice to a gas that has played an enormous role in
the development and sustainability of all life on this
wonderful Earth. Mother Earth has clearly ruled
that CO_2 is not a pollutant." —Dr. Robert Balling,
professor at the school of Geographical Sciences
and Urban Planning at Arizona State University
and director of ASU's Office of Climatology from
1989 to 2004, acknowledged recipient of funds
from Exxon and the Heartland Institute, he backs
CO_2 as a beneficial gas for plants.[372]

For a long time, pollution just meant spilling chemicals, oil or heat,
sound or radioactivity in our air or water or food sources faster than
it dispersed on its own. Significant pollution media events colored the
public's perception of pollution. The book mentioned earlier, *Silent
Spring* by Rachel Carson, which was stimulated by the pollution effects
of DDT, was one such event. Another was the Cuyahoga River, near
Cleveland, which became so polluted in 1969 that when sparks from
the wheels of a passing train reached oil- soaked items floating on the
surface, the river actually caught fire.

Some river! Chocolate-brown, oily, bubbling with
subsurface gases, it oozes rather than flows.[373]

Erin Brockovich, later the subject of a major movie with her name
as the title, filed a successful lawsuit against PG&E for their polluting
release of hexavalent chromium, also known as chromium-6, into the
river at Hinkley, California. These cases concerned what we might call
traditional pollution. Perhaps most important, a large economic litera-

ture has highlighted the significant pollution and congestion costs that burning fossil fuels imposes on American society.

The climate issue is all about CO_2, which even for scientists has not been seen as a traditional pollutant initially, but, given its overall impact when released on its current scale, it is the mother of all pollutants. If fossil fuels bore the full cost that they impose on the economy, then the federal gasoline tax could be quadrupled, and coal could be taxed on the order of 200% [374].

Of course, the desirable impact of such a rise in prices of gasoline and coal would be that desperate efforts to replace gasoline and coal would settle on the increasingly economic and clean hydrogen, clean electricity from hydro, solar, wind, thermal, geothermal and any other of the myriads of clean energy sources and carbon-neutral biofuels. These historic pollution events led directly to the establishment of the Environmental Protection Agency (EPA) and the Clean Air Act. On December 2, 1970, President Richard Nixon's proposal for an agency to protect human health and the environment and enforce congressional regulations to that end became a reality. It has a headquarters in Washington, DC, as well as 10 regional offices and 27 laboratories. The U.S. Clean Air Act defined a pollutant as

> *any air pollution agent or combination of such agents, including any physical, chemical, biological, radioactive (including source material, special nuclear material, and byproduct material) substance or matter which is emitted into or otherwise enters the ambient air.*

This broad definition is helped along by the obligation of the EPA administrator to publish various air pollutants' *"emissions of which, in his judgment, cause or contribute to air pollution which may reasonably be anticipated to endanger public health or welfare."* This is very much not the case with the Trump EPA.

In 2007, the U.S. Supreme Court upheld that the Clean Air Act gives the EPA the authority to regulate exhaust tailpipe emissions of greenhouse gases. Two years later, the EPA issued an endangerment finding. They said,

*greenhouse gases in the atmosphere may rea-
sonably be anticipated both to endanger public
health and to endanger public welfare. The major
assessments by the U.S. Global Climate Research
Program (USGCRP), the Intergovernmental Panel
on Climate Change (IPCC), and the National
Research Council (NRC) serve as the primary scientific basis supporting the Administrator's endangerment finding.*

This meant that greenhouse gases, including CO_2, fit the Clean Air Act's pollutant definition and that the EPA was obligated to regulate them. Humans have increased the levels of CO_2 by 44% in the last 140 years, from levels of 280 ppm to 403 ppm, primarily by the addition of the CO_2 from fossil fuels to the atmosphere, from where it is absorbed in the carbon sinks already operating, the oceans, geology and biomass. The increase in CO_2 has been observed as a proportion of the air, as we have already discussed, and can also be seen using spectroscopy, measuring changes in the electromagnetic spectrum offered by atmospheric gases. Climate scientists have also quantified the amount of warming you ought to see from a given amount of new CO_2 added to the atmosphere from human activities. Observations are matching expectations. When all is fully homogenized, we will see a 2.5°F (1.4°C) increase in temperature from preindustrial times. Additionally, we have measured and seen that the new carbon in the released CO_2 is carbon-12 as opposed to the radioactive version, carbon-14, which gives it a fossil fuel signature and, therefore, a thoroughly human origin as well.

The excess carbon pollution is warming the planet. A reasonable target number as illustrated by Bill McKibben in his July 2012 *Rolling Stone* magazine article[375] as 585 gigatons of CO_2. Burning only this amount would effectively keep the planet below its 3.6°F (2.0°C) temperature rise by midcentury, the target period for a transition to sustainability. The problem is that most people still have to realize that this is a "thing" and react appropriately by becoming more sustainable themselves. It's also the case that the expected oil production from the Canadian tar sands alone will result in exactly that scale of CO_2 production. Equally, the viscous tar sands in Venezuela also have this much pollution. If only one of these can produce, which should it be? All the other oil production in the world dwarfs this quantity of CO_2 in any case, so something has to break in the next decade or so. Luckily, renewable energy is beginning to have an impact on the amount of CO_2 released as its capacity replaces CONG electricity capacity and as we start to drive electric vehicles and as we replace our lightbulbs with LEDs and use energy-frugal fridges, and fuel cells and explore hydrogen production using renewable energy and water. Also, replacing oil and gas heating systems with air-sourced heat pumps with the new technology offering massive efficiency (Coefficient of Performance [COP] 6) and replacing fossil fuels with ag-waste cellulose-based carbon-neutral biofuels.

The story of fossil fuel leakages has countless entries for major pollution events, including the better-known events such as the Torrey Canyon oil tanker in Europe, the *Exxon Valdez* oil tanker spill, the spills from flaming Iraqi oil wells at the end of the first Iraq war, the BP Deepwater Horizon Gulf of Mexico oil leak, and the California, the Aliso Canyon or Porter Ranch natural gas leak, the worst gas leak in U.S. history and larger than the

BP Gulf of Mexico leak. The leak vented natural gas at the rate of 1.6 million pounds per day, the same as the emissions from 4.5 million cars a day, and this is likely to be the case from October 2015 to March 2016, with a potential total of 292 million pounds. In fact, we learn with natural gas that the methane, which is 21 times more of a greenhouse gas than CO_2, leaks all the time. Another leak from New Mexico that claimed to be even larger is the San Juan basin coalbed methane leak[376] whose existence was only revealed after satellites equipped with special cameras spotted the emissions. This leads us to believe that natural gas is a relatively uncontrolled gas and might have a much larger role to play in greenhouse gas pollution than previously expected. The industry tells us that methane leaks at a rate of 3% of the gas eventually sold, which is bad enough. Where gas is plentiful and cannot be piped to consumers, it is often burned in flares to get rid of it, generating CO2, far preferable to simply allowing the gas itself, methane or CH4, to leak. In fact, the percentage of leaked methane is about 10%, which is a serious amount of lost revenues and greenhouse gas to be adding to the atmosphere.

Figure 101: The relatively common site of an oiled bird after an oil spill, but this obvious damage is the tip of the iceberg when it comes to the wider ongoing impact of CONG on the ecosystem. Source: Pauk. Black Sea Oil Spill.12 November 2007.

This is probably a good point to mention again the damage to the ecosystem that is being wrought by CONG impacts. In chapter 2 we mentioned the impacts of renewable energy on birds being roasted by solar flux or chopped by wind turbine blades. It turned out that the numbers were miniscule and didn't even show up on a chart that compared death causes such as cats, building windows, moving vehicles, and so forth. It has to be remembered that the impact of CONG on ecosystems is much more insidious.

Physically, oil on feathers changes the insulation of the bird, making it vulnerable to the cold. The extra weight destroys the aerodynamic character of the wings and the bird can no longer evade predators, if they want an oily lunch. As it preens, its bill picks up oil, which it then ingests, affecting its internal organs with dehydration and metabolic imbalances. It will likely die if the oil cannot be removed by humans very early on.

It's not just the image of a bird black with oil all over its feathers but the impact of chemical pollution on the life cycle of birds that ends up altering fertility, chick growth and egg numbers, and this is just birds. Amphibious creatures have been through a terrible phase and act as the canary in the coal mine in terms of sensitivity to change in the environment, and they have higher barriers to survival in a polluted world. It's not the large animals, either, that are the key part of the die-off.

It's the tiny creatures in the marine environment, the source of life for the ecosystem, that are impacted significantly. Larval fish, plankton, seaweed, mussels, oysters, dolphins, whales, sea lions, sharks, turtles, algae and fish are all immensely impacted by oil spills. Oil on the water surface prevents sunlight from reaching photosynthesizing organisms, decreasing the amount of flora food available and not even by chemical means! Oxygen in the water decreases, resulting in a disruption in faunal populations.

Oil spills are temporally and geographically local in the damage they inflict, though. Even if we don't talk about individual oil spills, the background chemical impact on the environment of the wider human economy, which currently runs on oil and gas and coal, produces significant ongoing effects that dwarf individual oil spills. Waste and plastics at the other end of the consumer cycle also have a huge effect.

The Pacific and Atlantic gyres have been in the news for about a decade now but are not diminishing. Photographs of magnificent seabird carcasses on remote Pacific islands with brightly colored particles of plastic where their stomachs once were, suggesting that these morsels appeared like food to the birds but were deadly and then physically contributed to the bird's death. Ongoing pollution from CONG and its products and subsequent climate change can be blamed for effectively potentially destroying the ecosystem for 28 species of bird in North America alone.

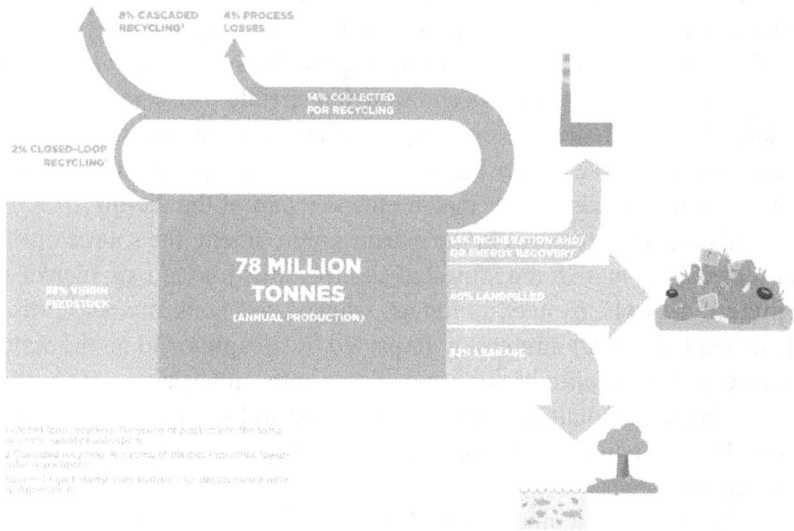

Figure 102: A diagram extracted from a report written by the Ellen MacArthur Foundation on the life cycle of plastics. Source: Ellen MacArthur Foundation.

In the very big picture we are speaking of the first steps in the sixth extinction,[377] the fastest extinction event in known history, which includes five other such extinction events of global species destruction. Bird mortality from fossil fuel pollution and climate change represents a far higher risk than that posed by any renewable energy solutions.

Science magazine published a study called "Methane Leaks from North American Natural Gas Systems"[378] that looked at over 200 earlier studies and concluded that natural gas leakage rates were 5.4%. More recent observations by the satellite have shown 9–10% leaks in methane from east Texas and North Dakota oil and gas basins. There was a hope that accessing nonconventional resources and using less coal and more natural gas for power generation would be a way to reduce emissions, but these observations put paid to that hope.

This actually means that fracking is speeding up pollution and therefore climate change, and this on the basis of the methane leaks alone without taking into account the emissions from burning the oil and gas produced. A key point of the study was that if natural gas electricity production was leaking at 5.4%, it would take 50 years before there was a net benefit.

If the data from the satellites is true, however, it will be more than a century before any benefit kicks in. When you consider that much of the new natural gas generating capacity is also replacing nuclear power, renewable energy and energy efficiency, there is a strong reason to suspect that climate change is being accelerated by the move to natural gas in the United States. When you consider that fracking also impacts drinking water supplies, there is a reason to stop spending billions on it and start supporting the energy capacity that does no harm (as far as we know).

To be fair, because CONG fossil fuels are in the Earth's crust, occasionally that crust will break open and oil and gas will be released by nature itself. The BP Gulf of Mexico oil spill revealed that Mother Earth is already prepared for the danger if the spill is in a tropical location. Deepwater Horizon's oil spill was digested by hungry microbes turning the oil into energy, water and CO_2. However, in Prince William Sound, where the temperatures are much lower, below the living range for these microbes, digestion was much slower, and evidence of the spill is still visible after decades.

Another index of how bad CONG leaks can be is the size of the fine paid to the EPA for a violation. Atlanta-based Colonial Pipeline pumps some of the largest volumes of refined petroleum products in the world and serves 50 million people on the East Coast of the United States. It has settled with the federal government over violations of the Clean Water Act, with oil spills in five states in the 1990s. In 2003, it was obliged to pay $34 million[379] as a civil penalty, the largest EPA fine in history, for a 1.4-million-gallon pipeline oil spill. In mid-September 2016, the company was responsible for a quarter of a million gallons of gasoline being leaked into rural Georgia.

If you want to mix the worst of both worlds, you need do no more than look at cold Russia, where there are constant pipeline leaks from old Cold War pipeline equipment, 60% of which has deteriorated and that has not been updated. An article[380] by Gleb Paikachev in the *Guardian* on August 5, 2016, spelled out how the northern Russian province of the Komi Republic is a beautiful remote spot with lakes and rivers, where the sun never sets, but also home to the Usinsk oilfields operated by Lukoil. Oil production from this spot dates from 1745 but was majorly developed in the 1960s and 1970s. In 2014, it produced 133 million barrels of oil, and plans exist to increase output to 182 million barrels by 2019. Dripping oil pumps, gas leaks and huge ponds of black sludge leave their mark in dying trees and undergrowth.

Individually small, they accumulate into a large part of the 1.5 million tons of oil spilled in Russia annually. This is more than twice the size of the spill from BP's Deepwater Horizon in 2010. For Russia, oil and gas amount to about half if its GDP, but there is a huge cost to the environment too. In 2014, Greenpeace recorded 11,709 oil pipeline breaks in a period when Canada, for example, had only 5 incidents. Employment in the region is very scarce, and the prevailing opinion about oil is that they don't want the business to depart, but they do want to be able to live in their home without pollution's threat to their health. Russia helps its oil industry greatly with subsidies for equipment but also turns a blind eye to the need to clean up oil spills too. The cost of drilling everywhere must reflect the value of a pristine environment. This is hardly achieved anywhere, but humankind is waking up to the challenge.

In the modern world, China has a reputation for building and operating lots of coal-fired power stations. China, though, is pressing its economy to make itself more energy efficient, clean up pollution and reduce the quantity of carbon emitted into the atmosphere. In 2015, China's carbon emissions were estimated to have fallen for the first time in 17 years, and 2016 was the third year in a row that coal consumption fell, with a goal to hit peak coal consumption by 2029. Slowing economic activity helps, but wind and solar are increasing their impact steadily. In 2016, China ordered a three-year suspension in the building of new coal-fired power plants. For some, this has almost the same quality of unlikeliness as the falling of the Iron Curtain in 1989, but a close read of the 13th Chinese 5-year economic plan[381] shows that there is plenty of intent to clean up their act and install renewable energy. Specifically, they lay out a plan to reduce emissions per unit of GDP by 40–45% by 2020 compared to 2005 levels. They also want to increase the share of renewable energy to 15% by 2020. In the case of a centrally planned government, a plan can be enacted efficiently, such as building a straight railway line between two cities for a fast train or directing scientists to develop health cures based on stem cells. No resistance is permitted. In democratic nations, however, this is not quite so straightforward.

The Ellen MacArthur Foundation has produced a study, "The New Plastics Economy: Rethinking the Future of Plastics,"[382] on the full life cycle of one of the major products of oil: plastics. This is important because about 6% of oil production is used to manufacture eight different forms of plastic.

As shown in the very revealing chart in figure 102, 86% of all the oil that is made into plastics is not recycled and forms part of a one-way system from ground to refinery, to plastic feedstocks, to packaging and products, and then 86% goes to incineration, landfills and, worse, into the wider environment as leakage. This is the part that is now ending up in the large oceanic gyres, which have appeared without apparent solution. The MacArthur Foundation champions a "closed-loop" economy, where feedstocks can be recycled endlessly, causing a huge improvement in our optimization of these resources. Currently, only 2% of the plastic that is recycled is put back into the manufacturing stage, and the remainder is not used again. The one-way economy with feedstocks from a nonrenewable resource, it goes without saying, is a nonsustainable attribute of our current economy that has external costs we need to address. Our generation is the one that is succeeding with solutions.

The extent of the penetration of waste plastic into our environment is astounding. One of the major components of this plastic pollution is a thing called a *nurdle*, of which 27 million tons are manufactured annually in the USA. Plastic manufacturers transport plastic resins from factory to factory in the form of huge bags full of nurdles, or resin pellets, that are less than 5 millimeters in diameter. These are also referred to, poignantly, as "mermaid's tears." A 2001 study[383] of Orange County, California, beaches found that 98% of the plastic debris found was composed of these nurdles. As if human predation on fish populations was not enough, especially with its toll of bycatch with dolphins, whales, sharks, turtles and sea lions killed by being caught by nets underwater and drowned. Plastic particles in the ocean interrupt life cycles. Physically, they choke animals even as small as krill, are mistaken as eggs for food, clog the digestive tracts of sea life and emit chemicals harmful to creatures' metabolism such as the ones that we humans are constantly telling ourselves not to consume in plastic containers, such as phthalates, bisphenol, polychlorinated biphenyls (PCBs) and dichlorodiphenyldichloroethylene (DDE).

A group named the 5 Gyres Institute[384] put out a study after a research voyage to every ocean in which they have found plastic everywhere they went. Every surface sample they took on a 2,500-nautical-mile trip across the North Atlantic from Bermuda to Iceland, for example, contained microplastic particles. They have completed similar trips across the five subtropical gyres, or "garbage patches." They have also found plastic on ocean floors. A previous expedition found that plastic "microbeads" the size of sand grains, which are an ingredient in cosmetic exfoliation products, also end up in the ocean. Success followed this campaign because certain states, as of January 23, 2016, led by Illinois and including California, Colorado, Connecticut, Iowa, Massachusetts, New Jersey and Wisconsin, have banned the production and sale of products containing microbeads. The Netherlands is the first country to announce that it will be free of microbeads. In many cases, biodegradable microbeads are still permitted.

We are innovating ourselves out of using petroleum as a feedstock for plastics and moving inexorably toward replacement polymers that can stand in for all eight forms of plastic. Some work really well and then biodegrade once they have been used. Instead of remaining in the environment for hundreds or thousands of years, they get wet and then have just months before they degrade into water and CO_2. The problem of environmental plastics is huge, but scientists are working on solutions, and like many others, the best working solutions seem obvious after the fact.

Boyan Slat,[385] a Dutch engineering student, saw more plastic than fish when on a diving trip to Greece in 2011. He came up with a design to effectively collect the plastic that is currently lost in the oceans when he was only 17 and won several awards. He reasoned that because it's difficult to go out and actively pick it all up, and the plastic is buoyant, and the oceans are moving, you can place a large collecting boom in a single spot and collect it all. Within a decade, this passive cleanup mechanism can clean up 50% of the damage. He put the idea on a crowdfunding campaign and within 100 days, 38,000 investors provided $2 million. This sent the team to the Pacific and Atlantic gyres, where they collected evidence and tested smaller prototypes of the design. The main idea is to have 100-km-long collecting booms that extend 3 meters into the top of the water where the majority of the plastic is situated. Periodically, about every month and a half, the accumulated plastic can be easily gathered and put into containers. It turns out this is an ideal feedstock for a plastics-to-oil plant, which can be conveniently based where the plastic is collected and where there is a market for the oil. There is enough plastic for them to collect 50 container loads a day, which has a huge value as fuel. This also helps to pay for the boom equipment and on paper at least renders the project profitable. There has been a lot of criticism, but Boyan is the "man in the arena" actively doing something about it while the armchair critics have contributed to the problem but have no better plan.

Chapter 9: Subsidies

Subsidies

"I would do away with these incentives that we give to wind and solar. I'd let them stand on their own and compete against coal and natural gas and other sources and let utilities make real-time market decisions on those types of things as opposed to being propped up by tax incentives and other types of credit that occur, both in the federal level and state level." —EPA administrator Scott Pruitt, in October 2017 at a Kentucky Farm Bureau event, inconceivably forgetting to mention the much higher subsidies already provided for CONG resources.

In this chapter, I look at four studies done in the last decade that discuss the government support for energy of all types and how renewable energy does in this competitive field. First, we look at the Brookings Report to highlight the importance of subsidies as an issue. Another report by DBL Investors, *What Would Jefferson Do?* explores the history of U.S. government subsidies for energy during the initial 15 years of that energy regime. It quantifies how the current federal support for renewable energy relates to previous energy support. I also look at a report prepared for the nuclear industry by Management Information Services Inc. (MISI), a Washington, D.C.–based research group that looks at subsidies offered over the period of 1990–2010. Then I look at another examination of the subject prepared by the Environmental Law Institute in 2009 for the years 2002–2008.

The upshot? They all show that renewable energy gets no respect.

Finally, just for flavor, we take a look at a study done to determine the cost of keeping the oil supply lines open. We must bear in mind that only a fraction of America's oil comes from the Middle East. This number is high and intimidating but paid anyway.

Subsidies are an important tool to provide certainty for economic actors in a strategically important industry. Subsidies remove barriers to product availability. In wartime, it's great for a government to subsidize armaments, energy and food supplies. In peacetime, those three industries are still strategic, but the pressure is always on energy to be reliable and available. Subsidies there make a real difference.

We all occasionally spoil ourselves with an expensive car or new phone, a camera, pair of shoes, a tie or a suit, to feel good about ourselves.

A country spending a bit more to ensure a supply of a strategic good is only doing the same. Anybody who's held military- or aviation-grade technology in their hands knows that a premium is already paid for quality design and reliability when lives and national strategic interests are on the line. Subsidies "are the expensive shoes in the wardrobe" and the extra money the government will spend to protect the economy.

Today those strategic interests have changed, and fossil fuels no longer represent the same level of strategic importance. In a peace-time world, something much more desirable is happening in energy that will hopefully push humanity up yet another notch in living standards and make the economy that much more resilient. In April 2012, the Breakthrough Institute,[386] with the Brookings Institution and the World Resources Institute, released *Beyond Boom and Bust*,[387] a report illustrating the need to reform cleantech subsidies to drive innovation, improve performance and reduce costs. The *New York Times* editorial board endorsed the report's recommendations, and Breakthrough staff gave Senate testimony on the findings.

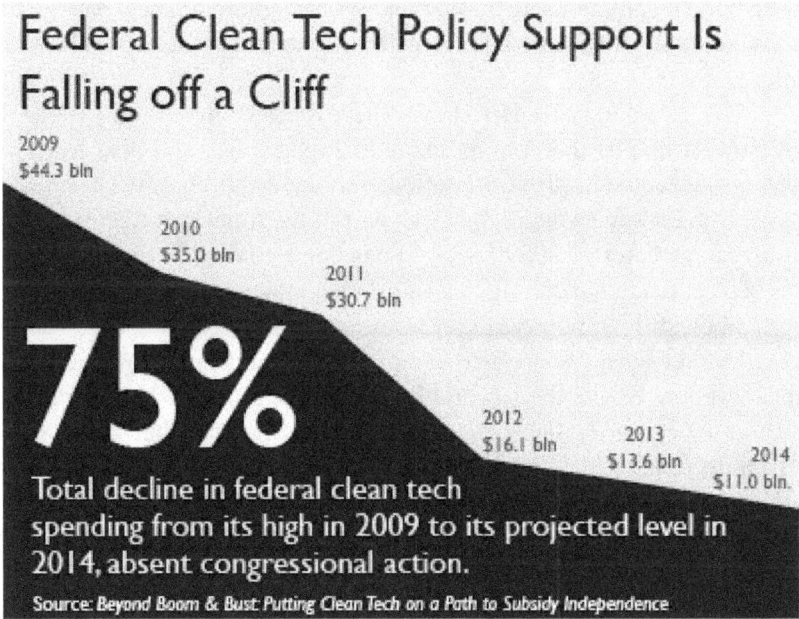

Federal Clean Tech Policy Support Is Falling off a Cliff

2009
$44.3 bln

2010
$35.0 bln

2011
$30.7 bln

75%

2012
$16.1 bln

2013
$13.6 bln

2014
$11.0 bln

Total decline in federal clean tech spending from its high in 2009 to its projected level in 2014, absent congressional action.

Source: *Beyond Boom & Bust: Putting Clean Tech on a Path to Subsidy Independence*

Figure 103: Policy support falling off a cliff since 2009. Source: Breakthrough Institute and Brookings Institution and World Resources Institute.

There is no integrated energy policy or effort to reform it in the United States. All changes are effectively being driven by private innovation and entrepreneurs and the available public support in the form of tax credits, grants and loan guarantees. All cleantech installations in the United States and elsewhere in the world are decreasingly dependent on supportive government policies, but these do not reflect the fact that cleantech cleans up so many other global problems that it is more than a strategic necessity.

This subsidy externality chapter underlines the fact that fossil fuels still obtain more subsidies than renewable energy, but the group interest is in having only cleantech as quickly as possible. Politics get in the way. Many of the existing U.S. renewable energy subsidies work for just a short time before expiring or receive a short-term extension, as has happened with the investment tax credit (ITC). These irregularities make it a hardly credible system that actually prevents the private entrepreneur from making long- term plans. Wind installations in the United States fell off a cliff in 2013 due to the end of the PTC subsidy. If this subsidy had offered a 10- or even 20- year scope, there would have been many more wind installations. In 2015, there was an optimistic new dawn with a 5-year extension, until 2022, of the investment tax credit (ITC) for solar and wind.

The Brookings Report shows the expected impact on cleantech investment due to the coming changes to subsidies and also looks at reforms that could improve the outlook for renewable energy. The heightened clean energy spending that was part of the American Recovery and Reinvestment Act of 2009 (ARRA) has come to a close. Now the ITC has been extended, resulting in a significant maintenance of federal support, albeit still temporary.

Market subsidies are also being cut in several European countries, and Germany, Spain, the UK and Italy all come to mind. The staggering success of countries like Germany to co-opt the electricity subscriber citizen to pay for the feed-in tariffs made Germany the world's leading solar and wind power per capita in a short amount of time and gave the country a huge relief from finding resources for 30% of its energy capacity. The German support succeeded due to a solid 20-year support period. This is a great opportunity for subsidy reform and finding a way to avoid the boom and bust cycle that often accompanies renewable energy subsidies and for optimizing use of public funds to provide long-term competitiveness. In the United States, the federal government has driven support for fossil fuel energy and innovation with subsidies for over 200 years.

Background

Economies use technology to turn feedstocks into finished products. This technology experiences a life cycle. Initially, a visionary sees how the economy could be improved with the technology. Then the innovator draws on personal resources to prove that this is the case. Eventually, a prototype is produced at a cost that dwarfs the incumbent technology, often inducing undue and premature condemnation. If a government sees that the technology is useful for strategic reasons, that is, it can be weaponized, then there is likely to be early subsidies. Every technology deserves support initially. CONG technologies got it and now clean energy needs it. Such a transition is complicated when the incumbent energy resources are financially golden and sitting on cash flows that are historically fabulous. You would expect exactly what you see in the real world: misinformation campaigns, political pressure and calls to end subsidies to the new competitors. A research report by OCI shows that during the U.S. election cycle of 2015–16, CONG companies spent a total of $354 million in campaign contributions and received an extra $29.4 billion in federal subsidies over those same years, an 8,200% return on investment.

Ironically, subsidies reduce the price of energy for the consumer. Arguments about subsidies talk about the absolute amounts of money spent in this effort but fail in general to include the extra cost of externalities. If you include the cost of the externalities of coal, oil, nuclear and gas energy in the total cost, then the CONG "cost of energy" is unsupportably high. We need warmth, cooling, work, sound, vision and mobility from energy and we worry about the cost of these services. CONG currently provides most of these services. CONG energy resources were adopted because they were plentiful, convenient, easy to transport and very dense forms of energy. Over time, CONG's negatives have quietly erupted. I say *quietly* because the extent of the damage is little known. I say *erupted* because the impact is becoming critical. I am suggesting that our general will to become sustainable would be transformed and increase significantly if we realized the extent to which CONG is hurting us today. The question is, are renewable energy sources better (cheaper, more effective, with fewer consequences) overall for us than CONG and if they are, should we be providing more public subsidies to make that happen or will it happen as a result of private activity anyway?

This book argues that the combination of externalities visited upon us by CONG means that any subsidies we also pay to CONG companies is compounding the damage. Conversely, the externalities of renewable energy are relatively so low that supporting sustainable ways of obtaining energy will expedite efficiencies, economies, savings, sustainability, health, pristine environments and practicality.

Subsidies ensure that incentives exist to promote a reliable, continuous supply of energy to the economy. Maintaining the economy is a fundamental security principle. Losing energy, even temporarily, for whatever reason, invites the prospect of dire economic dislocations followed by the Malthusian discomforts of hunger, a breakdown in law and order, predation, disease and ultimately a return to a potentially more barbaric existence. To avoid this, we all have an interest in maintaining our energy, food and water supply. Normally, this would be the domain of a government that stays on top of intelligence and creates the conditions necessary for resource security to thrive. I believe that our current CONG energy supply is fraught with dangers to our economy that make it progressively and insupportably expensive. We need to transition to sustainable energy as quickly as possible to avoid worse externalities and looming events called tipping points, from which there is no return to a world we call normal.

Renewable energy not only has lower externalities but also simply cancels out externalities from fossil fuels. In the discussion about the climate, for example, there is no better response to greenhouse gas emissions than to simply cancel them by replacing CONG, kilowatt for kilowatt, with some of the sustainable tried and tested technology choices until there is no more negative impact. The result of this would be a higher quality of life and greater quantity of energy available. When the energy source is effectively infinite, it is almost impossible to "waste" it.

The answer to the question, "Should subsidies be preferentially given to sustainable energy resources?" is unquestionably yes, and to a large extent this is already happening. Support for renewable energy still remains well below support for CONG. Bearing in mind the CONG subsidies, the answer to the question, "Should renewable energy compete on a level playing field?" is certainly yes, when all expenses, external and internal, are considered. In this circumstance, there is no question in my mind that we would accelerate the deployment of sustainable solutions, and this, of course, is what is actually happening globally. Let's hope it's fast enough!

Predictably, and for economic and political reasons, subsidy arguments support both sides. Competing energy sources do not enjoy an even playing field. Energy markets are anything but equally competitive. Costs of energy have been underwritten by global governments to different levels and both governments and private industry market their book and distort the picture and our perceptions of the issues. This is part of the complexity of the situation. The extent of the added value of renewable energy is still shrouded in obfuscation by incumbent CONG players and not understood by government, meaning there is no rational strategic approach or policy.

In a demonstration of how different voices cloud this issue, a November 2013 *Forbes* article, "Renewables Get 25 Times the Subsidy That Fossil Fuels Do," made the point that, globally, CONG subsidies are paid by many governments who are anxious not to help the energy companies but to help their citizens. There are 12 countries on this list, that, according to the International Energy Agency (IEA), spend $523 billion on CONG, 75% of the global total, versus the $88 billion spent globally to support renewable energy, including support from countries in the Middle East that are concerned with internal political security. In the West, Venezuela spends twice as much subsidizing gasoline as it does on health care.

On the other side of the argument, Bloomberg New Energy Finance summarizes the subject in an earlier July 2010 article called "Fossil Fuel Subsidies Are Twelve Times Renewables' Support," which focuses on a similar gross global subsidy figure from the IEA ($557 billion in 2008). When we stand back from the political fray, it's clear that something desirable, like clean energy, deserves to be supported to help bring down initial costs. The declines in cost per kilowatt of equipment have been significant and very helpful, whereas the CONG businesses are mature and very unlikely to provide similar efficiency and cost improvements.

In peace and war, energy is a strategic commodity. A guaranteed supply is required to run a modern economy. Underdeveloped economies frequently have power outages or no power at all. Billions of human beings live lives without the ability to cool food, recharge a cell phone, run a computer or read after sunset.

Whether this is a good thing or not, we are so set on the stimulation offered by modern goods and services brought to us by CONG that every part of our economy needs a guaranteed supply of power. In 1979, in Nigeria, the Nigerian Electricity Power Authority (NEPA) was unable to offer 100% generation. As a consequence, many services that depend on electrical power were cut off, along with GDP growth. The NEPA acronym was altered humorously to "No Electrical Power Available."

Our GDP growth depends on a continuous supply so that myriads of processes, manufacturing, transportation and communications can all help businesses run flawlessly and on time at any time. Ensuring this supply of energy from CONG means identifying resources and putting in place the entire chain of extracting, refining, combusting, distributing and delivering in such a way as to provide what amounts to a guarantee that this chain will not be broken. After well over a century of using increasing amounts of CONG, and the mastery of the logistics in developed nations to supply this 100% energy service, the volumes of resource used, and the risks are now so great that this guarantee can no longer be deemed reliable.

Today, the perception about renewable energy is that it exists solely on government subsidies at the current time and that it is uneconomic, lacks energy density and is a great white hope for the future. The question, "When are renew-

Interestingly, the use of sperm whale oil resulted in the derivation of the first unit of energy in 1860, when the Metropolitan Gas Act was passed in the British Parliament to regulate the operation of gas supply companies in London. The unit was the candlepower, which described the light emitted from a standardized whale spermaceti wax candle weighing one sixth of a pound and burning at a rate of 120 grains per hour. It was replaced in 1948 by the candela. Originally, one candlepower was equal to 0.981 candela, but today it is deemed to be effectively 1:1. Judging the number of candelas from any light source was very subjective at first, as they adjusted the distance of a comparison light source until it had the same subjective brightness and, using the inverse square law and the distance, were able to infer the relative amount of candela of the source. One candela is equivalent to today's 12.6 lumens. A common 60-watt incandescent bulb emits 800 lumens.

ables going to be profitable without government help?" appears to be a defensible, normal interrogation. The political divide is huge, and the conversation is complex enough that it's ripe for propaganda. Perception dominates in a voting world, and it's very easy to take a position that backs fossil fuels to the hilt and makes renewable energy look bad. I am hoping that this book will help to redress this impact and have an educational value for intellectually honest readers.

In coal mining, owners of mines can reclassify income normally subject to income tax as royalties, subject to a reduced tax rate. This measure was originally enacted during the Korean War when the U.S. Government was desirous of encouraging additional coal production and to that end they allowed a low tax rate which today still stands at just 15%.

Economic expansion has gone hand in hand with the initiation of new types of energy. Subsidization has encouraged all of these. In most developed countries, there has been a slow transition from wood and small mechanical hydro for milling and sawing. Then railways were built, and the industrial revolution happened, fostering the transition to coal. Use of oil was preceded by smelly whale oil. Vegetable oils, like peanut oil, powered Otto Diesel's compression ignition engine and meant that the whales were safer. Demand then grew for kerosene, which was refined from mineral oil and progressively to gasoline. Large hydroelectric projects emerged during the Depression, and nuclear power appeared after the Second World War.

Since the 1970s, at the first oil shock, renewable energy resources have begun to be used in increasing amounts. Attitudes have evolved alongside. Initially, there was a sort of new age wishful thinking that tied in with a 1960s-era liberalism. This then changed to cynicism as early mathematics suggested that renewable energy would always be too expensive and not conveniently energy dense. Today, cost is being resolved and energy density is addressed by scaling. In fact, the absence of a feedstock fuel in sustainable energy systems heralds the birth of resilience in the economy. This development underscores the emergence from the brief Anthropocene, the period of a couple of hundred years where CONG both accelerated us to a new quality of life and also became too dangerous to continue using, to a more mature, sustainable era. If this is true, then reaching this next stage will be to see many environmental challenges literally evaporate. We will be better able to focus on more quality-of-life issues, such as housing, healthcare, transportation and communications infrastructure.

One example close to home is the extent to which the Argentinian continental shelf is likely to be a repository for huge resources of oil and natural gas in the same way that the Gulf of Mexico and Brazilian offshore resources have been. Thirty years after the Falklands War, Cristina Fernandez de Kirchner, the Argentine president, nationalized YPF, the Argentinian oil company, in an emergency decree, releasing it from parent Spanish Repsol. Also available for exploration in view of the success of fracking for natural gas in the United States are the Vaca Muerta shale fields, but few are interested because of the dearth of incentives available from the Argentine government. British exploration of the Falklands, which aroused Argentinian indignation, occurred precisely because of British subsidies to the exploration companies. Few remember that a deal was struck shortly after the Falklands War to share 55% of the profit of every barrel of oil extracted from Falklands' waters with the Argentinians.

The correct approach of the Argentinians, given this situation, would have been to incentivize their own people and work with or without the British to locate every barrel of carbon fuel possible, especially while they could still be a boon to the Argentine economy. In view of the harsh effects of carbon on the climate, the failure by Argentina to develop its carbon fossil fuel resources is perhaps an unintended blessing. Failing to develop carbon resources and capping those in production may actually be necessary if we are rushing to avoid climate impacts.

In colonial days, British ships would come to American ports with free coal in their hulls. The ships used it as ballast to keep them stable in the ocean. A 10% tariff was placed on coal in early 1789 to subsidize U.S. local production and was rescinded in 1842. Government protection for local work and production was an economic reality. When coal was discovered in Pennsylvania, the state exempted it from taxation and incentivized its consumption. This practice spread across the United States, and by 1837, 14 U.S. states had commissioned geological surveys that mapped out mineral resources and cut the expense of exploration significantly.

Today we see this being emulated somewhat for renewables when some countries like the UK offer port facilities to ocean power producers testing prototypes for connection to the grid. In the United States, the

Interior Department identified sites for large-scale solar power installations on public lands, some of which have gone ahead.

The timber industry was aided by federal land grants, which also helped the railroads. The expansion of the rail network caused an exponential increase in the demand for coal and other goods. It was so successful that the price of coal soared, and this resulted in many more market participants. States granted special rights allowing vertical integration and further coal production.

U.S. energy subsidies derived from various sources:

- The tax code: Exemptions, allowances, deductions and credits
- Regulation: Federal mandates and controls enable so-called dangerous (nuclear) energy sources and pricing
- R&D: Funding for research and demonstration programs
- Market activity: Direct action on the marketplace
- Government services: Provided without a direct charge, that is, provision of ports and inland waterways as free public highways, deepening channels for large ships. Use of the military to protect shipping lanes and markets
- Disbursements: Direct grants such as those to construct and operate oil tankers

Over the decades of the twentieth century and often due to wartime activity, much infrastructure was installed in the United States that has subsequently become a key boost to fossil energy companies. The Big Inch and Little Inch oil pipelines were constructed during the Second World War in an effort to bring crude oil to the East Coast and avoid the risk of German submarines. Today those pipes bring natural gas east and represent a considerable savings to the industry.

The U.S. Defense Department spent almost half a billion dollars a year developing gas turbines that were taken from aircraft design and made to work on the ground to generate electricity. Billions were spent between the 1970s and 1980s as utilities sold off generating capacity and simply charged for the wiring they had installed. Energy service companies (ESCOs) sprang up, and many began to use gas turbines to provide electricity.

NASA needed solar power for its satellites, so almost $1 billion was spent from 1950 to 2006 on this technology, and although this is a much smaller amount than the sum spent on the turbines, it was critical in the pathway to eventual solar panel commercialization. A similar pathway was enjoyed by fuel cells.

The extent of the "subsidies" of all sorts that are enjoyed by the energy community is significant and the difficulty is drawing the line. These numbers reflecting Department of Energy funding are really only the R&D funding part of the equation. The considerable sums spent from other sources do not figure highly in this. The DBL Partners' research augmented the DOE numbers by adding in the many tax subsidies, like the percentage depletion provision. Just as drilling for oil and gas was accelerating in the United States, the Intangible Drilling Costs (IDC) deduction was introduced for recovering capital costs on dry wells, permitting the IDCs to be fully deducted in the first year. Then there is the percentage depletion allowance (PDA). Coal, oil and gas get used up over time as they are mined or drilled. The capital equipment and land rights needed to extract them can be repaid, a bit like depreciation, by allowing for depletion of the resource over its lifetime. In the United States, the IRS code (IRS Publication 53) has a "depletion allowance" permitting an owner to account for the depletion of reserves as product is sold. There are two approaches and tax payers can choose the option with the larger deduction: "Cost Depletion" allows the owner to recover up to the total capital investment by deducting a part of the capex over the life of the project. "Percentage Depletion Allowance" is 15% of revenues based on the average daily production of the coal, oil or gas, up to the total quantity of coal, oil or gas in the ground. This method can allow cumulative deductions that are larger than the total capex of the project. Imagine if a solar farm was able to deduct from its taxable income 15% of the revenues of the electricity it was selling. You can't deduct more than 100% of your taxable income, or 65% of your income from all sources, although you can accumulate any excess for future tax returns. Effective tax rates prior to 1986 on other industries averaged about 28% compared to rates on oil companies, which range from 6% to 24% under pre-1986 law.

In 2010, a group called the Green Scissors Campaign, representing interested entities including Friends of the Earth, produced a report[388] in which they characterized the start of 1932 as the start of coal subsidies in the form of lower tax rates to recover initial capital investment. This chapter has already explained that these subsidies go back much further than that. Renewable energy is competing against an energy incumbent,

CONG, that has already depreciated all its assets and is mature and at the top of its learning curve.

The Grantham Research Institute report,[389] *Unburnable Carbon 2013: Wasted Capital and Stranded Assets*, was written in partnership with Lord Nicholas Stern, the chairman of the Grantham Research Institute in the UK. It recognizes that operators in the financial markets have a very low expectation that government efforts to reduce greenhouse gases will be effective. This dysfunctionality reflects the responsibility right back to the private sector. The report recognizes the huge stakes that, for example, pension funds have in maintaining their asset values in an environmental scenario that currently looks very high risk.

Some groups are opposed to subsidies for renewable energy, and as expected they are constituted by the mature CONG industries. The CEOs of fossil fuel and utility companies such as Eni in Italy, GDF Suez in France and E.ON in Germany have called for a halt to renewable energy subsidies, but this can, of course, be disqualified due to their bias. For utilities, integrating intermittent energy with grid distribution is a significant challenge. Much renewable energy is wasted along with CONG energy as backup by keeping spinning reserves running just in case the wind dies suddenly.

Subsidies in the Initial 15 Years of an Industrial Growth Cycle

There is a point at the start of a new era of technological adoption when the new technology is more expensive and yet promising. At this point, subsidies are provided to test and implement the technology and reduce the cost until it becomes competitive at which point commercialization brings it to the mass market. Wood, coal, gas, nuclear all had support at the start. Much of that support became permanent in the tax code.

Throughout the world, renewable energy is benefiting from a cornucopia of tax breaks, feed-in tariffs, grant programs, and other support for the purchase and installation of different types of renewable energy. This situation makes it appear that renewable energy only exists because of government subsidies of one sort or another. Nancy Pfund of DBL Investors and Ben Healey wrote a study[390] on the support by government of new energy regimes in the early years of their transition and they confirm that this perception is 180 degrees distorted.

In mid-February 2013, I attended the launch of the Bloomberg New Energy Fact Book in Bloomberg's gorgeous New York building on 59th Street next to Bloomingdales. An audience question came up, set up as a contention between *Wall Street Journal* attitudes and the *New York Times* point of view, which is more permissive to renewables: "Why do renewables get such large subsidies?" This is tangibly emulated by the financial media such as CNBC and of course Fox News, where it's just mistakenly and egregiously accepted that renewable energy is forever more expensive than fossil fuels and will remain uneconomical.

Historical Average of Annual Energy Subsidies:
A Century of Federal Support

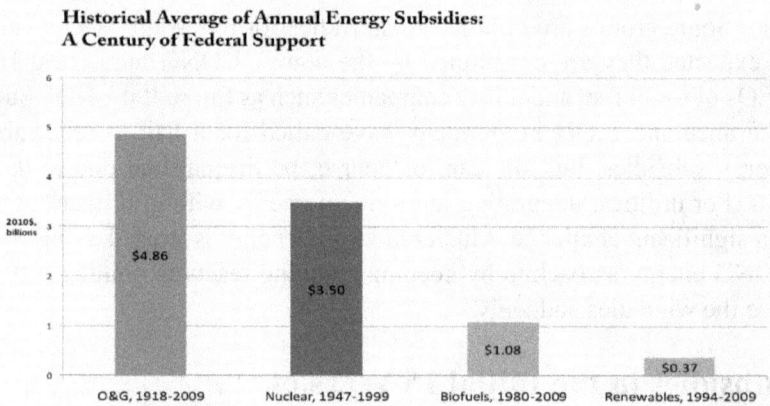

Figure 104: DBF chart showing average annual subsidies over the time period (at bottom of each column) showing that fossil fuels and nuclear have far outpaced renewable energy. Source: DBF Study, "What Would Jefferson Do"?

Since renewable energy is currently trying to get traction, and is in its early years, a comparison of renewables subsidies to those of CONG in the past when they were also just starting out makes sense.

Support for energy is a principal strategy to drive growth and has been for over 400 years. Pfund and Healey examined the extent of federal and state support for emerging technologies and targeted those 15 early years of each to get a good comparison. They were the first to quantify the federal commitment to renewable energy compared to support for energy types in earlier transitions. They found, for example, that nuclear power received funding that amounted to more than 1% of the federal budget over its first 15 years, while oil and gas made up 0.5%

of the federal budget. Renewable energy only represented 0.01%—one-tenth of 1 percent.

This means that oil and gas were supported at a level that was five times greater in the first 15 years of the subsidies' life, while nuclear was coddled at 10 times the public support. Subsidies for the "traditional" energy sources were many times what we are spending on renewables today.

If the federal government was to get rid of subsidies for all types of energy, including renewable energy now that the renewable energy market has reached a point where it is almost as economic, then all energy types would have to compete on their own merits. Established energy producers hardly innovate because they are complacent after receiving all the structural government support for so long, but it is important also to allow maturity in the renewables before doing so.

Provisions of the U.S. Tax Code that Subsidize Fossil Fuel Extraction

Tax Provision	10-year revenue score (billions of dollars)
1. Expensing intangible drilling costs	$13.9
2. Domestic manufacturing tax deduction for oil and gas	$11.6
3. Percentage depletion for oil and gas wells	$11.5
4. Percentage depletion for hard mineral fossil fuels	$1.7
5. Increase geological and geophysical expenditure amortization for independents	$1.4
6. Expensing of coal exploration and development costs	$0.4
7. Capital gains treatment for royalties	$0.4
8. Domestic manufacturing tax deduction for coal	$0.3
9. Deduction for tertiary injectants	$0.1
10. Exception for passive loss limitations for working interests in oil and gas properties	$0.1
11. Enhanced oil recovery credit	$0
12. Credit for oil and gas produced from marginal wells	$0
Total	$41.4

Source: OMB (2012).

Note: The last two provisions in this table are not expected to have a revenue impact because they phase out at oil prices below the levels expected over the ten-year scoring window.

Figure 105: Some tax code items are "hard-wired" to offer CONG a price break not allowable to renewable energy. Source: OMB (2012).

Permanent subsidies are an invitation to kill innovation and inject inertia lack in adapting to changing needs. Subsidies have been very helpful, but they need to adapt to changing circumstances.

In addition to the tax subsidies, Oil Change International (OCI) listed in a study they made the following observations about further ways the CONG community benefit:

- Master Limited Partnership tax exemptions
- Last-in, first-out (LIFO) accounting
- Lost royalties from onshore and offshore drilling
- Low-cost leasing of coal production in the Powder River Basin
- Taxpayers cover insolvent CONG companies for obligations to their communities and workers
- Inadequate industry fees recouped to cover the Abandoned Mine
- Land Grant Fund
- Inadequate industry support to cover worker health impacts
- Deduction for oil spill penalty costs
- Tar sands exemption from payments into the Oil Spill Liability Trust Fund
- Enhanced Oil Recovery Credit
- CO_2 Sequestration Credit
- Insurance costs for nuclear plants
- Exemption from compliance with the Safe Drinking Water Act
- Low Income Home Energy Assistance Program
- Subsidies for overseas fossil fuel projects

There are no permanent subsidies for renewable energy as there are woven into the tax code, since 1900, for oil and gas. In fact, the subsidies that do come are often temporary and renewed at the last moment, interrupting planning and financing for the new energy type.

Joseph Aldy, assistant professor of public policy at the Harvard Kennedy School of Government and a former special assistant to the president for energy and environment, noted in paper written for the Brookings Institution in February 2013 that the more than $4 billion in annual subsidies for the oil and gas industry, highlighted by the Pfund report above, does not benefit the economy anymore. Rather, it simply adds to the national debt and prolongs the country's dependence on a finite and damaging resource.

He noted that some of the tax benefits the CONG industries profit from have been in existence for more than 100 years, since a time when drilling was considered a much riskier activity that deserved federal support to support private investment. Today's calculus demands an adjustment of these tax preferences in view of the context of an evolved world where oil prices are supported by developing nations' growing appetites and put at risk by political developments in the Middle East.

Joseph Aldy's proposal calls for the elimination of 12 tax provisions that subsidize CONG energy and lead to the more level playing field so often claimed by the critics of renewable energy. This would further promote efficiency in allocating capital among all energy providers, including renewables. The removal of the CONG subsidies will have a small impact on production and will not affect pricing, reduce employment or weaken U.S. energy security. This proposal also complements other efforts to simplify the corporate tax code and help project the statement that large economies like that of the U.S. can phase out their dependence on CONG and it's a valuable strategy for all developing countries as well. It would put the United States' money where its mouth is with regard to a global energy policy in the most impactful manner.

The subsidies paid to CONG companies reduce the cost of activities in the upstream oil and gas markets, which alters the risk of such investments effectively distorting decision making and resulting in capital flowing away from more beneficial applications. Another study, completed by Resources for the Future in 2009, shows that removing these tax breaks cuts U.S. oil production by less than 0.3%. Today we are benefiting from the implementation of these tax breaks as the country moves toward becoming the largest oil and gas production source on Earth, even bypassing the Saudis by the year 2020. Aldy states that even if only three tax provisions were eliminated—expensing intangible drilling costs, section 199 domestic manufacturing tax deduction for oil and gas, and percentage depletion for oil and gas wells—it would yield 89% of the total potential economic benefit of his study. At the G20 meeting of world economies in Pittsburgh in 2009, world leaders recognized that fossil fuels were being subsidized to the tune of half a trillion dollars. Sending the message that eliminating subsidies for fossil fuels was helpful would have a good chance of having an impact on global fossil fuel consumption subsidies. This would reduce global demand by about 5%, or 4 million barrels per day. It would also impact pricing as well as the associated carbon emissions by about −7% by 2020 and −10% by 2050, to the benefit of all. The current tax breaks do not help new technologies or pollution-reducing technologies.

Federal Subsidies from 1950 to 2010

Subsidies to the mature fossil fuel energy industry are mainly written permanently into the tax code, and some of those are available to energy companies in general. By complete contrast, renewable energy subsidies are short-term legislation implemented via energy bills whose expiration dates cause havoc with capital allocation and planning. This has caused a significant sensitivity about food-for-fuel biofuel production elsewhere in the world. Asian countries such as China and Indonesia burn down primary forest to make way for palm oil plantations. The visual impact of the subsidy imbalance is well presented in the chart in figure 106.

A report written for the Nuclear Energy Institute was prepared by Management Information Services Inc. (MISI) as an independent analysis of subsidies provided by the federal government between 1950 and 2010. MISI is a Washington, D.C.–based research group with expertise in energy and technology that has Fortune 500 companies as customers along with not-for- profit groups, foundations, funds, academic research institutions and state and federal government agencies. The goal was for the nuclear industry to gain a toehold on how subsidies have changed over time and their effect on the nation's energy supply.

The conclusions, visible in figure 106, show that the largest beneficiaries of subsidies have been the oil and gas industries in the form of R&D programs. In the decade after 1997, federal spending on R&D for coal and renewable energy exceeded nuclear spending.

TYPE OF INCENTIVE	ENERGY SOURCE							SUMMARY	
	Oil	Natural Gas	Coal	Hydro	Nuclear	Renewables[2]	Geothermal	Total	Share
Tax Policy	194	106	35	13	-	44	2	394	47%
Regulation	125	4	8	5	16	-	-	158	19%
R&D	8	7	36	2	74	24	4	153	18%
Market Activity	6	2	3	66	-	2	2	80	10%
Gov't Services	34	2	16	2	2	2	-	57	7%
Disbursements	1	-	7	2	-18	2	-	-6	-1%
Total	369	121	104	90	73	74	7	837	
Share	44%	14%	12%	11%	9%	9%	1%		100%

Figure 106: Summary of U.S. federal energy incentives, 1950–2010. Source: Management Information Services Inc. 2011. All figures are in 2010 dollars. Rrenewables are primarily solar and wind.

It is clear that the renewable energy industry has been vanishingly small for much of the earlier part of this study, and consequently would show up as only a small absolute recipient of funding in any case. The MISI study shows that government priorities have changed over time. These subsidies represent a de facto energy policy where no official policy exists. While CONG energy sources have benefited from the lion's share of the subsidies, renewable energy has started to get traction. At their start, every energy resource depended on subsidies.

Now CONG resources no longer need subsidies to survive. Renewable energy on the other hand still does, but the trend is toward lower costs that will persist as our ability to innovate and save expense comes to bear, as it demonstrably does in every technology. Subsequent to the publication of this document, the MISI group, led by its president, Roger Bezdek, wrote a column for *World Oil Magazine*, which tends to support the oil industry. Inevitably, Bezdek's commentary embraces the way in which modern subsidies have ignored oil and gas for renewable energy.

Specifically, he points out that subsidies per watt are out of proportion with renewable energy, which is a fair comment but one that has to be qualified with the response that renewable energy is a desirable end point of global energy policy because it combats all the externalities our society is currently experiencing with CONG. The common criticism that "renewable energy only accounts for a tiny fraction of our energy, so why support it" takes no account of the fact that it is the fastest-growing piece of our energy capacity and represents a desirable end point for all of our energy production precisely because it has the clean and endless supply, resilient features we need.

Even in the leading early industrial nation to embrace renewable energy, Germany, the forces of traditional energy confounded the argument by claiming that the renewable energy subsidy there costs more than it actually does. Mr. Weber, director of Germany's Fraunhofer Institute for Solar Energy Systems, based in Freiburg, explains that the government spends only €40–€60 billion ($44.5–$66.6 billion) in annual subsidies for renewable energy. "If we're willing to burden the population with 180 billion euros of support for a dying industry, why do we worry about taking one-third of this to make Germany the world leader in photovoltaic technology?"

In May 2015, the International Monetary Fund released a study by Chris Hope and colleagues at the University of Cambridge, Judge Business School, on global fossil fuel subsidies[391] that revealed that global fossil fuel subsidies were greater than the combined global spending on healthcare, at an estimated $5.3 trillion, representing 6.5% of global GDP. An article in the *Guardian* newspaper on May 18, 2015, explained that this meant every minute $10 million of government money was being pumped into the fossil energy collective. This is $14.5 billion a day, or $600 million an hour, or $168,000 every second. The IMF uses the words "shocking" and "robust" to describe the conclusions of the report. The huge sum includes an appreciation for the pollution aspect of fossil fuels and the harm it does to human populations as well as to the increasing weather-imposing floods, droughts, storms from climate change. Lord Nicholas Stern at the London School of Economics commented on the study, saying that it was an understatement and that a complete accounting for the damage caused by climate change would produce an even higher number. The study says that ending direct subsidies to fossil fuels would result in a cut of about 20% in carbon emissions, a significant improvement in an area where little has been achieved. It also mentions that premature deaths from air pollution could be prevented in the amount of 1.6 million people per year.

Also, the resources freed by ending fossil fuel subsidies could drive economic growth, reduce poverty and allow investment in infrastructure, health, education and cutting taxes, which would promote growth. If fossil fuels suddenly cost what they should cost, inadequately subsidized renewable energy would suddenly become more competitive. Coal benefits from just over half of the total subsidies but has a reputation as a fuel that brings good economies. Oil attracts about one-third of the subsidies, and gas benefits from the remainder. The Chinese are the source of $2.3 trillion of the subsidies, with the United States responsible for $700 billion; Russia, $335 billion; Japan, $157 billion; and the European Union accounting for $330 billion. The climate change impact is about $1.27 trillion annually, or about 25% of the total subsidies. This very controversial number was carefully calculated using an official U.S. government estimate of $42/ton of CO_2 in 2015 dollars.

The UN's IPCC believe this figure is very likely an underestimate of the true damage. Direct subsidizing for consumers by government accounts for 6% of the total. Reduced sales taxes and cost of traffic congestion and accidents make up the remainder of the huge number. Now the cry is to get off fossil fuel subsidies rather than get off foreign oil! Just as education still rings with an unaltered Victorian authoritarianism and rote learning is still surviving somewhere in the curriculum, the global energy subsidies are also a blatant anachronism, with many subsidies, such as those in the United States, locked into the law itself. As we are increasingly aware that proven carbon resources must be left in the ground, many of the subsidies support the discovery of new CONG resources, and this is a match of wills being set up for a clash in the near future.

The *Guardian* newspaper has a sophisticated website that discusses many aspects of the risks and opportunities implied by CONG and renewable energy that is well worth a visit.[392] One interesting point that arises from this discussion is that the IMF numbers mean that the "subsidies" provided by absorbing the externalities exceed all the profits made by the world's top 20 oil and gas companies, with the notable exception of Exxon Mobil. Coal company subsidies were even larger than their revenues, at between $2 and $9 per $1 of revenues. By definition, this is unsustainable. As the impacts of climate change begin to really damage our civilization, governments are beginning to be serious about putting an internalized price on carbon. We often say that if a community was to experience automation and many jobs were to be lost, it would be good to compensate the workers who lost their livelihoods. Well, it appears that there is already such a subsidy in energy. Unfortunately, these subsidies are not actual cash, but a price on carbon would make them so.

The Environmental Law Institute
Subsidy Study of 2002–2008

This report did not look at nuclear subsidies.[393] It found that during this six- year period, the vast majority of federal energy subsidies supported energy sources responsible for emissions of greenhouse gases. Figure 107 gives a great visual representation of the scale of government support to fossil and renewables with a dimension for climate.

The journal *World Development* published a study[394] from the IMF in the summer of 2017 that revealed the extent of subsidies to fossil fuels. In 2013, subsidies amounting to $4.9 trillion rose to $5.3 trillion in 2015. The authors recognized that paying for the subsidies is an extra cost as well as an opportunity cost for not investing in faster, better, cheaper and sustainable. The authors highlight the fact that the real costs should also include all the environmental damage and deaths from air pollution that I have been underlining. In 2013, 6.5% of global GDP was spent supporting the fossil fuel industry. The study found that the top three subsidizers are China, the USA, and Russia. The EU comes in at a little less than half of the U.S. subsidies. Fossil fuel subsidies are enormous and unnecessary. Everyone pays in one form or another. Eliminating them would have resulted in cutting carbon emissions by 21% in 2013 alone. It would have reduced pollution deaths by 55% and increased global revenues by 4% of GDP. These subsidies lie there under the camouflage of taxes or complex pricing systems and are underappreciated by us all. The paper was written with this point in mind, to highlight that fossil fuels were being priced well below their true social costs and to generate an informed policy debate and accelerate a time when we can reap the huge benefits of going sustainable.

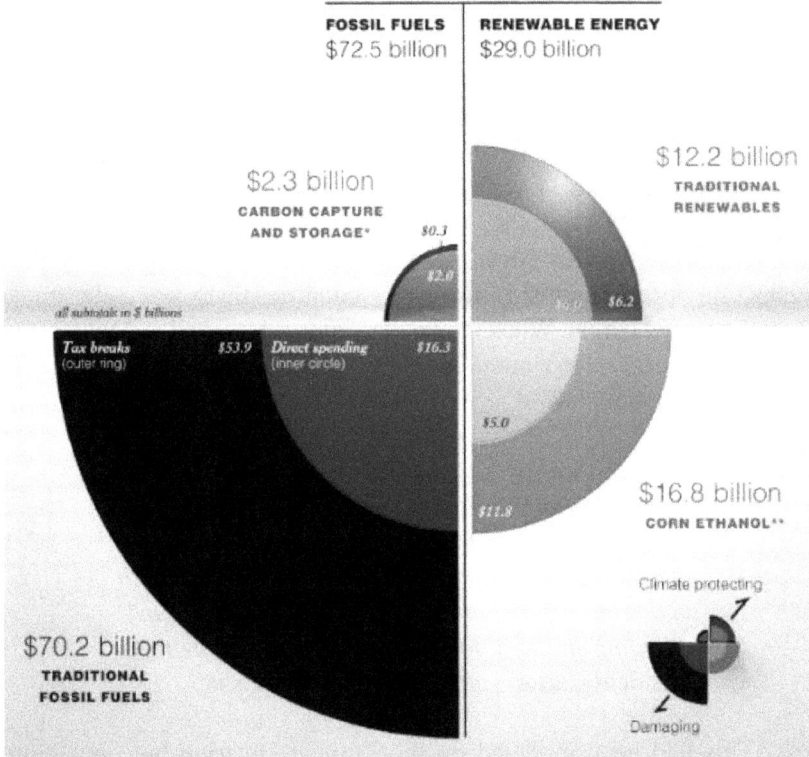

Federal Subsidies *(2002-08)*

FOSSIL FUELS	RENEWABLE ENERGY
$72.5 billion	$29.0 billion

$2.3 billion
CARBON CAPTURE AND STORAGE*

$12.2 billion
TRADITIONAL RENEWABLES

$0.3

$2.0

all subtotals in $ billions

Tax breaks (outer ring) $53.9

$6.2

Direct spending (inner circle) $16.3

$5.0

$16.8 billion
CORN ETHANOL**

$11.8

$70.2 billion
TRADITIONAL FOSSIL FUELS

Climate protecting

Damaging

Figure 107: The trend toward the USA being a low-carbon economy is resisted by an institutional tax code that has preferred the fossil fuel industries as strategic necessities in peace and war. This represents a considerable inertia for the new and sustainable entries into the market, so necessary, if for nothing else than to replace depleting fossil resources, to overcome. Source: The Environmental Law Institute, 2009.

Military Protection Expense of Oil Supply Lines

> *"... establishment of a naval base or fortified port in the Persian Gulf by any other power would be seen as a grave menace to British interests, and we should certainly resist it with all the means at our disposal"* —UK foreign secretary Lord Landsdowne, 1903.

The above quote comes from a period of diplomacy when drawing a line in the sand by military powers was relatively common to protect their policies and interests. British military policy became the foundation example for doctrinal foreign policy in the USA from President Monroe all the way to the current day. In 1943, President Roosevelt stated, "The defense of Saudi Arabia is vital to the defense of the United States."[395] After the Second World War, the Truman Doctrine was announced on March 12, 1947. Essentially, it was a warning to the Soviet Union not to make threats against Greece and Turkey. It was a pledge of support against Soviet communist influence for countries that were under threat from the Soviet Union. When the Soviet Union intervened in Afghanistan in December 1979, the United States became protective of its own activities in the Persian Gulf.

> *"Let our position be absolutely clear: An attempt by any outside force to gain control of the Persian Gulf region will be regarded as an assault on the vital interests of the United States of America, and such an assault will be repelled by any means necessary, including military force."* —Zbigniew Brzezinski, President Carter's national security advisor.

This had been modeled on the Truman Doctrine but bore more resemblance to the British declaration 76 years earlier concerning the same part of the world, except that now, the interests being defended were transparently those of energy supplies.

This particular subsidy is quite amazing. It stands to reason in a world where fossil fuels are unequally distributed that some nations will make strange bedfellows. A report[396] by Roger J. Stern entitled *United States Cost of Military Force Projection in the Persian Gulf, 1976–2007*, written in 2009, investigated the cost of the military forces fielded by the United States in the Persian Gulf, ostensibly to protect the U.S. supply of Middle Eastern oil. To field a single aircraft carrier continuously in the Persian Gulf requires a total of eight aircraft carriers. For the entire period, the cost was $6.8 trillion, amounting to $500 billion for 2007 alone (in 2008 dollars). In the report, he mentions that this cost is justified by U.S. energy insecurity and described it as "the loss of economic welfare that may occur as the result of a change in the price or availability of energy."

These costs dwarf the traditional tax subsidies, and if they were reinvested in developing and commercializing competitive, light renewable energy technologies or fourth (now fifth)-generation, safe nuclear power, we might be able to reduce our energy costs, increase our living standards and reverse the CONG externalities that are now consumed at such high volumes that the annual damage is debilitating. Not to mention that this would pitch our best talents at technology development to the service of our export industry and help the world achieve a higher standard and quality of life.

Payback

One question to ask of renewable energy is, given that it's received some subsidies, has it given taxpayers a return? Nature Energy published an analysis[397] that showed that it has provided a return. The author, Dev Millstein of Lawrence Berkeley National Laboratory, discovered that there were between 3,000 and 12,700 avoided premature deaths due to the replacement of fossil fuels by wind and solar energy between 2007 and 2015, including all the people who didn't get sick or claim sick days due to fossil fuel pollution and also the part of climate change mitigation already under way caused by the subsidies. The savings were between $35 billion and $220 billion. In this period, the U.S. government spent between $50 billion and $80 billion on wind and solar power subsidies, and even if we compare the lower end of the benefits with the higher bracket of subsidies, the climate and health benefits returned half of the subsidies. If subsidies stopped today, those benefits would continue to accrue. On top of this, new jobs and improvements in technology add significantly.

Renewable energy can generate the strategic power a nation requires for less money, minimal externalities and in future with no load to the tax code, thus returning billions to the taxpayer and making it possible for the economy to grow faster.

Chapter 10: Waste

Waste

"Put on a sweater." —President Jimmy Carter; on February 2, 1977, the newly inaugurated president appeared on television clad in that sweater and asked Americans to take a simple step to save energy: turn down the thermostat [for winter] and put on a sweater.

Energy cannot be destroyed or created. This is essentially the first law of thermodynamics. However, different types of energy can be transformed from one type to another. In this transference, much of it is lost (to use). In a car, chemical energy is transferred to the car as gasoline, which is turned by small explosions into thermal energy and then kinetic energy, forcing the pistons to cycle, which is transferred to the wheels, via a gearbox, to get you from home to school, then to work and via the shops back home again. In the process, more than 80% of the gasoline's energy is wasted, and air resistance and friction are forces that only add to the challenge. In addition, and as you will notice on the freeway, often only one person is driving, meaning that individual is accelerating and decelerating that ton and a half of metal and rubber just to transport her additional 125 pounds from A to B.

In a lamp, the electrical energy is converted into light energy, but not all the energy put in is useful. Only part of the energy is actually used directly for the purpose. The remainder is wasted as heat. In a world where energy costs money and is finite, there are huge incentives not to burn unnecessary energy. In our bulk, centralized power to grid system, there is a stack of inefficiency, starting with the combustion part where only about 40% of the energy in the coal or gas is extracted as power, then there are line losses in grids where the friction in the wires heats up the atmosphere. This process has become a complete waste. Sometimes as little as 10% of the energy in the coal will reach your house. Also, where the CONG energy you burn has externalities, then burning as little as possible is also a must. No modern appliance is 100% efficient, but at least over time, we have become increasingly able to engineer better efficiency into our appliances.

$$\text{Efficiency} = \frac{\text{Useful Energy}}{\text{Total Energy Input}}$$

In the business of investing in energy technologies, we hear a lot that whatever the innovation is, it cannot break the laws of thermodynamics. Today there are new technologies, tried and tested and commercial, that are busting out with far higher efficiency. Light-emitting diodes use only 15% of the electricity of incandescent because they have overcome the need to have incandescent metal. Now we can just produce the light instead of also making the heat. In extreme circumstances, this means that even air conditioners (the old kind still) can use less power to keep a space cool.

Electric cars have arrived! This is significant. First, you can get 3 miles for just 1 kWh (Tesla's 90-kWh battery is good for 270 miles[398]). This means you can actually refuel it with a garden- or roof-based solar array, or just put coal electricity in there and STILL get more miles per pound of released carbon than a gasoline car. The average daily distance driven is 21 miles. This only requires 7 kWh of electricity, easily within the ability of a person with a roof or garden to produce: 5 hours of sunshine per day can produce 7 kWh on a 2-kilowatt array. At $4 per watt (and the DOE Sunshot program[399] is aiming to reduce this to $1 by 2020, a level already reached at utility scale), that's only $4,000 to install. The goal is also to reduce the levelized cost of energy of a kilowatt-hour to just 6 cents. This means that daily trip will be done for 42 cents against the consumption of a gallon of gas at $3.00. Now we just need the electric vehicle (EV) to get cheaper. Also, the electric motor enjoys something that we needed gearboxes for in gasoline cars. The reason that hybrid electric vehicles do so well with fuel efficiency is that they use the motor to accelerate the car from 0 to 25 mph and then the gasoline internal combustion engine takes over. The electric motor has the same twisting power (torque) at low speeds (low rpm) as it does at any speed. The internal combustion engine has very low torque at low rpm, which is why we need to rev it up or increase its rpm to get going.

AVERAGE ESTIMATED FUEL ECONOMY BY MODEL YEAR

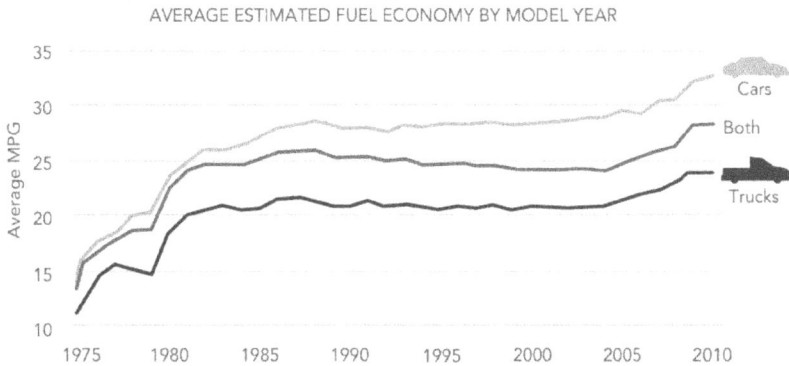

Figure 108: Illustration of the increased mileage efficiency of vehicles in the USA since 1975, the time of the first oil shock. Source: the Pew Environmental Group. www. pewenvironment.org/cleanenergy.

These three areas of energy use alone are about to make a huge impact on energy consumption while keeping the units of GDP growing fast. Light bulbs and car transportation make up a huge amount of electricity demand; in fact, cars switch the demand from liquid transportation fuels to electricity. Another demand appliance for electricity is the air conditioner. New designs that use only 30% of the power for the same cooling effect are now being tested and are only months away.

Other appliances and tools that we use regularly, such as cooking ranges, where resistance electric ranges have been very unpopular, could be transformed by the arrival of induction ranges that generate heat directly and efficiently in the bottom of the steel or iron saucepan or skillet, leaving the range surface cool to the touch. Even if we go back to transportation, there are now electric aircraft being tested by Airbus[400] and NASA,[401] and energy storage will make all the difference here too. A single 300-watt solar panel is plenty to power a modern flat-screen TV, which only draws about 100 watts. We are moving fast toward efficiency today, and the implications are huge.

Humans evolved from using muscle power that was fueled by food, mainly plants, that were in turn creating carbohydrates using the sun. However, the sun-to-muscle pathway is very inefficient.

Over time we started using fire and then chemical fuels, the CONG variety, which also originally came from the sun. The arrival of engines, such as Thomas Newcomen's steam engine, and then James Watt's improvements made it possible to suddenly have a textile industry, mass production of weapons and efficient mass transportation. We were able to use the machines to exert more leverage on the environment.

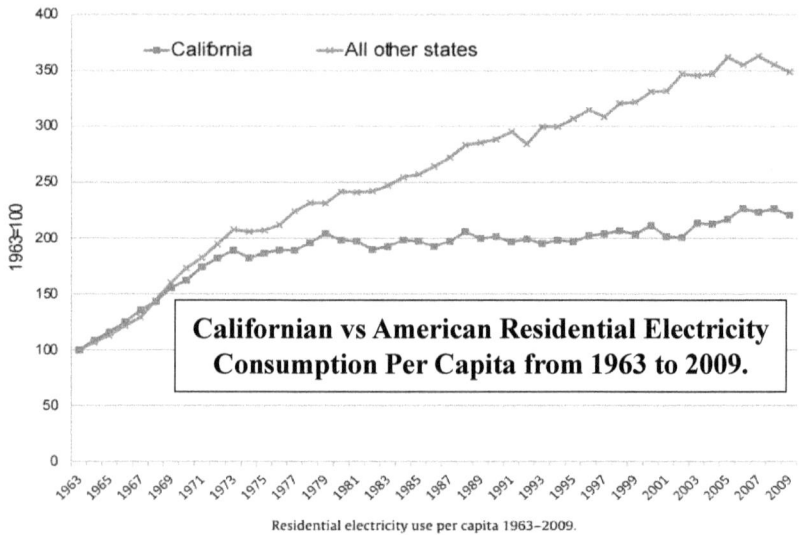

Figure 109: Comparison of the Californian residential electrical energy consumption per capita compared to all other states in the United States. After the oil shock of the 1970s, energy conservation measures took effect in California. Source: A Levinson / Journal of Economic Behaviour & Organization 107 (2014).

Then electricity joined the mix, and via electric motors, it powered machines and allowed better illumination, which just meant that working around the clock was then possible, and of course human beings made other human beings do just that. Coal was originally mainly used to generate electricity but moving to oil permitted more efficiency for ships and cars. B Alfred Crosby, author of *Children of the Sun*,[402] wrote that the internal combustion engine in trains, trucks, cars and tractors has been the most influential machine we have invented. By the end of the 20th century there were 500 million vehicles helping to burn 70 million barrels of oil a day. This is the anthropogenic part of climate change, and the century ended with coal and oil being dominant energy sources and burned very inefficiently.

Electricity is becoming our number one kind of energy. We use it to do everything, and increasingly things like ships, cars, trains and even planes[403] are moving to it. Humanity's energy use has gone up 20 times since 1850 and more than 5 times since 1950. Each person consumed

2,000 kWh of electricity a year in 1950 and 32,700 kWh in the year 2000. However, almost half of humanity, mostly living in the tropics, still rely on wood for fuel.

All of this energy consumption has gone along with an improvement in the efficiency with which we use it. Looking at cars, today we are familiar with the way that gas-guzzling vehicles were very expensive to drive when oil prices climbed. Originally, the first vehicles had very poor mileage. I always remember hearing in the military that a main battle tank used 4 gallons of diesel fuel per mile. Imagine the amount of fuel saved if today's efficiency could have been used before 1975 and the fuel that is being saved today. Having said that, the Model T Ford could travel 21 miles on a single gallon of gasoline, not much change from today.

Fuel efficiency was not the main issue between the two great wars and it was common to see vehicles then and even today that could only get 10 mpg. Today the smart car and the electric hybrid vehicles can obtain 60 miles to the gallon, and hybrid vehicles that use gasoline to generate power within the vehicle are obtaining over 100 miles to the gallon. Internal combustion engines, whose only function is to generate electricity to charge up a battery, are much more effective because they can always operate at their optimum settings and efficiency. We are discussing the waste in fuels that are finite in quantity. CONG fuels are one-use only and come from a finite resource that already has a "sell-by" date on it.

A paper[404] published by the DOE's Oak Ridge National Laboratory in October 2017 reveals the wasted fuel from just driving. It analyzes the impact of aggressive driving on fuel efficiency. We already know that the most inefficient part of the acceleration of an internal combustion engine car is the part from 0 to about 25 miles per hour, but fuel efficiency can be cut by 10–40% in stop-and-go traffic or 15–30% at highway speeds in small vehicles. This translates to a value between $0.25 and $1.00 per gallon of fuel that is essentially wasted by a common behavior, or hundreds of billions of dollars a year for years on end. It's clear that over the last few decades, the combination of inefficient engines and aggressive driving patterns has released more heat and pollution to the atmosphere than rubber to the road. Interestingly, the study also underlines that the increased efficiency of the electric motor and the ability to recover power from regenerative braking serve to partially forgive an aggressive stance.

There are two reasons we need to get a rein on the current situation. The first is because we cannot burn more than a limited amount of carbon if we want to avoid the worst effects of climate change, and the second, because in the volumes that we use gasoline today, we cannot continue to pump it at this rate for ever. It is for this reason alone if nothing else that we are transitioning away from fossil fuels for vehicles. The climate push has also accelerated this and the move to electric vehicles has suddenly brought us great improvements in efficiency.

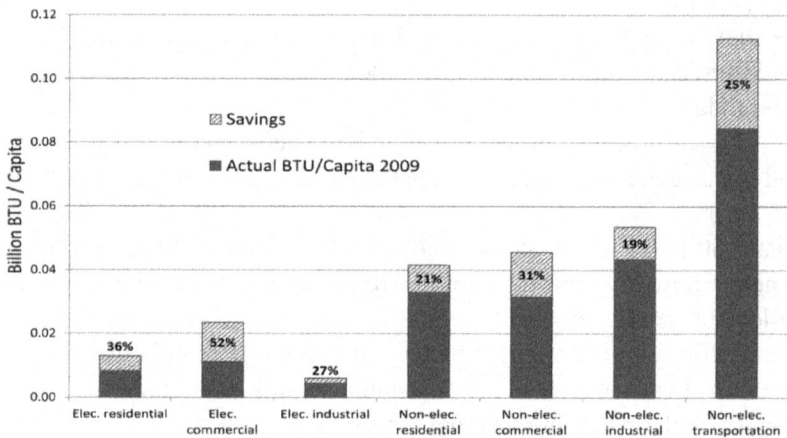

California energy savings 1963–2009.

Figure 110: Drop in consumption in BTU per capita of all the different energy types in California between 1963 and 2009. Source: A. Levinson / Journal of Economic Behavior & Organization 107 (2014).

Instead of looking at the internal combustion engine's improvements, we can look at the energy efficiency of a country or a state. In the United States, California enjoys a lovely mild climate. It's often not hot enough for air- conditioning in the summer (well, okay, in LA), and the winters are very mild and not much heating is required. This means the average household consumption of energy is very low at only 6.9 megawatt-hours per year.

As the seventh-largest global economy, California could be considered as a separate country. The state has 18 oil refineries with a combined capacity of 2 million barrels a day as of January 2015. Overall per capita state energy production was 48th highest of America's 50 states.

Much of the energy in California is renewable. They enjoy the fourth- largest hydroelectric capacity, second in wind and first in solar and geothermal. They are also 16th in nuclear but closed down the large San Onofre nuclear power plant near San Diego in June 2013. This last is a shame because, despite the imperfections of the light water reactor, it is nonetheless an important source of emissions-free electricity and could easily have continued to operate to help us all transition to cleaner sources of power. All this information is available on the EIA's residential energy consumption website.[405]

GDP vs. Energy Efficiency
(Top 40 Economies by GDP)

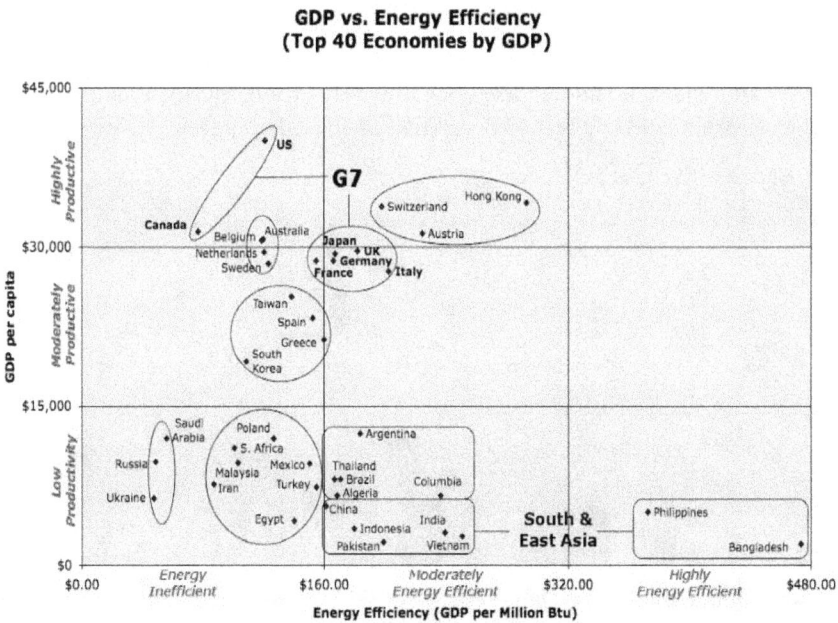

Figure 111: Illustration of the energy intensity of the top 40 global economies in 2005. Source: Peter Corless, 30 September 2005.

For 40 years, California has had an energy cost and logistical advantage over the rest of the United States and many countries worldwide, resulting in billions of dollars of savings.

The California Energy Commission (CEC) set the nation's first standards for energy efficiency for appliances and buildings. The California Public Utilities Commission (CPUC) also led the country in decoupling utility profits from sales of electricity and natural gas.

In 2012, California was the number one state for appliance standards and equal first with five other states for its building codes. Figure 109 shows that residential energy use grew by 120% in California during the period 1963– 2009 but by 245% in the other 49 states. The population of California also grew in this period, which means there were more "capitas" to dilute the energy number by.

California's energy is also sold at a higher price than in many other states, which gives it an advantage similar to that of Europe, where gasoline prices are over twice as high as those of the United States. The same is true of Japan, and cars in these countries are much smaller and more efficient. It's not for nothing that the smart car comes from there.

It wasn't just residential electricity consumption that showed this effect. In figure 110, we see the savings in each type of energy since 1963 and where it stands today in California. Total energy consumption in California actually fell 23% relative to the other states.

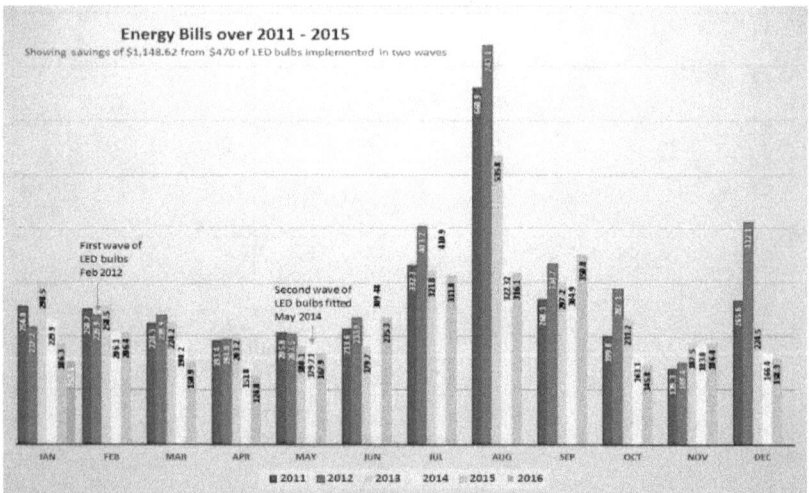

Energy Bills over 2011 - 2015
Showing savings of $1,148.62 from $470 of LED bulbs implemented in two waves

Figure 112: Five years of electricity bills in a New York apartment showing a 25% kWh consumption drop when 100% of lighting comes from LED bulbs. The leap in the summer months is related to air-conditioning consumption. New more efficient air conditioners also contributed to the reduction in energy intensity for the same comfort level. Source: NEF Advisors, LLC.

If the whole nation was experiencing the same drop in energy consumption the U.S. would be following the Obama administrations goal of greenhouse gas reductions to 17% below 2005 levels by 2020. If we look at any community, it uses a certain amount of energy but also generates a certain amount of GDP. The relationship between the two is called *energy intensity*. In general, historical energy intensity was poor and over time has improved.

Detailed studies are few, but there were some in 2004 and 2005 that give us the chart in figure 111. It shows energy intensity by comparing GDP per capita versus the energy efficiency in GDP per million BTU.

Any house, village, town or city that uses less energy but maintains the same living standards has become more efficient. In the past, early machines, internal combustion engines, electric motors, bulbs used more energy per unit of work than they do today. One of the more common changes in energy in recent years has been the arrival of a new kind of light bulb that uses technology called light-emitting diode, or LED, and that would cause Thomas Edison to turn in his grave as well as save a lot of money.

The incandescent bulb's time is over! In my apartment in New York City, I have 42 bulbs, a mixture of 13-watt (100W) and 9-watt (60W) bulbs; the figure in parentheses is the old incandescent wattage rating. The old bulbs were incandescent bulbs and used a modern variant of Thomas Edison's filament that glows white hot when an electric current passes through it. The tungsten filament heats up to 347°F (175°C). I had 60-watt and 100-watt bulbs with an A19 screw fitting, and in the course of an average month, I would burn almost 1,000 kilowatt-hours.

On achieving LED religion, however, I bought some LED bulbs and found an 8% overall savings in kilowatt-hours. In May 2014, I decided to do 100% of my apartment and experienced a full 25% reduction in my electrical consumption. The bulbs cost me $469.91, which is about 1,000% more than the incandescent version, and by end 2015, I had saved over $1,000, over twice my cost and the bulbs are still expected to last another 23 years!

If an apartment building or a village or small town or even a huge city like New York was to switch out all of its lighting for the new LED technology, it would save 85% of the electricity it normally uses on lighting. This would be a significant move toward efficiency. If the power used is CONG-based, it means savings in all the associated externalities, for example, in emissions, as well as leaving carbon in the ground. This sort of leap ahead also has another effect. If the power needed to go sustainable is now suddenly significantly reduced, sustainability is much closer. The LED revolution is shortly to be followed by several other revolutions in electricity usage. First is the air conditioner, which will cut cooling energy costs by over 50% and make humid summers more economically tolerable. The next is the transition to electric cars, which cuts the energy required to travel a mile to just 0.3 kWh per mile as opposed to 1/30th of a gallon of gasoline per mile.

There is a light bulb, still "switched on," in Livermore, California, that was manufactured in 1890 by the Shelby Electric Company.[406] The bulb in question was fitted to a fire department building. It was originally in a hose cart building, then was moved to a downtown garage used by the fire and police departments. Later it was moved to a newly constructed City Hall. In each location, the bulb's purpose was 24/7 illumination. Mike Dunstan, a local journalist, noticed the bulb's longevity in 1972. He wrote "Light Bulb May Be World's Oldest" for the *Tri-Valley Herald*. Then the Guinness Book of World Records, Ripley's Believe It or Not and General Electric, which had acquired the Shelby Electric Company, all confirmed from the evidence that this was the oldest working light bulb in the world. The bulb was moved a final time in 1976 when the fire station moved again. This time they severed the cable holding the bulb rather than risk unscrewing it.

Figure 113: The Shelby light bulb in Livermore, California, which has been permanently switched on for 115 years. Source: Documentary.

Figure 114: Some of the credible universities, institutes, agencies and multi-billion-dollar companies that have been tempted to explore whether LENR technologies really can give us much more efficient power.

They took 22 minutes to transfer the bulb and Ripley's "Believe it or Not", allowed that this was not going to interrupt the extraordinary full-time illumination of the bulb which has now persisted for 115 years, albeit at only 4 watts. There are several other, almost as remarkable bulbs from this age still operational.

Early on, the industry embraced reliability and built it into their products. Then General Electric and the Pheobus cartel[407] decided to limit the life of the light bulb in order to encourage consumers to buy more. This works best when the manufacturer owns a large share of the market and customers can complete the circle and buy replacements from them. However, increase competition, and the reliability of products improves again. Originally, the idea was to stimulate industry and get the economy moving, but it quickly became the case that new goods were associated with shoddy goods that needed replacement.

It stands to reason that we would only get better at using energy more efficiently over time. There is another arena of energy resource that is currently emerging as a possibility. Completely unproven, but a fascinating possibility is that we can obtain the equivalent of nuclear power but without the Hinkley Point price tag. Low-energy nuclear reactors (LENR) could offer us a huge improvement in our energy efficiency, but this still lies tantalizingly out of our reach, even though many of the credible brand names we consider to be household names are putting money into confirming whether or not such energy is actually available (figure 114).

Wasted Water

All energy production consumes water, adding to the cost. Much of the water is lost as steam evaporation into the atmosphere, and much is returned but polluted by thermal energy and toxic to the local environment. Solar power often gets a clean bill of health when it comes to water, but the panels or mirrors do better, especially in desert conditions, when they are cleaned of dust. Solar thermal or concentrated solar thermal applications often use water as a thermal fluid medium and consequently use more. Predictably, hydroelectricity uses more water, but that water is replaced by rainfall and replenished by rivers, although there is a lot of evaporation from lake surfaces that otherwise would not happen.

Both coal and nuclear power use thermal energy from coal and nuclear fuels to heat water to steam to turn a Rankine cycle turbine and consequently use similar but vastly larger amounts to achieve that goal. Thermal energy resources often work better if the water is very cold and so turbines often have cooling water going through them as well. Geothermal resources have to pump water back down into the hot rocks, where the heat is collected, but this can be a circular tour as the same water is cycled again and again. The chart in figure 115 is taken from a study[408] completed by the River Network in 2012, called "Burning Our Rivers," and graphically demonstrates another reason why fossil fuels are to be avoided in the fuel mix. We already are short of water for other purposes, such as desperately needed agriculture, often under threat of droughts. Wasting it by generating power with fossil fuels is not a good idea.

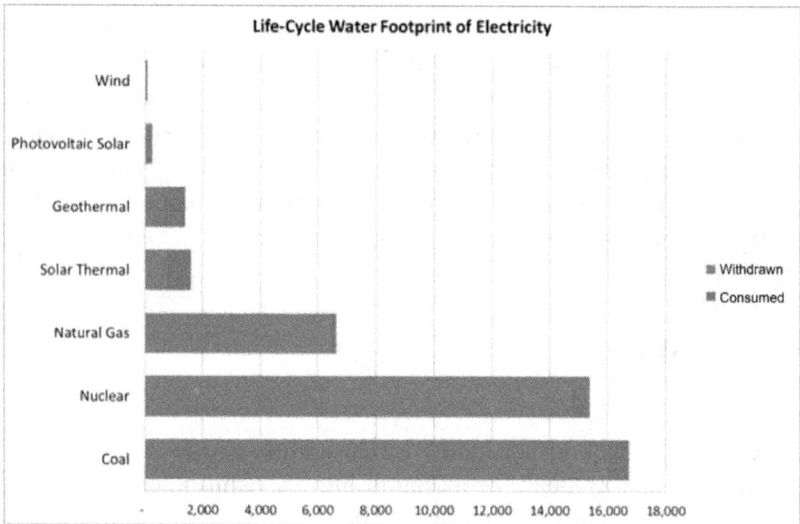

Figure 115: The different quantities of water consumed by the different major electricity production technologies. Source: "Burning Our Rivers: The Water Footprint of Electricity". River Network in 2012.

Blockchain

In 2008, Bitcoin became the first of the flood of cryptocurrencies now (March 2018) numbering 1,384 and growing fast. This extraordinary growth heralds a new method of establishing title that is not based like regular currency on the promise to the bearer of backing up that value, but by simple mathematics, cryptography and the reliability coming from knowing that your transaction is being monitored by many cryptocurrency "miners" around the world at the same time.

Blockchain is the name of the software ledger system that follows cryptocurrencies around. When Bitcoin started in 2008, the "proof of work" algorithm on which the blockchain ledger is based offered 50 bitcoins for every "block" of transaction detail that was added to the ledger by a miner. This number halves every 4 years and stands at 12.5 bitcoin per block. One new block is "awarded" to a miner every 10 minutes as a reward for getting the work done.

Energy consumption for the number crunching is intense. A recent Greentech Media article quoted the electrical consumption of Bitcoin alone as being 32 terawatt-hours, ahead of 159 countries, including Ireland. Given that a single Bitcoin transaction is equal to the energy consumption of 4,000 credit card transactions, it's not surprising that competitor Ethereum, the world's number two blockchain network, uses less power, at 11 terawatt- hours, but even this uses the same amount of energy as Zambia.

The value of Bitcoin has grown, and it is currently experiencing high volatility. There is a consensus that the underlying technology is here to stay because cryptocurrencies solve problems in myriads of industries. They offer speed, transparency, accuracy, reliability and trustworthiness born of being simultaneously calculated on many global sites. It means that value, title and provenance are all easier to come by, facilitating financing activity and improving the experience for all players.

Size of the Bitcoin blockchain from 2010 to 2016, by quarter (in megabytes)

125,000

100,000

75,000

50,000

25,000

0

Blockchain size in megabytes

Q3 Q4 Q1 Q2 Q3 Q4 Q1 Q2 Q3 Q4 Q1 Q2 Q3 Q4 Q1 Q2 Q3 Q4 Q1 Q2 Q3 Q4 Q1 Q2 Q3 Q4
'10 '10 '11 '11 '11 '11 '12 '12 '12 '12 '13 '13 '13 '13 '14 '14 '14 '14 '15 '15 '15 '15 '16 '16 '16 '16

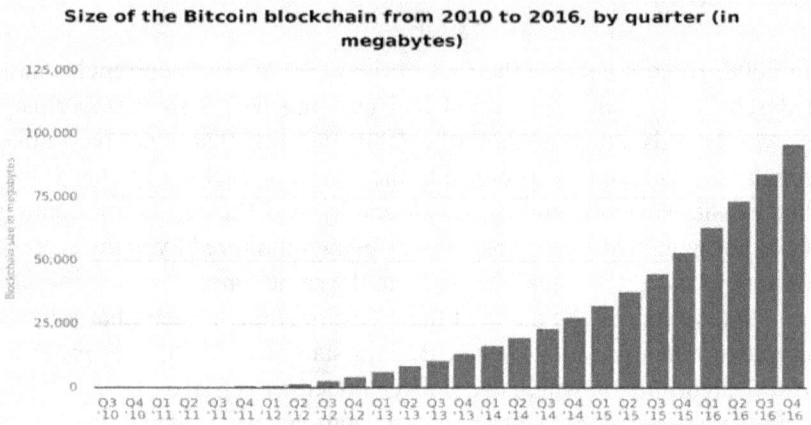

Figure 116: Size of the Bitcoin blockchain file from 2010 to 2016 in megabytes. End of 2017 it was 149,000 megabytes in size. Source: Statista 2017.

Renewable energy is one industry that is benefiting. One new cryptocurrency, for example, is called SolarCoin, which aims to transition the energy consumption of blockchain to carbon-free resources and consume a fraction of the power as well, making it thousands of times more energy efficient than Bitcoin.

Blockchain file sizes have been growing steadily too. Since 2010, the Bitcoin blockchain alone has increased in file size to 149 gigabytes. You can monitor this progress on a site called the Bitcoin Energy Consumption Index. Energy cash flows offer a perfect growth industry for blockchain, taking the industry from a cumbersome trading and clearing industry into a clean and efficient marketplace. Renewable energy, already the darling of the socially responsible and sustainable practices. These investment themes offer plenty of opportunity for blockchain. The many different forms of electron energy production all merge in the grid, but the blockchain ledger can tell you which ones came from which source—wind, solar, storage or another. Certificates, RIN credits, tax credits and other guarantees are all mechanisms that can be massively cleaned up with blockchain, which deals with the complex data flows and cash flows easily. As production thresholds are reached, credits are automatically generated and distributed. L03 Energy, a solar installer based in Brooklyn, NYC, has a joint venture with Siemens that uses a blockchain platform to timestamp small-scale trading of neighborhood power and microgrid electricity transactions for residents of Gowanus and Boerum Hill.

In Australia, a recent ICO welcomed a new cryptocurrency based on Ethereum called Power Ledger, designed to aid utilities in their peer-to-peer relationships with subscribers and other utility players and integration of distributed power generation. In the United States, another recent ICO that raised $40 million was Grid +, which is much more to do with distributing wholesale power with a retail platform in the Texas electricity market via existing wiring connections. Both systems are using tokens for actual cost price kilowatt-hours that are provided to the subscriber. The National Renewable Energy Laboratory is also working with blockchain and working with a blockchain web services provider called BlockCypher to explore supporting distributed energy markets. NREL sees blockchain as an enabling technology to facilitate continuous competitive pricing, low transaction costs and efficient matching of energy demand and supply and, most importantly, to leverage underutilized energy capacity.

It's only a matter of time before blockchain also revolutionizes electronic voting, and we can expect our democracies to be desperate to adopt these admirable characteristics in coming elections.

Chapter 11: Conclusion

Conclusion

Unlike renewable energy, CONG energy is laden with significant externalities. The climate is approaching several tipping points (such as methane release from warming permafrost tundra, natural gas leakage and white polar ice caps turning to dark water, keeping more of the sun's heat and causing a change in the Earth's albedo), and our geopolitical system hovers perilously close to conflict, partly because of the unequal distribution of CONG resources and nuclear know-how, which can make countries into difficult neighbors but which also stimulates other countries without resources such as Israel and Japan to compensate.

We suffer greatly from poor health and economic loss from lost work mostly because of pollution, but also due to the expense of oil and its availability. Days lost from health problems directly impact an economy. On a wider scale, the impact of chemical toxins in the environment costs us IQ points, adds to health costs and destroys the wider, valuable and necessary richness of our planet by destroying species and putting at risk ecosystem food chains, which are like biological tipping points. Such is the rate of extinctions that we are clearly in a sixth global extinction and there is no evidence of a cause outside the human cause.

Stultified tax codes are supportive of embedded tax breaks to strategic industries, while new industry growth is resisted by incumbent players, who will do everything they can to impede negative impacts on their bottom lines. We provide subsidies to the wrong things because profitable CONG industries pay lobbyists well enough to keep politicians' hands tied. Even knowing that all of this will come to an end as the supplies of CONG are exhausted is merely to confirm the disaster that will come with business as usual.

The cumulative consideration of all the externalities that stem directly from the impact of the use of CONG on the economy and our environment is that the actual cost is high, the secondary effects horrible and they are already significant today. We are going to have to do something about it and quickly. We have discussed the cumulative damage in the various chapters.

Considering all this "damage" in its entirety makes the case clearly that the transition needs to happen in as fast a manner as possible.

Summary of the CONG Externalities

Chapter 3: Climate. Tipping points, melting ice, wildfires, landslides, storms, hurricanes, tornadoes, accelerating sea levels, increasing CO_2 and other greenhouse gases, complexity caused by denier politics, resistance to change from the fossil fuel industry, the sixth global extinction, droughts, global security made more vulnerable, climate and war refugees, disease ranges spreading, drunken forests, release of mercury and methane, Arctic warming faster than equatorial warming, land-based ice presenting possibility of rapid sea level rise in Greenland and the Antarctic.

Chapter 4: Depletion. The massive current rates of CONG energy consumption mean limited resource life cycle, importance of replacement energy sources which we have, the importance of treating CONG as a bridge to sustainability as rapidly as possible. New technology releases new CONG resources, fracking and tar sands, but we can't afford to accumulate the carbon emissions, which stay in the atmosphere for hundreds of years. Also, subsidence and earthquakes from resource extraction cause considerable alarm and damage.

Chapter 5: Geopolitics. It's clear that political forces conspire against a sustainable planetary regime. The idea of countries in harmony and moving in aspirational ways to improve quality of life everywhere appears to be unachievable if you have a zero-sum game approach, which is apparently common in our politicians. Conflict over resources can go away with sustainability, but a mind-set trapped in CONG tries to resuscitate coal, an otherwise doomed industry. Clearly, rare, good, informed thinking is required, which is difficult to ensure. An equitable peaceful, progressive and happy world is possible, but not a common vision. The nuclear world remains locked until it can embrace new technology and progress. Proliferation remains a huge risk that has increased in criticality today. Interestingly, China, the big polluter, installing coal-fired power stations on a weekly basis, is now leading the world in cleaning up its act.

There is also now very good evidence that a highly organized and well-thought-out industrial and political effort, both in the United States and in other countries, is trying to muddy the waters of progress with misinformation about science. They make it appear that the science is incomplete and deny that the climate is warming on a scale that actually threatens our ability to do something about it. They take advantage of the complexity of the definition of science to sow doubt in the minds of voters. Whereas the difference in accepting a position based on good scientific evidence rather than on the evidence of "feelings" is very clear, it is nonetheless a battleground of the modern era, reinforced, incredibly, by this Trump administration.

Chapter 6: Health. Legacy CONG technologies have a massive global impact on health via atmospheric and waterway pollution. The cost of treating sick people accounts for a percentage of GDP and keeps them from working, as well. Common respiratory diseases, cancers and many other infirmities that can be reversed are visited on people around the world. Even when the Japanese casualties from the Second World War atomic bomb attacks are added to the nuclear impact on health, it is still a far lower risk to human health than polluting coal.

Chapter 7: Price Volatility. This externality causes significant cyclical and economic challenges to growing, global economies. In the United States, oil prices are arguably responsible for many recessions. Economic growth is often choked off in a growing economy as the commodity market inflates the price of CONG resources. In sustainable energy, there is a huge benefit to a growing economy not to have a price per ton of wind or barrel of sunshine.

Chapter 8: Pollution. CONG energy creates brownfield sites and has to be cleaned up and causes species extinctions and illnesses and death. Complex chemicals have impacts on insects and animals. The bee populations are at risk because there is little will to find an alternative to agricultural insecticides. CO_2 is a pollutant due to the damage it does. Health is a major problem due to pollution. In the United States, the Trump EPA is unwinding protections that will take years to reapply. Particulate matter is also a massive issue, and city dwellers end their lives with lungs that look like they were smokers, even if they were not. Despoliation of natural scenic beauty is the least of the problems but a significant one nonetheless. Pipeline pollution is significant. The oceanic plastic gyres in every part of the globe present a huge wildlife and pollution challenge. Luckily, more coal companies are folding in 2018 than during the entire first term of the Obama administration, despite the supportive words of the Trump administration because they haven't yet understood that it is the economics of cheaper natural gas that is forcing out coal and keeping the United States compliant with its reduction in carbon emissions without doing so deliberately.

Chapter 9: Subsidies. Baked into tax law in many countries, CONG tax breaks will be next to impossible to prize from the grip of conservative politicians. In the face of this, sustainable energy formats struggle to compete, with infrequent renewals of insufficient subsidies that are small in comparison. They have definitely helped but have slowed down progress.

Imagine if all subsidies were to be channeled to where the economic benefit is greatest, that is, the vision of this book, where sustainability grows the economy much more. The sustainable world would likely already be here. Different studies confirm that both in the United States and overseas, subsidy regimes for CONG hold everyone back and cause huge global problems.

Chapter 10: Waste. Given that CONG resources are finite, it's amazing more is not done to enhance efficiency, but oil companies exist to sell more oil, so efficient cars were out of the question in countries where gasoline fuel is cheap. In Europe and Japan, where gasoline is expensive, however, there are cars that can easily do 60 miles per gallon, and the winning aspect of the new electric vehicles is exactly that their operating costs per mile are more than competitive with gasoline and diesel. This lesson is that technology has guided efficiency over time and new standards are cutting the BTUs consumed per unit of GDP, more in some states, such as California, than in others. All appliances, HVAC, lighting, vehicles and electrical storage are evolving to become much more efficient currently, but the infrastructure buildout implied is still massive. This could be good news for the billions of human beings who still subsist on wood or dung energy every day, but geopolitics could still make their lives miserable. Built-in obsolescence may hit the EV or LED at some juncture if permitted by regulation. Water is a precious commodity, too, and linked directly to energy production. Some sustainable forms of energy production such as wind and PV use very little water.

Vision of the Sustainable Alternative

The good news is that we have the means to do something about it. It's a fix, however, that's made difficult by incumbent players who are unwilling to allow the changes to occur. Nonetheless, the transition is something that is happening at an accelerating rate anyway, because of the sensibilities of private players and voters in response to the immense opportunity of change and economic gain. The transition to sustainability, and not merely in just the energy sphere, is the greatest business opportunity and transition in human history, and it's happening today.

When you look at all of this together, the daily current extraction of 96 million barrels of oil and 10 million tons of coal and production of natural gas at a rate of 1.3 million barrels of oil equivalent is mostly combusted, releasing huge quantities of CO_2 and causing all these other externalities. It is overwhelming, and we don't see it clearly. Some are blind to it and conflicted on how to operate with this elephant in the room.

I see it clearly as a multi-trillion-dollar opportunity. We can cut fossil fuel emissions by using the clean, green alternatives that are proven and increasingly cheaper. We can install air-sourced heat pumps and avoid emitting CO_2 for the entire winter but still stay toasty and warm. We can recharge our electric cars at home with solar and wind and electrical storage. We can replace fossil liquid fuels with cellulose-derived biofuels. We can work on the project to take CO_2 out of the atmosphere. We can work on replacing early nuclear technology with far friendlier fourth-generation technologies that can consume existing nuclear waste as fuel.

If you can see a world with all the overwhelming CONG externalities erased from the global economy, then you would see a transformation that also needs to be highlighted because not only is it possible but also it's very likely to be the case that a lot of good can come from this transition. Although this might appear a naïve, utopian vision, it nonetheless stems from exactly what would happen if you stop the damage from CONG externalities. Let's get visionary:

If we can transition from CONG to sustainable energy, then:

- The global climate impact of increasing CO_2 and other greenhouse gases won't stop until we extract these gases actively from the atmosphere, but at least we will have prevented the climate problem from getting worse so long as we are still this side of important tipping points.
- Solar, wind, hydro, geothermal and all the many other formats of sustainable energy resources do not have, by definition, any major limitations on them. They are not infinite, either, but have a replacement function or are so large in scope that

depletion is not an issue. If wound up with higher efficiency, then this issue is even more beneficial.

- Since sustainable resources are far more distributed than CONG resources, geopolitical pressures can be eased, and a more equitable global stability can be achieved. The "haves" will be more widespread. This might even go so far as to help cut defense budgets and move toward a world where resource competition is almost nonexistent. To those who would furrow their brows and say that this bodes badly, I would point out that grim determination would be best served by championing the new sustainability and obtaining this vision.
- Health outcomes will improve markedly, which will have a double- edged impact on the economy both by more employment and by lower medical costs.
- Since a ton of wind and a barrel of sunshine have no commodity value, there is nothing to prevent a growing economy from continuing to grow faster. CONG economies are "governed" by this unfortunate effect, creating economic cycles that last often decades and interfere with other forms of progress.
- Transition of subsidies away from CONG is a desirable outcome that will only accelerate sustainability and end up reinforcing the strategic energy security issues for which CONG subsidies existed in the first place.
- Since you can't waste sunshine, there is a different definition of waste in a sustainable economy.

What an agreeable picture this paints and what a likely picture. This is not whimsical or pie in the sky, merely a straight-line progression from CONG to sustainable with all its benefits.

It remains to be seen whether modern humans, with their vibrant modern economies and the ability to glimpse our universe in all its glorious electromagnetic spectrum—to the earliest rays of light that come from the origins of the Big Bang—can pull this off in time. We are seriously preparing for a presence on the moon and Mars and clearly have the ability to get there. (Already this administration is canceling a huge infrared telescope that represents our ability to see our surroundings in a way that would reflect massive benefits on humankind. Those who cancel such a thing see no link between an advanced telescope and the benefits it confers in the human mind, which is a form of blindness in its own right and which brings up the phrase *the blind leading the blind*.)

I have portrayed a vision of a new world that is rapidly happening as a result of private activities, regardless of any impeding forces. It's exciting and reveals something that is truly awesome about humanity's future.

Let's hope the forces of darkness can be overcome and a more rapid advance made on this vision.

Appendix

Units of Energy and Temperature

Any numbers I produce in this book are backed by at least one source, but conclusions are often my own—I would like to hope that they are clear and rational. In general, I use kilowatt-hours (kWh) to describe units of energy and British thermal units (BTUs) to describe thermal energy. One BTU is the equivalent of the heat from a single lit match but is officially a measure of the energy required to heat 1 pound of water by 1°F. There are 3,412 BTUs in a single kilowatt-hour. Wherever I quote a temperature, it is in Fahrenheit, the American fashion, with the Celsius equivalent in brackets.

Not all the information about renewable energy contained in this book is quantitative; I have collected much supporting material. Know that I have not consciously selected an idea and then cherry-picked data to fit it. Instead, I strive to understand the science behind issues and look at the evidence used by scientists to support their conclusions. Herein, I attempt to articulate this information in a simple way so that concepts like anthropogenic climate change become better explained. This book represents an exploration of these externalities, and I quote the experts and their attempts to quantify them.

Just in case I'm interpreted as an idealist allowing wishful thinking to overcome my rationale, I am representing that my position comes from an impartial consideration of facts before drawing any conclusions. I reference all energy figures from resources such as BP, the National Renewable Energy Laboratory (NREL), and the International Energy Agency (IEA), among others, which have reputations to maintain.

The lack of an energy policy in the United States, where energy is critical for the well-being of the world's largest global economy, is a signal gap in the U.S. approach to this essential aspect of the economy. The unwritten policy appears to be "If it ain't broke, don't fix it" and "If it's profitable, defend it to the last." The United States is still the largest economy in the world, with China catching up. In the United States, just 5% of the world's population makes 25% of its wealth—and a proportionate quantity of pollution and waste in the process, the "throwaway society." It seems somehow a wasted opportunity for a nation with all this innovative capacity not to stick with its real technological advantages and implement a new era of cheaper sustainable living—but the vision is not there yet, or trusted.

Instead, the United States stands as an example of a country that questions the need to become sustainable while committing all the regressive fossil fuel sins because there is not enough widely known about the issues and the opportunity being wasted.

Notes

1   Al Gore's optimistic 2016 TED Talk. Climate Reality Project website. Available at: https://www.climaterealityproject.org/blog/ ideas-worth- spreading?gclid=CM3DmrfBss0CFdgYgQodPkcACQ

2   Martin Heidegger. Being and Time. 1927. His main work was the elucidation of the existential subjectivity of being which led to insights about the huge changes in the human condition and what it means to be a human being. A documentary was made, called "Being in the World". Available at: http:// beingintheworldmovie.com/

3   Theodore Kaczynski's manifesto, submitted to be published by the New York Times and the Washington Post, which was accomplished on September 19, 1995. Available at: http://editions-hache.com/essais/pdf/kaczynski2.pdf

4   A group calling itself the Olga Cell of the Informal Anarchist Federation International Revolutionary Front has claimed responsibility for the non-fatal shooting of a nuclear-engineering executive on 7 May in Genoa, Italy. Nature International Weekly Journal of Science. Available at: http://www. nature.com/news/anarchists-attack-science-1.10729

5   Hans Rosling. TED Talk. The magic washing machine. TEDWomen 2010 · 9:15 · Filmed Dec 2011. Available at: http://www.ted.com/talks/ hans_rosling_and_the_magic_washing_machine?language=en

6   The White House, in a report to Congress, has put the probability at 83% that a worker making less than $20 an hour in 2010 will eventually lose their job to a machine. Even workers making as much as $40 an hour, face odds of 31 percent. Available at: https://www.whitehouse.gov/sites/default/files/ docs/ERP_2016_Book_Complete%20 JA.pdf whitehouse

7   Rachel Carson. "Silent Spring". 1962. Available at: https://www.amazon.com/Silent-Spring-Rachel- Carson/ dp/0618249060?ie=UTF8&*Version*=1&*entries*=0

8   Mohamed Yunus earned a Nobel Peace Prize in 2005 for positive aspects of capitalism. Available at: https://www.nobelprize.org/nobel_prizes/peace/ laureates/2006/yunus-bio.html

9   Paul Hawken. "The Ecology of Commerce". 1993. Commentary about the "Take, make, waste" industrial habit stealing from future generations. Available at: https://www.amazon.com/ Ecology-Commerce-Revised-Declaration- Sustainability/dp/0061252794

[10] Ray Anderson: The business logic of sustainability Filmed February 2009 at TED2009. Available at: https://www.ted.com/talks/ray_anderson_on_the_business_logic_of_sustainability?la nguage=en

[11] Kirsten Korosec. In U.S., there are twice as many solar workers as coal miners. Fortune Magazine. January 16, 2015. Available at: http://fortune.com/2015/01/16/solar-jobs-report-2014/

[12] Pascala, S. and Socolow, R. "Stabilization Wedges: Solving the Climate Problem for the Next 50 Years with Current Technologies", Science, 13 August 2004, Vol. 305, No. 5686, pp 968-972. Available at: http://mae.princeton.edu/people/faculty/socolow/Science-2004-SW-1100103- PAPER-AND-SOM.pdf

[13] Amory Lovins, Reinventing Fire: Bold Business Solutions for the New Energy Era (2011) p. 251 ISBN 978-1-60358-371-8. Available at: http://www.rmi.org/Content/Files/RMI_SolutionsJournal_Fall09.pdf

[14] In 2010 and 2011, to strengthen strategic Chinese export industries the Chinese government added to the global industry's struggles by awarding their own solar companies $40.7 billion. Available at: http://mercomcapital.com/loans-and-credit- agreements-involving-chinese-banks-to-chinese-solar-companies-since-jan-2010

[15] The Economist. The cost del sol. Sustainable energy meets unsustainable costs. Jul 20th, 2013. Available at: http://www.economist.com/news/business/21582018-sustainable-energy-meets- unsustainable-costs-cost-del-sol

[16] Dr. Ben Gaddy, Director of Technology Development, Clean Energy Trust. Dr. Varun Sivaram, Douglas Dillon Fellow, Council on Foreign Relations; and Dr. Francis O'Sullivan, Director of Research and Analysis, MIT Energy Initiative. "Venture Capital and Cleantech: The Wrong Model for Clean Energy Innovation." July 2016. The report details the state of venture capital investment in cleantech startups since the sector's boom in 2006, its bust in 2009, its current status in 2016 and its future. Available at: http://energy.mit.edu/publication/venture-capital- cleantech/

[17] Jeremy Grantham. Unburnable Carbon. 2013. Available at: http://www.lse.ac.uk/GranthamInstitute/wp-content/uploads/2014/02/PB-unburnable-carbon-2013-wasted-capital-stranded-assets.pdf

[18] Ross Koningstein and David Fork were two Google engineers who revealed the thinking behind dropping the RE<COAL initiative in a November 2014 Greentech Media article. Available at: http://www.greentechmedia.com/articles/read/google- engineers-explain-why-they-stopped-rd-in-renewable-energy

[19] The partial pressure of CO_2 from the Keeling Curve, calculated by readings made in Hawaii. Available at: https://www.co2.earth/

[20] Professor Klaus Lackner. The Center for Carbon Negative Emissions. Available at: https://engineering.asu.edu/cnce/klaus-lackner/

21 Professor Klaus Lackner's CO2 absorption technology. Nova. March 4, 2015. Available at: https://www.youtube.com/watch?v=i70QI_ezDbw

22 Peter Harrison. Air Fuel Synthesis. This is a YouTube video produced by CNN's Dan Rivers, in the UK. November 28th, 2014. Available at: https://www.youtube.com/watch?v=aKPLYhlECww

23 Berkshire Hathaway's Warren Buffet's utility, NV Energy paid only 3.87 cents per kilowatt hour for solar electricity from First Solar's Nevada based 100-megawatt solar installation. Available at: http://www.bloomberg.com/news/articles/2015-07-   07/buffett-scores-cheapest-electricity-rate-with-nevada-solar-farms

24 Thomas Lansdall-Welfare, Saatviga Sudhahar, James Thompson, Justin Lewis. Content analysis of 150 years of British periodicals. FindMyPast Newspaper Team and Nello Cristianinia, Intelligent Systems Laboratory, Department of History, University of Bristol, School of Journalism, Media and Cultural Studies, University of Cardiff and FindMyPast Newspaper Archive Limited. www.britishnewspaperarchive.co.uk. Edited by Kenneth W. Wachter, University of California, Berkeley, CA, and approved November 30, 2016. PNAS. Available at: http://www.pnas.org/content/early/2017/01/03/1606380114.full.pdf

25 Frankfurt School—UNEP report. Global Trends in Renewable Energy Investment 2016. Prepared by the Frankfurt School-UNEP, Collaborating Centre for Climate & Sustainable Energy Finance and Bloomberg New Energy Finance, the United Nations Environment Programme (UNEP) and Bloomberg New Energy Finance. Available at: http://fs-unep-centre.org/publications/global-trends-renewable-energy-investment- 2016

26 Thomas Friedman. Hot Flat and Crowded. Why We Need a Green Revolution - and How It Can Renew America. November 2009. ISBN: 978-0-312-42892-1. Available at: http://www.amazon.com/Hot-Flat-Crowded-Revolution- America/dp/0312428928

27 Craters thought to be enormous bubbles of emerging methane which have left holes in the ground in Russia. There is also some thought that they may simply be melting water from ice "Pingoes" which are also present. Available at: http://news.nationalgeographic.com/news/2015/02/150227-siberia-mystery-holes- craters-pingos-methane-hydrates-science/

28 Electrification named as the 20th Century's greatest achievement by the U.S. National Academies of Science. April 2012. Available at: http://www.greatachievements.org/?id=2949

29 Marjolein P. Baar, Renata M.C. Brandt, Diana A. Putavet, Julian D.D. Klein, Kasper W.J. Derks, Benjamin R.M. Bourgeois, Sarah Stryeck, Yvonne Rijksen, Hester van Willigenburg, Danny A. Feijtel, Ingrid van der Pluijm, Jeroen Essers, Wiggert A. van Cappellen, Wilfred F. van IJcken, Adriaan B. Houtsmuller, Joris Pothof, Ron W.F. de Bruin, Tobias Madl, Jan H.J. Hoeijmakers, Judith Campisi, Peter L.J. de Keizer10,'Correspondence

information about the author Peter L.J. de Keizer. Targeted Apoptosis of Senescent Cells Restores Tissue Homeostasis in Response to Chemotoxicity and Aging. Cell, Volume 169, Issue 1, p132–147.e16, 23 March 2017. Available at: http://www.cell.com/cell/fulltext/S0092-8674(17)30246-5

[30] Henry Lee and Grant Lovellette. Will Electric Cars Transform the U.S. Vehicle Market? An Analysis of the Key Determinants. Belfer Center Discussion Paper 2011-08. July 2011 Available at: http://www.belfercenter. org/sites/default/files/legacy/files/Lee%20Lovellette%20Ele   ctric%20 Vehicles%20DP%202011%20web.pdf

[31] McKinsey & Company. Report with Bloomberg New Energy Finance on Electric vehicles. An integrated perspective on the future of mobility. October 2016. Available at: https://data.bloomberglp.com/bnef/sites/14/2016/10/ BNEF_McKinsey_The-Future- of-Mobility_11-10-16.pdf

[32] Peter O'Connor is a Kendall Science Fellow at the Union of Concerned Scientists researching the integration of solar power and electric vehicles into the electricity system.Available at: http://blog.ucsusa.org/peter-oconnor/ what-electric-vehicle-sales- in-2016-mean-for-the-future-of-transportation

[33] The Half the Oil Plan (2013), a plan to cut 11 million barrels per day of oil consumption in the U.S. by 2035. Available at: http://www.ucsusa.org/clean-vehicles/clean-fuels/half-the-oil-how-it-works#.WH6lU_krJaQ

[34] Edison Power, subsidiary of Sunvault (ticker: SVLT) has contracted with Robert Murray Smith for a graphene battery for a vehicle, made with units the size of 500 leaf paper reams. Available at: https://www.youtube.com/ watch?v=P266pdT71tI

[35] Alexander J. Epstein. The Moral Case for Fossil Fuels. Penguin. 2014.ISBN 978-0- 698-17548-8

[36] Alexander J. Epstein, Bill McKibben. The Ethics for Fossil Fuel Use" a debate at The Program in Values and Ethics in the Marketplace. Duke University. 2012. Available at: https://www.youtube.com/watch?v=0_a9RP0J7PA

[37] Professor Geoffrey Heal of Columbia University in the City of New York, wrote a National Bureau of Economic Research working paper. Available at: http://www.nber.org/papers/w22525

[38] The EU has reduced its carbon emissions already by almost 25% since 1990 and at the same time increased its economy by almost 50%. Available at: http://ec.europa.eu/clima/policies/international/negotiations/paris/ index_en.htm

[39] Jason M. Breslow. Investigation Finds Exxon Ignored its Own Early Climate Change Warnings. September 16, 2015. Available   at: http://www.pbs.org/wgbh/frontline/article/ investigation-finds-exxon-ignored-its-own- early-climate-change-warnings/

[40] H.R. Brannon, Jr., A.C. Daughtry, D. Perry, W.W. Whitaker, & M. Williams. Radiocarbon Evidence on the Dilution of Atmospheric and Oceanic Carbon by Carbon from Fossil Fuels. Production Research Division, Humble Oil

and Refining Company, P.O. Box 2180, Houston 1, Texas. October 1957. Volume 38. Number 5. The American Geophysical Union. Available at: http://agupubs.onlinelibrary.wiley.com/hub/journal/10.1002/(ISSN)2324-9250/

41 Author: E. Robinson, & R.C. Robbins. Sources, abundance, and fate of atmospheric pollutants. Written for the American Petroleum Institute. Stanford Research Institute. February 1968. Available at: https://www.smokeandfumes.org/documents/16

42 Inside Climate News, a website and Pulitzer Prize winning Media group who, with Columbia University's School of Journalism, wrote about Exxon Mobil's climate pivot that happened in the 1980's. Available at: https://insideclimatenews.org/news/25112015/exxon-deep-cuts-climate-change- research-budget-1980s-global-warming

43 E2. An independent and non-partisan group of business leaders, investors and advocates for smart economic policies. Available at: http://www.e2.org/economic- benefits-californias-climate-leadership/

44 California Senate Bill 32 (SB32). Available at: https://leginfo.legislature.ca.gov/faces/billNavClient.xhtml?bill_id=201520160SB32

45 Pope Francis' encyclical letter "Laudato Si". Available at: https://w2.vatican.va/content/dam/francesco/pdf/encyclicals/documents/papa- francesco_20150524_enciclica-laudato-si_en.pdf

46 The Economist. Puffs of hope. Renewable energy. August 1st, 2015. Available at: http://www.economist.com/news/leaders/21660124-wind-and-solar-energy-are- increasingly-competitive-lot-has-change-they-can-make

47 Go100%. A website that covers global projects to obtain 100% renewable energy. Available at: http://www.go100percent.org/cms/index.php?id=17

48 The Sierra Club campaign to support 100 cities contemplating renewable energy to sign on to make themselves 100% renewable. Available at: http://www.sierraclub.org/ready-for-100

49 Carbon Disclosure Project. Available at: https://www.cdp.net/en/cities/world- renewable-energy-cities

50 Hans Rosling. The Health and Wealth of Nations. A video of a chart showing the growth in both over the last 200 years. Available at: http://www.gapminder.org/

51 World Population was 7.47 on December 14th, 2016 in the opinion of this website: http://www.worldometers.info/world-population/

52 Jeremy Rifkin. The Hydrogen Economy. Pages 61, 62. Jeremy P. Tarcher/Putnam. 2002.

53 The Declaration of Independence". National Archives and Records Administration.

54 Jens Alber, Jan Delhey, Wolfgang Keck and Ricarda Nauenburg. First European Quality of Life Survey 2003. Social Science Research Centre (WZB), Berlin.

55 Naomi Oreskes & Erik M. Conway. Merchants of Doubt. How a Handful of Scientists Obscured the Truth on Issues from Tobacco Smoke to Global Warming. Bloomsbury Press. ISBN: 978-1-60819-394-3.

56 Cristin E. Kearns, Dorie Apollonio, Stanton A. Glantz. Sugar industry sponsorship of germ-free rodent studies linking sucrose to hyperlipidemia and cancer: An historical analysis of internal documents. Available at: https://doi.org/10.1371/journal.pbio.2003460

57 "Scientists' Report Documents ExxonMobil's Tobacco-like Disinformation Campaign on Global Warming Science". Union of Concerned Scientists. 3 January 2007. "Fairness and accuracy in Myron". The Myron Ebell Climate.

58 "Exxon Mobil softens its climate-change stance". Post-gazette.com. 2007-01-11. Retrieved 2011-08-25.

59 Richard Seager. "Causes and Predictability of the 2011 to 2012 California Drought, 2014. The claim is made that the California drought, the most severe as of the summer of 2014 in 1,200 years, is not related to climate change.

60 The Department of Energy website shows the ramp up in production of U.S. oil and the reduction in dependence on foreign oil. Available at: http://energy.gov/science-innovation/energy-sources/fossil

61 Thomas Friedman. "Hot, Flat and Crowded". 2008. Farrar, Straus and Giroux, NY. P 21.

62 Available at: http://www.bp.com/en/global/corporate/about-bp/energy-economics/statistical-review-of-world-energy/review-by-energy-type/natural- gas/natural-gas-prices.html

63 Stephen &Donna Leeb. The Oil Factor: Protect Yourself-and Profit-from the Coming Energy Crisis. February 12, 2004.

64 Matt Ridley is a climate change denier but accepts that human caused climate change is happening. He just thinks it takes much longer for any of the alarmist predictions to manifest. He has a blog at: http://www.rationaloptimist.com/blog/my- life-as-a-climate-lukewarmer.aspx

65 Germany Reaches New Levels of Greendom, Gets 31 Percent of Its Electricity From Renewables. Bloomberg Business Week. Caroline Winter. August 14, 2014.

66 Karl R. Haapala; Preedanood Prempreeda. Comparative life cycle assessment of 2.0 MW wind turbines. International Journal of Sustainable Manufacturing, 2014 DOI: 10.1504/IJSM.2014.062496.

67 The Alternative Fuels Data Center compares many liquid fuels and gases. Available at: http://www.afdc.energy.gov/fuels/fuel_comparison_chart.pdf

68 The European Nuclear Society has a page on fuel comparison. Available at: https://www.euronuclear.org/info/encyclopedia/f/fuelcomparison.htm

69 Alliance Bioenergy Plus. Available at: http://www.amgrenewables.com/

70 Clinton Campbell, Robert Margolis, Paul Denholm and Garvin Heath. "Land-use Requirements for Solar Power Plants in the United States",

National Renewable Energy Laboratory report. July 30, 2013. Available at: www.renewableenergyworld.com/rea/blog/post/print/2013/08/how-much-land-does- solar-take-not-so-much-nrel-finds

[71] "Land-use Requirements and the Per-capita Solar Footprint for Photovoltaic Generation in the United States." NREL.

[72] Fthenakis, V.; Kim, H.C. (2009). "Land Use and Electricity Generation: A Life- Cycle Analysis." Renewable and Sustainable Energy Reviews (13); pp. 1465–1474.

[73] Professor Alexander Mitsos and Corey J. Noone. MIT. Manuel Torrilhon. RWTH Aachen University in Germany. Heliostat Field Optimization: A New Computationally Efficient Model and Biomimetic Layout. December 22, 2011.

[74] BACKGROUNDER. –A Comparison: Land Use by Energy Source— Nuclear, Wind and Solar.

[75] Aramco World, Arab and Islamic cultures and connections. September/ October issue 1981, Volume 32, Number 5. Available at: https://www.saudiaramcoworld.com/issue/198105/saudi.arabia.and.solar.energy.a.spe cial.section.htm

[76] The New Economy, an Australian website for clean energy news. Available at: http://reneweconomy.com. au/2013/citigroup-how-solar-module-prices-could- fall-to- 25cwatt-41384

[77] Updated Capital Cost Estimates for Utility Scale Electricity Generating Plants. April 2013. U. S. Energy Information Administration.

[78] Huadian Gansu Zhangye Solar Thermal Power (1 X 6MW) Project Feasibility Study Stage. Feasibility Study Report. F7221K-A01-01. NWEPDI of CPECC. April 2010.

[79] The Desert Tortoise relocation project at Ivanpah. Available at: https://www.youtube.com/watch?v=A--1eRAcQd0

[80] Frank R. Spellman. Environmental Impacts of Renewable Energy. Taylor & Francis Group. 2015.

[81] Available at: http://www.solaripedia.com/259/3256/Ivanpah+SEGS+%28California+USA%29.htm l

[82] Available at: http://www.solaripedia.com/13/31/solar_one_and_two_%28now_defunct%29.html

[83] U.S. Fish & Wildlife Service. Available at: http://www.fws.gov/mountain-prairie/pressrel/11-64.html

[84] Susan Kraemer. One Weird Trick Prevents Bird Deaths At Solar Towers. Cleantechnica. April 16th, 2015. Available at: http://cleantechnica.com/2015/04/16/one-weird-trick-prevents-bird-deaths-solar- towers/

[85] Vasilis M. Fthenakis, Hyung Chul Kim, And Erik Alsema. Emissions from Photovoltaic Lifecycles. NREL. Environ. Sci. Technol. 2008, 42, 2168–2174. Available at: http://pubs.acs.org/doi/pdf/10.1021/es071763q

[86] "An yll wynde that blowth no man to good, men say." John Heywood's A dialogue conteinyng the number in effect of all the prouerbes in the Englishe tongue, 1546. John Heywood (1497–1580) was an English writer known for his plays, poems, and collection of proverbs.

[87] Wind energy is considered a disaster responding to the hoax of climate change in this vociferous website which of course also discusses wind turbine syndrome. Available at: http://www.windturbinesyndrome.com/wind-turbine-syndrome/what-is- wind-turbine-syndrome/

[88] The Caithness Windfarm Information Forum. Available at: http://www.caithnesswindfarms.co.uk/

[89] RenewableUK. A leading renewable energy trade association. Available at: http://www.renewableuk.com/en/events/conferences-and-exhibitions/renewableuk- 2015/

[90] Risø National Laboratory for Sustainable Energy. Available at: http://orbit.dtu.dk/en/organisations/risoe-national-laboratory-for-sustainable-energy%2869f3623e-9f3f-48aa-8b46-4b4fb2abab7f%29.html

[91] Available at: http://www.liveleak.com/view?i=3cd_1383772851#Opj3eWpLL6Co282t.99

[92] David Wahl, Philippe Giguere. Ice Shedding and Ice Throw—Risk and Mitigation. Wind Application Engineering. GE Energy. Available at: http://www.cbuilding.org/sites/cbi.drupalconnect.com/files/ger4262.pdf

[93] Cattin et al. Wind Turbine Ice Throw Studies in the Swiss Alps. EWEC 2007. Based on studies of a 600 kW Enercon E-40 at 2,300 mASL in Swiss Alps

[94] Summary of Wind Turbine Accident Data to 30 September 2014. PDF. Caithness Windfarm Information Forum.

[95] Payback time for renewable energy. NREL factsheet. Available at: http://www.nrel.gov/docs/fy13osti/57131.pdf

[96] Sibley and Monroe. 1992.

[97] Kevin J. Gaston and Tim M. Blackburn. April 1997. How many birds are there? Available at: http://link.springer.com/article/10.1023%2FA%3A1018341530497

[98] K. Shawn Smallwood, "Comparing bird and bat fatality-rate estimates among North American wind-energy projects", Wildlife Society Bulletin, 26 Mar. 2013. Available at: http://onlinelibrary.wiley.com/doi/10.1002/wsb.260/pdf

[99] Wallace P. Erickson, Gregory D. Johnson and David P. Young Jr. A Summary and Comparison of Bird Mortality from Anthropogenic Causes with an Emphasis on Collisions. USDA Forest Service. PSW-GTR-191. 2005. Available at: http://www.fs.fed.us/psw/publications/documents/psw_gtr191/Asilomar/pdfs/1029- 1042.pdf

[100] Erickson WP, Wolfe MM, Bay KJ, Johnson DH, Gehring JL (2014) A Comprehensive Analysis of Small-Passerine Fatalities from Collision with

Turbines at Wind Energy Facilities. PLoS ONE 9(9): e107491. doi:10.1371/journal.pone.0107491

[101] State of North America's Birds 2016. North American Bird Conservation Initiative. Available at: http://www.stateofthebirds.org/2016/overview/results- summary/

[102] BirdLife International (2014) We have lost over 150 bird species since 1500. Presented as part of the BirdLife State of the world's birds website. Available from: http://www.birdlife.org/datazone/sowb/casestudy/102

[103] Sovacool, Benjamin K., 2009. "Contextualizing avian mortality: A preliminary appraisal of bird and bat fatalities from wind, fossil-fuel, and nuclear electricity," Energy Policy, Elsevier, vol. 37(6), pages 2241-2248, June. Available at: https://ideas.repec.org/a/eee/enepol/v37y2009i6p2241-2248.html

[104] PLos Biology published a report in 2011 which was written by the Census of Marine Life scientists. Available to: http://journals.plos.org/plosbiology/article?id=10.1371/journal.pbio.1001127

[105] Bernie Krause. A recorder of natural sounds in many global habitats. Available at: http://onpoint.wbur.org/2015/08/28/the-worlds-disappearing-natural-sound

[106] Bernie Krause. TED Talk. The voice of the natural world. TEDGlobal 2013 · 14:48 · Filmed Jun 2013. Available at: https://www.ted.com/talks/bernie_krause_the_voice_of_the_natural_world?language=en

[107] John Bakeless. America As Seen by Its First Explorers: The Eyes of Discovery. Dover Language Books & Travel Guides. Paperback—January 20, 2011. Available at: http://www.amazon.com/America-Seen-Its-First-Explorers/dp/0486260313

[108] Bald Eagle seriously injured by wind turbine. Available at: https://www.youtube.com/watch?v=jwVz5hdAMGU

[109] The White Throated Needletail death on YouTube. Geobeats news service. July 1, 2013. Available at: https://www.youtube.com/watch?v=IO5LjgIEofg

[110] Rare swift killed by Scottish wind turbine. Available at: http://www.scotsman.com/news/scotland/top-stories/birdwatchers-see-rare-bird- killed-by-wind-turbine-1-2980240

[111] Exxon Mobil pleads guilty to bird deaths. Available at: http://abcnews.go.com/Business/story?id=8322081

[112] BP and Pacificorp pay fines for killing birds. Available at: http://www.businessinsider.com/obama-eagle-death-wind-farm-oil-energy-epa-2013- 5

[113] Pawel Plonczkier and Ian C. Simms. Journal of Applied Ecology. 2012. Available at: http://onlinelibrary.wiley.com/doi/10.1111/j.1365-2664.2012.02181.x/epdf

[114] Mike Barnard. 10 August 2012. Want to save 70 million birds a year? Build more windfarms. RenewEconomy. Available at: http://reneweconomy.com.au/2012/want- to-save-70-million-birds-a-year-build-more-wind-farms-18274

[115] Erin F. Baerwald, Genevieve H. D'Amours, Brandon J. Klug and Robert M.R. Barclay. Barotrauma is a significant cause of bat fatalities at wind turbines.

[116] "NREL Study Finds Barotrauma Not Guilty", November 27, 2012. Available at: http://www.nrel.gov/wind/news/2013/2149.html

[117] Germany has 74% of its power supplied by renewable energy. 2014. Available at: http://gas2.org/2014/05/27/for-one-hour-germany-was-powered-by-74-renewables/

[118] Information supplied by Agora Energiewende, a research institute in Berlin, showed that Germany's demand for electricity was almost 100% supplied by renewable energy including a large amount of wind on the 15th May 2016. Available at: http://www.bloomberg.com/news/articles/2016-05-16/germany-just-got-almost- all-of-its-power-from-renewable-energy

[119] Posthumous pardons of First World War shellshock victims. Available on: http://www.telegraph.co.uk/news/1526437/Pardoned-the-306-soldiers-shot-at-dawn- for-cowardice.html

[120] Information Paper: Evidence on Wind Farms and Human Health. February 2015. PDF. National Health and Medical Research Council. Available at: http://www.nhmrc.gov.au/_files_nhmrc/publications/attachments/eh57a_information_paper.pdf

[121] Ian Clark, William N. Alexander, William J. Devenport, Stewart A. Glegg, Justin Jaworski, Conor Daly, and Nigel Peake. "Bio-Inspired Trailing Edge Noise Control", 21st AIAA/CEAS Aeroacoustics Conference, AIAA AVIATION Forum, (AIAA 2015-2365). Available at: http://dx.doi.org/10.2514/6.2015-2365

[122] UK Renewable Energy Roadmap. Available at: https://www.gov.uk/government/uploads/system/uploads/attachment_data/file/48128/2167-uk-renewable-energy-roadmap.pdf

[123] Positive environmental impacts of offshore wind farms. European Wind Energy Association. Available at: http://www.ewea.org/fileadmin/files/members- area/information-services/offshore/research- notes/120801_Positive_environmental_impacts.pdf

[124] Kristopher B. Karnauskas, Julie K. Lundquist & Lei Zhang. Southward shift of the global wind energy resource under high carbon dioxide emissions. Nature Geoscience 11, 38–43 (2017)doi:10.1038/s41561-017-0029-9. Published 11 December 2017. Available at: https://www.nature.com/articles/s41561-017-0029- 9#auth-1

[125] Lake Powell receives enough silt to fill 1,400 container ships a day. Available at: http://www.lakepowellviewestates.com/lake-powell-page-arizona-real-estate-   blog/bid/199806/The-Damn-Glen-Canyon-Dam

[126] The Tarbela Dam. Available at: http://www.wapda.gov.pk/htmls/pgeneration-dam-tarbela.html

[127] Tarbela Dam silting. Available at: http://www.internationalrivers.org/sedimentation-problems-with-dams

[128] The Cerron Grande Dam in El Salvador. Available at: http://www.cathalac.org/lac_atlas/index.php?option=com_content&view=article&id=34:cerron-grande-el-salvador&catid=1:casos&Itemid=5

[129] The Colorado River does not reach the ocean. Available at: http://www.smithsonianmag.com/science-nature/the-colorado-river-runs-dry- 61427169/?no-ist

[130] Lindzen, Richard S., Ming-Dah Chou, and Arthur Y. Hou; 2001: "Does the Earth Have an Adaptive Infrared Iris?" Bulletin of the American Meteorological Society, Vol. 82, No. 3, pp. 417-32

[131] Earthquakes and Dams. Zipingpu Dam in Sechuan Province, China. Available at: http://www.telegraph.co.uk/news/worldnews/asia/china/4434400/Chinese- earthquake-may-have-been-man-made-say-scientists.html

[132] Greenhouse Gases from Reservoirs Caused by Biochemical Processes. Rikard Liden. The World Bank. April 2013. Available at: https://openknowledge.worldbank.org/handle/10986/16535

[133] Three Gorges Dam Power Output. Available at: http://news.yahoo.com/chinas- three-gorges-dam-breaks-world-hydropower-record-004111862.html

[134] Thayer Watkins. "The Catastrophic Dam Failures in China in August 1975". San Jose State University. Retrieved 2013-11-25.

[135] Marc F. Bellemare. "Rising Food Prices, Food Price Volatility, and Social Unrest." June 26, 2014. The Oxford University Press. An article published in the January 2015 issue of the American Journal of Agricultural Economics, mentions this study.

[136] Thomas Friedman. "WikiLeaks, Drought and Syria". January 21st, 2014. New York Times. Available at: http://www.nytimes.com/2014/01/22/opinion/friedman- wikileaks-drought-and-syria.html?_r=0

[137] DeCicco, J.M., Liu, D.Y., Heo, J. et al. Climatic Change (2016). doi:10.1007/s10584-016-1764-4. Carbon balance effects of U.S. biofuel production and use. Available at: http://link.springer.com/article/10.1007%2Fs10584-016-1764- 4

[138] Dupont Biofuels. "DuPont to the Senate Ag Committee: 'Drive by the RFS, we have completely re-imagined how we fuel our planet.'" Available at: http://biofuels.dupont.com/cellulosic-ethanol/nevada-site-ce-facility/single-blog- post/blog/posts/show/post/dupont-to-the-senate-ag-committee-driven-by-the-rfs-we- have-completely-re-imagined-how-we-fuel/

[139] Amanda Peterka, E&E reporter. Biofuel-obsessed U.S. seen blowing 'historic opportunity' in nascent green chemical market. Friday, May 23, 2014. Available at: http://renmatix.com/wp-content/uploads/2014/06/greenwire.pdf

[140] Marc F. Bellemare. July 27, 2012. "Drought, Extreme Temperature, and the Consequences of High Food Prices". University of Minnesota.

[141] Aki Kortelainen et al. Real-Time Chemical Composition Analysis of Particulate Emissions from Woodchip Combustion. Energy Fuels. January 15, 2015. DOI: 10.1021/ef5019548

[142] NRDC Issue Brief. Sami Yassa, Senior Scientist. syassa@nrdc.org. Think Wood Pellets are Green? Think Again. May 2015. Available at: https://www.nrdc.org/sites/default/files/bioenergy-modelling-IB.pdf

[143] Astronauts Anders, Borman and Lovell see the Earth rising above the Moon's horizon as they orbit the Moon. Reconstruction of the moment to celebrate the 45th anniversary of the event on Christmas Eve, 1968 available at: https://www.nasa.gov/content/goddard/nasa-releases-new-earthrise-simulation-video

[144] Galileo Galilei. 1632. Dialogue Concerning the Two Chief World Systems. Available at: http://law2.umkc.edu/faculty/projects/ftrials/galileo/dialogue.html

[145] Greg Craven. "The Most Terrifying Video You'll Ever See". Available at: https://www.youtube.com/watch?v=zORv8wwiadQ

[146] The World Health Organization. Climate change and human health - risks and responses. Available at: http://www.who.int/globalchange/summary/en/index12.html

[147] "We're the first generation to feel the impact of climate change, and the last generation that can do something about it." - Jay Inslee, governor of Washington State (since January 2013) in the 9-part documentary "Years of Living Dangerously" on climate change that premiered on April 13th, 2014 on the premium TV network Showtime.

[148] Paracelsus, aka Philippus Theophrastus Aureolus Bombastus von Hohenheim. John S. Rigden (2003). Hydrogen: The Essential Element. Harvard University Press. p. 10. ISBN 978-0-674-01252-3.

[149] F. J. Moore. A History of Chemistry, New York: McGraw-Hill (1918) pp. 34–36

[150] "Kepler's decision to base his causal explanation of planetary motion on a distance-velocity law, rather than on uniform circular motions of compounded spheres, marks a major shift from ancient to modern conceptions of science.... [Kepler] had begun with physical principles and had then derived a trajectory from it, rather than simply constructing new models. In other words, even before discovering the area law, Kepler had abandoned uniform circular motion as a physical principle." Peter Barker and Bernard R. Goldstein, "Distance and Velocity in Kepler's Astronomy", Annals of Science, 51 (1994): 59–73, at p. 60.

[151] Joseph Louis Legrange, who was an Italian initially but made many contributions in mathematics and astronomy. The Legrange point, where gravity between different celestial bodies is equally strong between them, was named for him.

[152] Antoine Lavoisier named hydrogen. "Hydrogen". Van Nostrand's Encyclopedia of Chemistry. Wylie-Interscience. 2005. pp. 797–799. ISBN 0-471-61525-0.

[153] L'Entreprenant was a reconnaissance balloon used by the French army in battles in Northern Europe between 1794—1799. They used hydrogen, originally generated by using sulphuric acid but then other means were found.

[154] Weeks, Mary Elvira (1932). "The discovery of the elements. IV. Three important gases". Journal of Chemical Education 9 (2): 215. Bibcode: 1932JChEd...9215W. doi: 10.1021/ed009p215.

[155] Jean-Antoine Chaptal, a high-ranking figure in Napoleon's government and a chemist.

[156] Joseph Priestley and Carl Scheele discovered oxygen independently. Cook & Lauer 1968, p.499 - 500

[157] The USGS Water Science School. Oxygen is 60% of our body weight. Available at: http://water.usgs.gov/edu/propertyyou.html

[158] Joseph Priestley discovered "fixed air" or carbon dioxide (CO_2).

[159] Fourier J (1824). "Remarques Générales Sur Les Températures Du Globe Terrestre Et Des Espaces Planétaires". Annales de Chimie et de Physique 27: 136–67.

[160] Eunice Newton Foote. At the 10th annual meeting of the American Association for the Advancement of Science (AAAS) in Albany in 1856, Joseph Henry of the Smithsonian Institution read a paper on behalf of author Mrs. Eunice Foote. AAAS, 1857, Proceedings of the American Association for the Advancement of Science. Tenth Meeting, held at Albany, New York, August 1856. Joseph Lovering, Cambridge, 258 p. Available at: http://www.searchanddiscovery.com/documents/2011/70092sorenson/ndx_sorenson.pdf

[161] John Tyndall discovered that methane and carbon dioxide are both opaque to long wave radiation and infra-red radiation. Available at: http://www.aip.org/history/climate/co2.htm. Tyndall's paper "On the Absorption and Radiation of Heat by Gases and Vapours, and on the Physical Connexion of Radiation, Absorption, and Conduction" in 1861, was the first to establish that CO_2, water vapor, and other gases had radiative properties, although others had speculated similarly before.

[162] The Tyndall Centre for Climate Change Research. Available at: http://www.tyndall.ac.uk/

[163] Chamberlin, T.C., 1899. An Attempt to Frame a Working Hypothesis of the Cause of Glacial Periods on an Atmospheric Basis, Journal of Geology, Vol. 7. pp. 575, 667, 751. Available at: https://archive.org/details/jstor-30056816

[164] Svante Arrhenius realized that changing carbon dioxide would also have a direct effect on surface temperature. Available at: http://earthobservatory.nasa.gov/Features/Arrhenius/

[165] Earth's population in 1800. United States Census Bureau. These figures include estimates from several sources. Available at: https://www.census.gov/population/international/data/worldpop/table_history.php

[166] The Gaia Hypothesis/Paradigm: Available at: http://www.gaiatheory.org/

[167] Milutin Milankovitch. Milankovitch cycles. Available at: http://www.ncdc.noaa.gov/paleo/milankovitch.html

[168] Milutin Milankovic. Contribution to the mathematical theory of climate. Belgrade. April 5, 1912. Available at: http://elibrary.matf.bg.ac.rs/bitstream/handle/123456789/3680/mm12F.pdf?sequence=1

[169] D.M. Etheridge, L.P. Steele, R.L. Langenfelds, R.J. Francey, J.-M. Barnola and V.I. Morgan. 1998. Historical CO_2 records from the Law Dome DE08, DE08-2, and DSS ice cores. In Trends: A Compendium of Data on Global Change. Carbon Dioxide Information Analysis Center, Oak Ridge National Laboratory, U.S. Department of Energy, Oak Ridge, Tenn., U.S.A.

[170] Peter Sinclair. Climate Denial Crock of the Week. Available at: https://www.youtube.com/watch?v=8nrvrkVBt24

[171] Layers in lake clays illustrate the 21,000-year Milankovitch cycles. Bradley, Wilmot H. (1929). "The Varves and Climate of the Green River Epoch." U.S. Geological Survey Professional Papers 158E: 85-110.

[172] Milutin Milankovic. Mathematical Theory of Heat Phenomena Produced by Solar Radiation. Gauthier-Villars. Paris. 1920. Available at: https://www.amazon.fr/math%C3%A9matique-ph%C3%A9nom%C3%A8nes-thermiques-radiation-Milankovitch/dp/B001BW5600

[173] Vladimir Ivanovitch Vernadsky. Mankind was after all, capable as a life form of altering the biosphere. Available at: http://biospherology.com/vernadsky.htm

[174] Gilbert N. Plass, James Rodger Fleming, Gavin Schmidt. Carbon Dioxide and the Climate. American Scientist. 1956. Available at: http://www.americanscientist.org/issues/feature/2010/1/carbon-dioxide-and-the-climate

[175] Gilbert Norman Plass. Made three predictions in Time Magazine in 1953 about CO2 and the climate and was almost spot on in each instance. Available at: "Science: Invisible Blanket - TIME". Time. May 25, 1953.

[176] Carbon Dating invented in the late 1940's by Willard Libby who earned a Nobel Prize. (1946). "Atmospheric helium three and radiocarbon from cosmic radiation". Physics Review 69 (11–12): 671–672.

[177] Time Magazine. "One Big Greenhouse". May 28, 1956. Available at: http://content.time.com/time/magazine/article/0,9171,937403,00.html

[178] Benjamin Franta. The Guardian. Available at: https://www.theguardian.com/environment/climate-consensus-97-per-cent/2018/jan/01/on-its-hundredth-birthday-in-1959-edward-teller-warned-the-oil-industry-about-global-warming

[179] National Academy of Sciences. It was possible to change climate without meaning to—something they called "inadvertent modifications of weather and climate"—and they specifically cited carbon dioxide as a contributing factor. Available at: http://www.smithsonianmag.com/science-nature/ talking-about-climate-change-how- weve-failed-and-how-we-can-fix-it-180951070/#z8srgz7QcsGhlZOl.99

[180] Revelle and Keeling wrote a paper for the President's Science Advisory Committee called, "Restoring the Quality of Our Environment. November 1965. Available at: http://dge.stanford.edu/labs/caldeiralab/Caldeira%20 downloads/PSAC,%201965,%20 Restoring%20the%20Quality%20of%20 Our%20Environment.pdf

[181] James Hansen. Target Atmospheric CO_2: Where Should Humanity Aim? National Aeronautics and Space Administration. Goddard Institute for Space Studies. 2008. Available at: https://arxiv.org/ftp/arxiv/ papers/0804/0804.1126.pdf

[182] Peter Gwynne. The Cooling World. April the 28th, 1975. Newsweek, Science section. Available at: https://www.scribd.com/doc/225798861/ Newsweek-s-Global- Cooling-Article-From-April-28-1975

[183] Peter Gwynne. My 1975 'Cooling World' Story Doesn't Make Today's Climate Scientists Wrong. Wednesday, December 2014. Inside Science. Available at: https://www.insidescience.org/news/my-1975-cooling-world-story-doesnt-make- todays-climate-scientists-wrong

[184] In 1979 the first "World Climate Conference" organized by the World Meteorological Organization (WMO) expressed concern that "continued expansion of man's activities on Earth may cause significant extended regional and even global changes of climate". It called for "global cooperation to explore the possible future course of global climate and to take this new understanding into account in planning for the future development of human society." Available at: https://www.wmo.int/pages/themes/climate/ international_background.php

[185] The Jason Committee. A government advisory group comprising younger scientists who were not of the post war national laboratory background, who were able to provide a new perspective and help the defense establishment. Gordon J.F. MacDonald et al." The Long-Term Impact of Atmospheric Carbon Dioxide on Climate". Jason Report. April 1979. SRI International.

[186] Easter Island. The Rapa Nui of Easter Island were long thought to have devolved into war as they consumed the last of their timber forest, putting fishing and travel beyond their capabilities. Jared Diamond used the story in his book, "Collapse". New data suggest that the Moai, the Easter Island statues could have been erected by far fewer individuals and that the population decline was more likely due to something that the indigenous American's had experienced, Small Pox and other

European diseases. Available at: http://arstechnica.com/science/2016/02/ new-evidence-easter- island-civilization-was-not-destroyed-by-war/

[187] Yann Arthus-Bertrand. Ted Talk. A wide-angle view of fragile Earth. TED2009 · 14:54 · Filmed Feb 2009. Available at: https://www.ted.com/ talks/yann_arthus_bertrand_captures_fragile_earth_in_wide_an gle

[188] Paul Bernstein and W. David Montgomery. The cost of controlling carbon dioxide emissions: final report. Prepared by W. David Montgomery. Washington, D.C. Charles River Associates Inc., 1991.1944. Available at: http://trove.nla.gov.au/version/33840524

[189] National Research Council (NRC). National Academy of Sciences (NAS). Surface Temperature Reconstructions for the Last 2,000 Years. 2006. ISBN: 978-0-309-10225-4. Available at: http://www.nap.edu/catalog/11676/ surface-temperature- reconstructions-for-the-last-2000-years

[190] The Global Carbon Project was established in 2001 in recognition of the large scientific challenges and critical nature of the carbon cycle for Earth's sustainability. Available at: http://www.globalcarbonproject.org/about/ index.htm

[191] Jeremy Grantham. Unburnable Carbon 2013: Wasted Capital and Stranded Assets. Available at: http://carbontracker.live.kiln.digital/Unburnable-Carbon-2-Web- Version.pdf

[192] Kim Naudts1, Yiying Chen, Matthew J. McGrath, James Ryder, Aude Valade, Juliane Otto, Sebastiaan Luyssaert. Europe's forest management did not mitigate climate warming. Science Magazine, 5th February 2016. Available at: http://science.sciencemag.org/content/351/6273/597

[193] Coco Liu, Scientific American. China's Great Green Wall Helps Pull CO2 Out of Atmosphere. April 24, 2015. Available at: http://www.scientificamerican. com/article/china-s-great-green-wall-helps-pull-co2- out-of-atmosphere/

[194] Scientists have managed to reach back for data as far as 800,000 years using the trapped bubbles in Antarctic ice

[195] Aradhna Tripati. Honored with the E.O. Wilson Award for her work to understand past climate change. December 17th, 2014. Available at: http://www.biologicaldiversity.org/news/press_releases/2014/wilson-award-12-17- 2014.html

[196] New York Times. Science Times. August 9, 2016. A Keeper of Carbon is at Risk. Peat bogs contain almost as much carbon as the atmosphere. Available at. http://www.nytimes.com/2016/08/09/science/climate-change-carbon-bogs- peat.html?_r=1

[197] Peter Mayhew, Gareth Jenkins. University of York. "Fossil Record Supports Evidence of Impending Mass Extinction." Proceedings of the Royal Society. 24 October 2007.

[198] Georg Feulner. Formation of most of our coal brought Earth close to global glaciation" has been published in the journal PNAS. vol. 114 no. 43. Available at: http://www.pnas.org/content/114/43/11333.abstract

[199] NPR Finally Stops Referring to Global Warming Deniers as "Skeptics". 11/17/2014 7:59 pm EST. Mark Boslough, Physicist. The Huffington Post.

[200] The BBC Trust issued a report on impartiality and scientific evidence. Trust Conclusions on the Executive Report on Science Impartiality Review Actions. July 2014. Available at: http://downloads.bbc.co.uk/bbctrust/assets/files/pdf/our_work/science_impartiality/tr ust_conclusions.pdf

[201] Rand Simberg. The Other Scandal in Unhappy Valley. The Competitive Enterprise Institute's conflation of a child molesting coach at Penn State University with Michael Mann's scientific achievement of providing color at long last on the likely evolution of annual temperatures over the last 1,000 years. Available at: https://cei.org/blog/other-scandal-unhappy-valley

[202] Mcintyre, S. and R. McKitrick (2005). "Hockey sticks, principal components, and spurious significance". Geophysical Research Letters.

[203] Dr. Michael Mann hitting back at deniers in the New York Times. If you See Something, Say Something. January 17th, 2014. Available at: https://www.nytimes.com/2014/01/19/opinion/sunday/if-you-see-something-say-something.html?_r=0

[204] The letter written by 255 scientists and published in Science Magazine. May 7, 2010. Volume 328, Issue 5979, pp. 689-690. Available at: http://science.sciencemag.org/content/328/5979/689.full

[205] Skepticalscience.com has an updated chart showing a common method used by denialists to say that there is no global warming. Available at: https://www.youtube.com/watch?v=xWdJuNYLTLs

[206] Peter Sinclair has a series of videos discussing the positions of climate deniers. Available at: https://www.youtube.com/user/greenman3610

[207] James Delingpole. "The last thing I would want is for Monbiot, Mann, Flannery, Jones, Hansen and the rest of the Climate rogues' gallery to be granted the mercy of quick release. Publicly humiliated? Yes please. Having all their crappy books remaindered? Definitely. Dragged away from their taxpayer funded troughs and their cushy sinecures, to be replaced by people who actually know what they're talking about? For sure. But hanging? Hell no. Hanging is far too good for such ineffable toerags." Available at: http://www.desmogblog.com/james-delingpole

[208] A leaked Heartland Institute memo. Confidential Memo: 2012 Heartland Climate Strategy. January 2012. Brendan Demelle and Richard Littlemore. Evaluation shows "Faked" Heartland Climate Strategy Memo is Authentic. Wednesday, February 22, 2012. Available at: http://www.desmogblog.com/evaluation-shows-faked-heartland- climate-strategy-memo-authentic

[209] Naome Oreskes. The American Denial of Global Warming. Video of a presentation. Available at: https://www.youtube.com/watch?v=cIMkIBdV2yw

[210] Naome Oreskes, Erik M. Conway. Merchants of Doubt. How a handful of Scientists Obscured the Truth on Issues from Tobacco Smoke to Global

Warming. Bloomsbury Press, 2010. ISBN: 978-1-59691-610-4. Website: http://www.merchantsofdoubt.org/
[211] Robert Kenner. Merchants of Doubt. A Sony Pictures Classic Release in March 2015. Based on the acclaimed book of the same name by Naome Oreskes and Erik M. Conway.
[212] Maxwell T. Boykoff, Jules M. Boykoff. Balance as Bias: global warming and the U.S. prestige press. Elsevier Ltd. 2003. Available at: http://www.eci.ox.ac.uk/publications/downloads/boykoff04-gec.pdf
[213] John Oliver's discussion of the consensus of climate scientists. Available at: https://www.youtube.com/watch?v=cjuGCJJUGsg
[214] "Trust Conclusions on the Executive Report on Science Impartiality Review Actions". The BBC Trust. July 2014.
[215] Joseph Bast and Roy Spencer. The Myth of the Climate Change '97%'. What is the origin of the false belief—constantly repeated—that almost all scientists agree about global warming? Wall Street Journal. May 26, 2014. Available at: http://www.wsj.com/articles/SB10001424052702303480304579578462813553136
[216] The Heartland Institute. A libertarian and conservative think tank founded in 1984 and based in Chicago. The primary US based originator of material supporting denialist claims about climate change and with a past based in the tobacco misinformation. Website: https://www.heartland.org/. The Heartland Institute is one of approximately 40 groups set up to misinform the public about climate change. A comprehensive list with relatively detailed activity logs is available at: http://www.fightcleanenergysmears.org/behind_the_smears.cfm
[217] David Michaels. Doubt is Their Product. How Industry's Assault on Science Threatens your Health. Oxford University Press. 2008. ISBN 978-0-19-530067-3. Available at: http://www.amazon.com/Doubt-Their-Product-Industrys- Threatens/dp/019530067X
[218] NASA's Dr. Gavin Schmidt goes into hiding from seven very inconvenient climate questions. Available at: https://wattsupwiththat.com/2015/05/19/nasas-dr- gavin-schmidt-goes-into-hiding-from-seven-very-inconvenient-questions/
[219] Freud, A. (1937). The Ego and the mechanisms of defense, London: Hogarth Press and Institute of Psycho-Analysis. Denial as a mechanism of the immature mind. Available at: http://d1nl9ryf5ceszm.cloudfront.net/book/the-ego-and-the- mechanisms-of-defence-international-psycho-analysis-library-_26u39w.pdf
[220] Ben Stein. Expelled: No Intelligence Allowed. Mostly about creationism, it complained about the treatment meted out to scientists on the fringe. 18th April 2008. An article in the Scientific American about this movie debunks all of its egregious misinformation tactics. Available at: http://www.scientificamerican.com/article/six- things-ben-stein-doesnt-want-you-to-know/

[221] CO2isgreen. Available at: http://co2isgreen.org/

[222] CO2isgreen's advertisement. Available at: https://www.youtube.com/watch?v=TxCQHn-w0Bw

[223] Exhaled CO2 comes from our food. https://www.skepticalscience.com/breathing- co2-carbon-dioxide.htm

[224] Lung metrics. Available at: http://www.anaesthetist.com/icu/organs/lung/Findex.htm#lungfx.htm

[225] Current World Population. Available at: http://www.worldometers.info/world- population/

[226] McGoogan, Ken (2002). Fatal Passage: The True Story of John Rae, the Arctic Hero Time Forgot. New York: Carroll & Graf. ISBN 0-7867-0993-6.

[227] The Albedo is the ability of the surface of the Earth to reflect shortwave energy back into space. Ice is almost perfect for this with a high albedo, while Open Ocean is almost the opposite, with its dark blue color. Available at: https://www.esr.org/outreach/glossary/albedo.html

[228] Pan-Arctic Ice Ocean Modeling and Assimilation System (PIOMAS) is a sea ice volume estimating model. Available at: http://neven1.typepad.com/blog/2015/03/piomas-march-2015.html

[229] NASA's Operation IceBridge. Available at: https://blogs.nasa.gov/icebridge/

[230] Lewis Gordon Pugh talks about his record-breaking swim across the North Pole. He braved the icy waters (in a Speedo) to highlight the melting icecap. Watch for astonishing footage -- and some blunt commentary on the realities of supercold-water swims. Available at: https://www.ted.com/talks/lewis_pugh_swims_the_north_pole

[231] Kendra Pierre-Louis and Nadja Popovich. Climate Desk New York Times. Of 21 Winter Olympic Cities, Many May Soon Be Too Warm to Host the Games. January 11, 2018. Available at: https://www.nytimes.com/interactive/2018/01/11/climate/winter-olympics-global- warming.html

[232] United States Geological Survey (USGS) article stating that human (anthropogenic) release of CO2 is 135 times greater than all the land and subsea volcanic degassing activity. Available at: http://volcanoes.usgs.gov/hazards/gas/climate.php

[233] BP Statistical Review 2018. Available at: https://www.bp.com/content/dam/bp/en/corporate/pdf/energy-economics/statistical- review/bp-stats-review-2018-full-report.pdf

[234] David King, Director. Martin Durkin, Producer. The Great Global Warming Swindle. Channel 4. 2006.

[235] Nicolas Caillon, Jeffrey P. Severinghaus, Jean Jouzel, Jean-Marc Barnola, Jiancheng Kang, Volodya Y. Lipenkov. Timing of Atmospheric CO2 and Antarctic Temperature Changes Across Termination III. Science 14 March 2003: Vol. 299 no. 5613 pp. 1728-1731. DOI: 10.1126/science.1078758. Available at: http://www.sciencemag.org/content/299/5613/1728.short

[236] Climate Denial 101x. University of Queensland. A really great initiative to explore the climate issues subject to denial. Available at: https://www.youtube.com/watch?v=dHozjOYHQdE&list=PL-Xgw8LFaM3D-IMt7lByYIX1T0mduh-5V&index=8

[237] The Petition Project. Available at: http://www.petitionproject.org/index.php

[238] NASA say that 97% of climate scientists agree that global warming is both anthropomorphic and dangerous for humanity. Available at: http://climate.nasa.gov/scientific-consensus/

[239] Richard S.J. Tol. Quantifying the consensus on anthropogenic global warming in the literature: A re-analysis Energy Policy, Volume 73, Issue null, Pages 701-705.

[240] James Lawrence Powell. Holds a Ph.D. in Geochemistry from the Massachusetts Institute of Technology and several honorary degrees, including Doctor of Science degrees from Berea College and from Oberlin College. Taught Geology at Oberlin College for over 20 years. Served as Acting President of Oberlin, President of Franklin and Marshall College, President of Reed College, President of the Franklin Institute Science Museum in Philadelphia, and President and Director of the Los Angeles County Museum of Natural History. President Reagan and later, President George H. W. Bush, appointed him to the National Science Board, where he served for 12 years. Asteroid 9739 Powell is named after him. He has written eleven books, the most recent of which is Four Revolutions in the Earth Sciences: From Heresy to Truth, published by Columbia University Press. In 2015, he was elected a Fellow of the Committee for Skeptical Inquiry (CSI). Cook et al concluded 97% of climate scientists agreed that human activities resulted in global warming. A closer look at the scientific literature resulted in Powell's 99.95% figure. Available at: http://www.jamespowell.org/

[241] Cook et al "Quantifying the consensus on anthropogenic global warming (AGW) in the scientific literature" (2013 Environ. Res. Lett. 8 024024).

[242] Consensus on consensus: a synthesis of consensus estimates on human-caused global warming. Cook, J., Oreskes, N., Doran, P. T., Anderegg, W. R., Verheggen, B., Maibach, E. W., ... & Nuccitelli, D. (2016). Consensus on consensus: a synthesis of consensus estimates on human-caused global warming. Environmental Research Letters, 11(4), 048002. Link to paper. Available at: http://dx.doi.org/10.1088/1748-9326/11/4/048002

[243] Richard A. Muller. The Conversion of a Climate-Change Skeptic. July 28, 2012. New York Times. Available at: http://www.nytimes.com/2012/07/30/opinion/the-conversion-of-a-climate-change-skeptic.html?_r=1&pagewanted=all

[244] Henrik Svensmark. Open Source Systems, Science, Solutions. OSS. Available at: http://ossfoundation.us/projects/environment/global-warming/myths/henrik-svensmark

[245] Jasper Kirkby. Effect of water vapor nucleation from naturally existing and industrial aerosols. Also investigated the impact galactic cosmic rays (GCR) on cloud formation using the CERN Synchrotron. TED Talk available at: http://ed.ted.com/lessons/cloudy-climate-change-how-clouds-affect-earth-s- temperature-jasper-kirkby#review

[246] Peter Laut. Climate Change: The Role of Flawed Science. November 2009. PDF. Available at: http://www.realclimate.org/wp-content/uploads/PETERLAUT- ANALYSIS-CLIMATE-CHANGE-CPN1.pdf

[247] U.S. Geological Survey of Carbon emissions from Mount Kilauea. February 15, 2007. Which produces more CO2, volcanic or human activity? A weekly feature provided by scientists at the Hawaiian Volcano Observatory. Available at: http://hvo.wr.usgs.gov/volcanowatch/archive/2007/07_02_15.html

[248] U.S. Department of Energy's Carbon Dioxide Information Analysis Center (CDIAC). Boden, T.A., G. Marland, and R.J. Andres. 2017. Global, Regional, and National Fossil-Fuel CO2 Emissions. Carbon Dioxide Information Analysis Center, Oak Ridge National Laboratory, U.S. Department of Energy, Oak Ridge, Tenn., U.S.A. doi 10.3334/CDIAC/00001_V2017. Available at: http://cdiac.ornl.gov/trends/emis/tre_glob_2014.html

[249] Gifford Miller, Áslaug Geirsdóttir, Yafang Zhong, Darren J. Larsen, Bette L. Otto-Bliesner, Marika M. Holland, David A. Bailey, Kurt A. Refsnider, Scott J. Lehman, John R. Southon, Chance Anderson, Helgi Björnsson, Thorvaldur Thordarson. Abrupt onset of the Little Ice Age triggered by volcanism and sustained by sea-ice/ocean feedbacks. Geophysical Research Letters. First Published: 31 January 2012. Available at: http://onlinelibrary.wiley.com/doi/10.1029/2011GL050168/abstract

[250] Norman A. Phillips. The General Circulation of the atmosphere: a numerical experiment. PDF. The Institute for Advanced Study, Princeton. 17th October 1955. Quarterly Journal of the Royal Meteorological Society. Vol. 82. No. 352. April 1956. Available at: http://www.phy.pku.edu.cn/climate/class/cm2010/Phillips_QJRMS_1956.pdf

[251] Sir Paul Nurse. Science Under Attack - BBC Documentary. Available at: https://www.youtube.com/watch?v=C3JEaigwAbg

[252] Mark Levin on his national radio show saying all the real, old scientists never mentioned climate change so that's the proof that its just been concocted by the left for political reasons. Available at: https://www.youtube.com/watch?v=lcNZ2LXE4kg

[253] Sea Level Rise, IPPC AR5 Working Group 1. Slides. Available at: http://onlinelibrary.wiley.com/doi/10.1029/2011GL050168/full https://www.ipcc.ch/pdf/unfccc/cop19/3_gregory13sbsta.pdf

[254] Strange craters, burning ice and 'drunken trees': Climate change is causing the planet to behave in mysterious was, scientists claim. Daily Mail. 6 August 2014. Available at: http://www.dailymail.co.uk/sciencetech/article-2717938/

Strange-craters-burning-ice-drunken-trees-How-climate-change-causing-planet-behave- mysterious-ways.html

[255] Alexander Nauels, Joeri Rogelj, Carl-Friedrich Schleussner, Malte Meinshausen and Matthias Mengel. Linking sea level rise and socioeconomic indicators under the Shared Socioeconomic Pathways Published 26 October 2017. Published by IOP Publishing Ltd. Available at: http://iopscience.iop.org/article/10.1088/1748-9326/aa92b6

[256] Melissa Allison. The Effect of Rising Sea Levels on Coastal Homes. Zillow Porchlight. 2nd August 2016. Available at: http://www.zillow.com/blog/rising-sea- levels-coastal-homes-202268/

[257] New York City's first steps to preventing the inevitable Hurricane Sandy storm surge of 14 feet above high tide by imagining walls and berms that prevent water flow. Available at: https://vimeo.com/117303273

[258] Bill McKibben. Rolling Stone. Global Warming's Terrifying New Math. July 19, 2012. Available at: http://www.rollingstone.com/politics/news/global-warmings- terrifying-new-math-20120719

[259] Carbon Tracker Initiative. Available at: http://www.carbontracker.org/

[260] Climate Consent. Summary of Carbon numbers with a daunting conclusion. Available at: http://www.climateconsent.org/pages/carbonmaths.html

[261] Steven C. Sherwood, Climate Change Research Centre, University of New South Wales, Sydney, New South Wales 2052, Australia. Matthew Huber, Purdue Climate Change Research Center, Purdue University, West Lafayette, IN 47907. An adaptability limit to climate change due to heat stress. Edited by Kerry A. Emanuel, Massachusetts Institute of Technology, Cambridge, MA. March 24, 2010. Available at: http://www.pnas.org/content/107/21/9552.full.pdf

[262] Hubert H. Lamb. "Climate: Present, Past & Future—Vol 2". 1977. Routledge.

[263] Luke Howard. The Climate of London. 1818-20. Landsberg, Helmut Erich (1981). The urban climate. Academic Press, New York, p.3.

[264] Colin P. Kelley, Shahrzad Mohtadi, Mark A. Cane, Richard Seager, and Yochanan Kushnir Climate change in the Fertile Crescent and implications of the recent Syrian drought. National Academy of Sciences. March 2014.

[265] Colin P. Kelley, Shahrzad Mohtadi, Mark A. Cane, Richard Seager, and Yochanan Kushnir. Climate change in the Fertile Crescent and implications of the recent Syrian drought. Proceedings of the National Academy of Sciences. (PNAS). Vol. 112, no 11. Available at: http://www.pnas.org/content/112/11/3241.abstract

[266] The military cost of fuel. National Defense Magazine. Available at: http://www.nationaldefensemagazine.org/archive/2010/April/Pages/HowMuchforaGa llonofGas.aspx

[267] Munich Re's annual assessment of global insurance events. Topics GEO. Available at: https://www.munichre.com/site/touch- naturalhazards/

get/documents_E1018449711/mr/Assetpool.shared/Documents/5_Tou ch/_Publications/302-08606_en.pdf

[268] National Association of Insurance Commissioners website with details on the climate surveys taken by larger insurance companies. Available at: http://www.naic.org/cipr_topics/topic_climate_risk_disclosure.htm

[269] Jeremy Grantham, Environmental philanthropist. "Unburnable Carbon". The world spends $674 billion annually on carbon fossil fuels but puts itself at huge risk. Explored in Chapter 5.

[270] Mercer. Investing in a Time of Climate Change. Available at: http://www.mercer.com/content/mercer/global/all/en/insights/focus/invest-in-climate-change-study-2015.html

[271] "The Human Universe". Presented by Professor Brian Cox. Produced by Gideon Bradshaw and Andrew Cohen. BBC. Jean-Jacques Hublin, director at the Max Planck Institute for Evolutionary Anthropology, Leipzig, Germany. Nature, Vol 546, Page 289. June 8th, 2017 Available at: http://www.nature.com/news/oldest-homo-sapiens- fossil-claim-rewrites-our-species-history-1.22114

[272] Bloomberg uses NASA satellite and other data to provide an amazing animated chart showing the impact of human and natural forces impacting global warming. Available at: https://www.bloomberg.com/graphics/2015-whats-warming-the-world/

[273] Airline flights in a day from Aviation Benefits Beyond Borders. Available at: http://aviationbenefits.org/media/26786/ATAG_AviationBenefits2014_FULL_Low Res.pdf

[274] Oil extracted from the ground globally in 2014. IEA. Available at: http://www.eia.gov/cfapps/ipdbproject/iedindex3.cfm?tid=5&pid=53&aid=1&cid=regions&syid=2010&eyid=2015&unit=TBPD

[275] Alliance BioEnergy Plus, Florida. A company that licenses a patent from the University of Central Florida that covers the efficient, cheap exploitation of cellulose, turning any source of the substance into sugars and lignin, with 100% conversion in 15 minutes using very little energy, no heat, no enzymes, no chemicals, and not even any liquids. A true global breakthrough in biofuels that results in a significant replacement of mineral fuels. Available at: http://www.alliancebioe.com/

[276] Michael E. Mann, Stefan Rahmstorf, Byron A. Steinman, Martin Tingley and Sonya K. Miller. The Likelihood of Recend Record Warmth. Nature.com. Scientific Reports 6, Article number: 19831. Published online: 25th January 2016. Available at: https://www.nature.com/articles/srep19831

[277] Gerrit Hansen & Daithi Stone. Assessing the observed Impact of anthropogenic climate change. Nature.com. Published online 21st December 2015. Available at: https://www.nature.com/articles/nclimate2896?foxtrotcallback=true

[278] Carson, Rachel (2002) [1st. Pub. Houghton Mifflin, 1962]. Silent Spring. Mariner Books. ISBN 0-618-24906-0.

[279] Hidden Costs of Energy: Unpriced Consequences of Energy Production and Use. 2010. Available at: http://www.nap.edu/download. php?record_id=12794

[280] Hubbert, M. King (June 1956). "Nuclear Energy and the Fossil Fuels" (PDF). Shell Oil Company/American Petroleum Institute. Retrieved 2014-11-10., Presented before the Spring Meeting of the Southern District, American Petroleum Institute, Plaza Hotel, San Antonio, Texas, March 7–8-9, 1956. Available at: http://www.hubbertpeak.com/hubbert/1956/1956.pdf

[281] Energy Information Administration. 2008. Annual Coal Report. Available at: http://www.eia.doe.gov/cneaf/coal/page/acr/acr sum.html

[282] R.E.H.Sims,R.N.Schock,A.Adegbululgbe,J.Fenhann,I.Konstantinaviciute, W. Moomaw, H.B. Nimir, B. Schlamadinger, J. Torres-Martínez, C. Turner, Y. Uchiyama, S.J.V. Vuori, N. Wamukonya, X. Zhang, 2007: Energy supply. In Climate Change 2007: Mitigation. Contribution of Working Group III to the Fourth Assessment Report of the Intergovernmental Panel on Climate Change. [B. Metz, O.R. Davidson, P.R. Bosch, R. Dave, L.A. Meyer (eds)], Cambridge University Press, Cambridge, United Kingdom and New York, NY, USA.

[283] William E. Mulligan. Air Raid! A Sequel. Captain Moci's daring bombing raid on the British in Bahrain. October 18, 1940. Aramco World. July/August 1976. Volume 27, Number 4. Available at: https://www.saudiaramcoworld. com/issue/197604/air.raid.a.sequel.htm

[284] Dr. Peter W. Becker. German Wartime Energy Innovation. The Role of Synthetic Fuel in World War II Germany. University of South Carolina. Available at: http://www.airpower.maxwell.af.mil/airchronicles/ aureview/1981/jul-aug/becker.htm

[285] The Bruce-Lovett report was mentioned in Peter Grose's biography, "Gentleman Spy: The Life of Allen Dulles". The CIA's own history department still could not find a copy by 1995. Even gross had not seen the report itself but used notes from history Professor Arthur M. Schlesinger who informed them that the report was in Robert Kennedy's papers before they were deposited in the John F. Kennedy Presidential Library in Boston. More at: http://www.nytimes.com/1987/07/22/opinion/washington-file-and-forget.html

[286] Gawrych, Dr. George W. (1996). The 1973 Arab-Israeli War: The Albatross of Decisive Victory. Combat Studies Institute, U.S. Army Command and General Staff College. "Intro" (PDF). Archived from the original (PDF) on June 10, 2007. "Part I" (PDF). Archived from the original (PDF) on June 8, 2011. "Part II" (PDF). Archived from the original (PDF) on June 8, 2011. "Part III" (PDF). Archived from the original (PDF) on June 8, 2011. "Part IV" (PDF). Archived from the original (PDF) on June 8, 2011. "Part V" (PDF). Archived from the original (PDF) on June 8, 2011. "Part VI" (PDF). Archived from the original (PDF) on June 8, 2011. "Part VII" (PDF).

Archived from the original (PDF) on June 8, 2011. "Notes" (PDF). Archived from the original (PDF) on March 19, 2009. Retrieved May 28, 2015.

[287] "U.S. Oil Production and Imports 1920 to 2005" by David Moe RockyMtnGuy - self-made using Excel from data published by the Energy Information Administration of the Department of Energy. Licensed under CC BY-SA 3.0 via Commons - https://commons.wikimedia.org/wiki/File:US_ Oil_Production_and_Imports_1920_to_2005.png#/media/File:US_Oil_ Production_and_Imports_1920_to_2005.png

[288] British troops were deployed to Kuwait in 1961. Available at: https:// paradata.org.uk/events/persian-gulf

[289] President Reagan's National Security Council (NSC) report on the situation concerning Iran and Iraq in 1984. Available at: http://www.wpainc.com/ Archive/Reagan%20Administration/WFM%20Papers%20fr om%20 Reagan%20Archives/Iran- Iraq/Presentation%20on%20Gulf%20Oil%20 Disruption%205-22-84.pdf

[290] The Organization of the Petroleum Exporting Countries (OPEC) is a permanent, intergovernmental Organization, created at the Baghdad Conference on September 10–14, 1960, by Iran, Iraq, Kuwait, Saudi Arabia and Venezuela. The five Founding Members were later joined by nine other Members: Qatar (1961); Indonesia (1962)— suspended its membership from January 2009-December 2015; Libya (1962); United Arab Emirates (1967); Algeria (1969); Nigeria (1971); Ecuador (1973)—suspended its membership from December 1992-October 2007; Angola (2007); and Gabon (1975) - terminated its membership in January 1995 but rejoined in July 2016. OPEC had its headquarters in Geneva, Switzerland, in the first five years of its existence. This was moved to Vienna, Austria, on September 1, 1965. Available at: http://www.opec.org/opec_web/en/about_us/24.htm

[291] National Energy Policy. Report of the National Energy Policy Development Group. May 2001. Available at: http://wtrg.com/EnergyReport/National-Energy- Policy.pdf

[292] Duffy, Michael; James Carney (13 July 2003). "A Question of Trust". Time. This article mentions that President Bush had heard from British Intelligence that Saddam Hussein had sought uranium yellowcake.

[293] Evan Abramson. "When the Water Ends". Yale Environment 360. Mediastorm. Tells the story of how drought conditions in East Africa have communities fighting and dying to control dwindling water and grassland supplies. Available at: http://e360.yale.edu/feature/ when_the_water_ends_africas_climate_conflicts/2331/

[294] Robert Cox, journalist and defender of the mothers of the disappeared, published articles in the Buenos Aires Herald, a newspaper for the English-speaking audience in Argentina. Time Magazine, Sunday, August 21st, 2010. Available at: http://content.time.com/time/world/article/0,8599,2007460,00. html

295 Horacio Verbitsky, The Flight. Confessions of an Argentine Dirty Warrior. ISBN-13: 978-1565840096. This is the account of Lieutenant Commander Adolfo Francisco Scilingo's participation in the death flights in the Dirty War. Available at: http://www.amazon.com/The-Flight-Confessions-Argentine-Warrior/dp/1565840097

296 "Spain tries Argentine ex-officer". BBC News. January 20, 2005. Available at: http://news.bbc.co.uk/2/hi/americas/4173215.stm

297 James Dao, the New York Times, Wednesday, August 21st, 2002. Available at: http://www.nytimes.com/2002/08/21/world/us-releases-1980-s-files-on-repression-in- argentina.html

298 History of the Falkland Islands. Available at: http://www.historyworld.net/wrldhis/PlainTextHistories.asp?historyid=ac51

299 Darwin, Charles (1839). Narrative of the surveying voyages of His Majesty's Ships Adventure and Beagle between the years 1826 and 1836, describing their examination of the southern shores of South America, and the Beagle's circumnavigation of the globe. Journal and remarks. 1832–1836. (The Voyage of the Beagle) III. London: Henry Colburn. pp. 149–150.

300 Alan Thornett, The Falklands Oil Rush and Thatcher's War. 2012. Available at: http://socialistresistance.org/3329/the-falklands-oil-rush-and-thatchers-war

301 David Vine. Base Nation. How U.S. Military Bases Abroad Harm America and the World. August 2015. Britain's hypocrisy. Chagossians had to depart their island but the Britons on the Falklands had self-determination rights. Available at: http://www.huffingtonpost.com/david-vine/the-truth-about-diego- gar_b_7585546.html

302 Kim Sabido, ITN reporter who went to the Falklands conflict. His version of the story available at: http://www.radionz.co.nz/audio/player?audio_id=2520788

303 The Treaty on the Non-Proliferation of Nuclear Weapons (NPT). Available at: http://www.nti.org/glossary/#nonproliferation-treaty

304 Meitner, L.; Frisch, O. R. (1939). "Disintegration of Uranium by Neutrons: A New Type of Nuclear Reaction". Nature 143 (3615): 239. Available at: http://www.nature.com/nature/journal/v143/n3615/abs/143239a0.html

305 Stephane Groueff. Dr. Eugene Wigner, Voices of the Manhattan Project. December 10, 1964. A phone interview from New York with Eugene Wigner. Concerning the letter of Szilard and Einstein to President Roosevelt. Available at: http://manhattanprojectvoices.org/oral-histories/eugene-wigners-interview-1964

306 Hoddeson, Lillian; Henriksen, Paul W.; Meade, Roger A.; Westfall, Catherine L. (1993). Critical Assembly: A Technical History of Los Alamos During the Oppenheimer Years, 1943–1945. New York: Cambridge University Press. ISBN 0- 521-44132-3. OCLC 26764320.

307 Eric Schlosser. Command and Control: Nuclear Weapons, the Damascus Accident, and the Illusion of Safety (8/18/13). Available at: http://

www.amazon.com/Eric-Schlosser-Damascus-Accident-   Illusion/dp/
B00HTJT5HY/ref=sr_1_3?s=books&ie=UTF8&qid=1444622611&sr=1-
3&keywords=eric+schlosser+command+and+control
[308] Daley, Ted, "Apocalypse Never: Forging the Path to a Nuclear Weapon-
Free World", Rutgers University Press, 2010. p. 240. Available at: http://
rutgerspress.rutgers.edu/product/Apocalypse-Never,3949.aspx
[309] Nuclear testing chart. Available at: https://commons.wikimedia.org/wiki/
File:Worldwide_nuclear_testing.svg#/media/Fil   e:Worldwide_nuclear_
testing.svg
[310] In latest test, North Korea detonates its most powerful nuclear device
yet. 6th September 3 2017. Available at: https://www.washingtonpost.
com/world/north-   korea-apparently-conducts-another-nuclear-test-south-
korea-   says/2017/09/03/7bce3ff6-905b-11e7-8df5-   c2e5cf46c1e2_story.
html?noredirect=on&utm_term=.721f60beb973
[311] "Big Ivan, The Tsar Bomba ("King of Bombs")". September 3, 2007.
Retrieved June 11, 2014. Available at: http://nuclearweaponarchive.org/
Russia/TsarBomba.html
[312] End the Nuclear Insanity. A letter written to take advantage of a UN Assembly
cycle to advance the effort of disarming the nuclear powers, something
those powers had signed onto with the Nuclear Proliferation Treaty in 1970.
Available   at: http://www.huffingtonpost.com/jose-ramoshorta/end-the-
nuclear- insanity_b_12436344.html
[313] "Proceedings into the fire at Windscale Pile Number One (1989 revised
transcript of the "Penney Report")" (PDF). UKAEA. 18 April 1989.
Available at: http://news.bbc.co.uk/2/shared/bsp/hi/pdfs/05_10_07_ukaea.
pdf
[314] Percentage of global nuclear energy vs energy consumption of all kinds in
2014. Taken from the BP Statistical Review of World Energy 2018.
[315] From an interview with Gerald Feldhamer who provided unpublished
photographs of the Manhattan Project physicists on his board. NEF Advisors,
LLC.
[316] Kenneth Chang. Alvin Radkowsky obituary. March 5, 2002. Available at:
http://www.nytimes.com/2002/03/05/world/alvin-radkowsky-86-developer-
of-a- safer-nuclear-reactor-fuel.html
[317] Jewish Press. Friday February 3rd, 1984. P. 41. New Power Technology To
Market Israeli Inventor's Nonproliferative Reactor.
[318] Professor Geraldine Thomas. Chair of Molecular Pathology at Imperial
College, London. She kept in touch with Russian scientists and kept data on
the Chernobyl accident and collected tissue samples from individuals who
died. Available at: http://www.imperial.ac.uk/people/geraldine.thomas
[319] Hayes, Daniel F. (August 1956). "A Summary of Accidents and Incidents
Involving Radiation in Atomic Energy Activities, June 1945 through

December 1955". Oak Ridge, Tennessee: U.S. Atomic Energy Commission. pp. 2–3.

[320] Zeilig, Martin (August–September 1995). "Louis Slotin and 'The Invisible Killer'". The Beaver 75 (4): 20–27.

[321] Martha Smith-Norris."Only as Dust in the Face of the Wind": An Analysis of the BRAVO Nuclear Incident in the Pacific, 1954 The Journal of American-East Asian Relations Vol. 6, No. 1 (SPRING 1997), pp. 1-34".

[322] Lieutenant Colonel Petrov, The Man Who Saved the World! Available at: bigstory.ap.org/article/1d519e7c53f649d7baa23e72b41c1754/russian-who-saved- world-recalls-his-decision-5050

[323] Robert M. Gates. From the Shadows: The Ultimate Insider's Story of Five Presidents and How They Won the Cold War. New York: Simon & Shuster, 1996, page 114. Available at: http://www.simonandschuster.com/books/From-the- Shadows/Robert-M-Gates/9781416543367/browse_inside

[324] Nixon's Nuclear Crisis. Extract from The Arrogance of Power: The Secret World Of Richard Nixon, by Anthony Summers, with Robbyn Swan, published by Gollancz. Copyright Anthony Summers 2000. The Guardian. 2nd September 2000. Available at: https://www.theguardian.com/weekend/story/0,3605,362958,00.html

[325] The story of the nuclear reactor coolant accident on the Russian submarine K-19. Available at: http://articles.latimes.com/1994-01-03/news/mn-8123_1_soviet-nuclear-submarine

[326] K-19 Submarine decommissioning and scrapping. Available at: http://www.nationalgeographic.com/k19/k19_html_scene10.html

[327] Vasili Alexandrovitch Arkhipov's refusal to agree to launch a nuclear missile, saved the world. Available at: http://www.theguardian.com/commentisfree/2012/oct/27/vasili-arkhipov-stopped- nuclear-war

[328] Lloyd, Marion (13 October 2002). "Soviets Close to Using A-Bomb in 1962 Crisis, Forum is Told". Boston Globe. pp. A20. Available at: http://www.latinamericanstudies.org/cold-war/sovietsbomb.htm

[329] J.M. Montmollin and W.R. Hoagland. Analysis of the Safety Aspects of the MK 39 MOD 2 Bombs Involved in B-52G Crash Near Greensboro, North Carolina. February 1961. Available at: http://nsarchive.gwu.edu/nukevault/ebb475/docs/doc%205%20AEC%20report%20G oldsboro%20accident.pdf

[330] Eric Schlosser. Command and Control: Nuclear Weapons, the Damascus Accident, and the Illusion of Safety. 632 pp. The Penguin Press.

[331] The Damascus Incident. An accident where a socket from a socket wrench dropped 80 feet down the side of a Titan missile, hitting the metal skin of the missile, causing a fuel leak, which eventually exploded. Available at: http://www.encyclopediaofarkansas.net/encyclopedia/entry- detail.aspx?entryID=2543

[332] NPR. In a special series on superstorm Sandy. Sandy Reveals Long Island Utility's 'Boondoggle' Past. November 17, 2012. Steve Henn.

333 LongIslandPress.com Nuclear Waste: 20 Years after Shoreham's Closure. A Look at What Long Island's Failed Nuclear Plant Could Have Been, By Timothy Bolger and Christopher Twarowski on June 11th, 2009.

334 A Tale of Two Reactors. The New American Magazine. Ed HIserodt, Sunday 6 July 2008.

335 The International Atomic Energy Agency (IAEA). Available at: https://www.iaea.org/

336 Glenn Kessler. "Did Iran's supreme leader issue a Fatwa against the development of nuclear weapons?".

337 Interview with Alexei Yablokov. "Comments on Russia's atomic suitcase bombs". Available at: http://www.pbs.org/wgbh/pages/frontline/shows/russia/suitcase/comments.html

338 Testimony of Dr. Alexei Yablokov to the House National Security Committee, in October 2, 1997. Available at: http://www.pbs.org/wgbh/pages/frontline/shows/russia/suitcase/yablokov.html

339 Video of Interview with Indira Ghandi talking about the developing problems in East Pakistan, soon to be Bangladesh. Available at: https://www.youtube.com/watch?v=_MATAqeiL-4

340 Abdul Qadeer Khan was head of Pakistan's enrichment project at Kahuta and claimed they had enough uranium 235 to detonate a bomb in 1984. Available at: http://www.nti.org/country-profiles/pakistan/

341 India rejected signing the the Treaty on Non-Proliferation of Nuclear Weapons because it was unfair. Available at: http://www.nti.org/country-profiles/india/

342 India's nuclear capability and relationship with the rest of the world. Available at: http://www.nti.org/country-profiles/india/

343 Jonathan Watts. China: the air pollution capital of the world. The Lancet. Volume 366, No. 9499, p1761–1762, 19 November 2005. Available at: http://www.thelancet.com/journals/lancet/article/PIIS0140-6736(05)67711-2/fulltext?version=printerFriendly

344 670,000 Chinese deaths from smog in 2012. Teng Fei, associate professor. Tsinghua University, Beijing. South China Morning Post. 5 November 2014. Available at: http://www.scmp.com/news/china/article/1632163/670000-deaths-year- cost-chinas-reliance-coal?page=all

345 Malcolm Moore. China's 'airpocalypse' kills 350,000 to 500,000 each year. Telegraph. 7 January 2014. Available at: http://www.telegraph.co.uk/news/worldnews/asia/china/10555816/Chinas- airpocalypse-kills-350000-to-500000-each-year.html

346 Freese, Barbara. Coal: A Human History. Hardback, Perseus Publishing. The paperback edition by Penguin Books. 2004. Available at: https://www.amazon.com/Coal-Human-History-Barbara-Freese/dp/0142000981

347 300,000 deaths annually from coal burning in China today and 5.5 million worldwide. The University of British Columbia's School of Population and

Public Health in Vancouver. Available at: http://news.ubc.ca/2016/02/12/poor-air-quality- kills-5-5-million-worldwide-annually/

[348] Aaron J. Cohen, H. Ross Anderson, Bart Ostro, Kiran Dev Pandey, Michal Krzyzanowski, Nino Künzli, Kersten Gutschmidt, C. Arden Pope III, Isabelle Romieu, Jonathan M. Samet and Kirk R. Smith. Urban Air Pollution. 2002. Available at: http://apps.who.int/publications/cra/chapters/volume2/1353-1434.pdf

[349] Paul R. Epstein, Jonathan J. Buonocore, Kevin Eckerle, Michael Hendryx, Benjamin M. Stout III, Richard Heinberg, Richard W. Clapp, Beverly May, Nancy L. Reinhart, Melissa M. Ahern, Samir K. Doshi, and Leslie Glustrom. 2011. Full cost accounting for the life cycle of coal in "Ecological Economics Reviews." Robert Costanza, Karin Limburg & Ida Kubiszewski, Eds. Ann. N.Y. Acad. Sci. 1219: 73–98. Available at: http://www.chgeharvard.org/sites/default/files/epstein_full%20cost%20of%20coal.pd f

[350] The EIA state that combustion of 1 short ton (2,000 pounds) of coal will generate about 5,720 pounds (2.86 short tons) of carbon dioxide. Available at: http://www.eia.gov/coal/production/quarterly/co2_article/co2.html

[351] Brightergy. Blog about the true cost of coal. Available at: https://brightergy.com/blog/the-true-cost-of-coal/

[352] Susan Kemp and Jeffrey Olsen of the U.S. Geological Survey. Elevated levels of Mercury Found in Fish in Western U.S. National Parks. 16 April 2014. Available at: http://pubs.usgs.gov/of/2014/1051/pdf/ofr2014-1051.pdf

[353] Michael Hendryx and Melissa M. Ahern Mortality in Appalachian Coal Mining Regions: The Value of Statistical Life Lost. Public Health Rep. 2009 Jul-Aug. This study found that the mortality of residents from all causes were highest in heavy coal mining areas in Appalachia and lowest in non-coal mining locales. Available at: http://www.ncbi.nlm.nih.gov/pmc/articles/PMC2693168/pdf/phr124000541.pdf

[354] Farmer Suicides. U.S. National Library of Medicine, National Institutes of Health. Available at: http://www.ncbi.nlm.nih.gov/pmc/articles/PMC2802368/

[355] Germanwatch. Keeps data on climate related disasters around the globe. Available at: http://germanwatch.org/en

[356] Barbara A. Maher, Imad A. M. Ahmed, Vassil Karloukovskia, Donald A. MacLaren, Penelope G. Foulds, David Allsop, David M. A. Manne, Ricardo Torres- Jardón, and Lilian Calderon-Garciduenas. Magnetite pollution nanoparticles in the human brain. Proceedings of the National Academy of Sciences (PNAS). Vol. 113 no. 39 Edited by Yinon Rudich, Weizmann Institute of Science, Rehovot, Israel, and accepted by Editorial Board Member A. R. Ravishankara July 25, 2016. Available at: http://www.pnas.org/content/113/39/10797.abstract

[357] EEA Report No 13/2017. Air quality in Europe - 2016. Published 11 Oct 2017. Available at: https://www.eea.europa.eu/publications/air-quality-in-europe-2017

[358] BBC News. The Dangers of mining around the world. 14 October 2010. Available at: http://www.bbc.com/news/world-latin-america-11533349

[359] David J. Blackley, DrPH; James B. Crum, DO; Cara N. Halldin, PhD; Eileen Storey, MD; A. Scott Laney, PhD. Resurgence of Progressive Massive Fibrosis in Coal Miners— Eastern Kentucky, 2016. Available at: http://www.npr.org/documents/2016/dec/blacklungreport121516.pdf

[360] Guxens, Mònica et al. Air pollution exposure during fetal life, brain morphology, and cognitive function in school-age children. Biological Psychiatry, Volume 0 , Issue 0. Available at: http://www.biologicalpsychiatryjournal.com/article/S0006-3223(18)30064-7/fulltext

[361] World Health Organization. Fact Sheet No. 313. March 2014. Ambient (Outdoor) air quality and health. Impact of indoor and outdoor fossil fuel burning on latent death rates. Available at: http://www.who.int/mediacentre/factsheets/fs313/en/

[362] Ailun Yang and Yiyun Cui. World Resources Institute. Global Risk Assessment. November 2012. Available at: http://www.wri.org/publication/global-coal-risk- assessment

[363] Greenpeace report. The Great Water Grab. How the Coal Industry is Deepening the Global Water Crisis. 2013. Available at: http://www.greenpeace.org/international/Global/international/publications/climate/2016/The-Great-Water-Grab.pdf

[364] Kirsten Korosec. In the U.S., there are twice as many solar workers as coal miners. Fortune Magazine. January 16, 2015. Available at: http://fortune.com/2015/01/16/solar-jobs-report-2014/

[365] Investor's Business Daily. War on Fossil Fuels: One Coal-ossal Mistake. January 2016. Politics Editorial. Available at: http://www.investors.com/war-on-fossil-fuel-a- mistake/

[366] Sovacool, Benjamin K. (2008). "The costs of failure: A preliminary assessment of major energy accidents, 1907–2007". Energy Policy 36: 1806.

[367] Petr Beckmann. Health Hazards of NOT going Nuclear. Apr 01, 1977. ISBN: 978- 0911762175. Available at: http://www.amazon.com/The-Health-Hazards-Going- Nuclear/dp/0911762175

[368] A Tale of Two Reactors. The New American Magazine. Ed HIserodt, Sunday 6 July 2008.

[369] Ralph Vartabedian, Crude-oil train wrecks raise questions about safety claims, LA Times, March 12, 2015

[370] National Fire Protection Association. Information on gasoline fires in service stations in the U.S. Available at: http://www.nfpa.org/research/reports-and- statistics/fires-by-property-type/business-and-mercantile/fires-at-us-service-stations

[371] Nick Watts and 63 fellow authors. The Lancet Countdown on health and climate change: from 25 years of inaction to a global transformation for public health. The Lancet. Published: 30 October 2017. Available at: http://www.thelancet.com/pdfs/journals/lancet/PIIS0140-6736(17)32464-9.pdf

[372] Dr. Robert Balling. DeSmogBlog article on him. Available at: http://www.desmogblog.com/robert-c-balling-jr

[373] Time Magazine. The Burning River that Sparked a Revolution. June 22, 1969. Available at: http://time.com/3921976/cuyahoga-fire/

[374] Parry, Ian W. H., and Kenneth A. Small. 2005. "Does Britain or the United States Have the Right Gasoline Tax?" American Economic Review, 95(4): 1276-1289.

[375] Bill McKibben. Rolling Stone. Global Warming's Terrifying New Math. July 12 2012. Available at: https://www.rollingstone.com/politics/news/global-warmings- terrifying-new-math-20120719

[376] The San Juan, New Mexico, coalbed methane leak. Available at: http://www.scientificamerican.com/article/the-biggest-methane-leak-in-america-is-in- new-mexico/

[377] Kolbert, Elizabeth (2014). The sixth extinction: an unnatural history. New York: Henry Holt and Co. ISBN 9780805092998. Available at: http://www.amazon.com/The-Sixth-Extinction-Unnatural-History/dp/0805092994

[378] A. R. Brandt, G. A. Heath, E. A. Kort, F. O'Sullivan, G. Pétron, S. M. Jordaan, P. Tans, J. Wilcox, A. M. Gopstein, D. Arent, S. Wofsy, N. J. Brown, R. Bradley, G. D. Stucky, D. Eardley, R. Harriss Science magazine. Methane Leaks from North American Natural Gas Systems. Available at: http://nature.berkeley.edu/er100/readings/Brandt_2014.pdf

[379] Colonial Pipeline is a company that serves 50 million on the Eastern seaboard of the U.S. They have had pipeline leaks. Available at: https://thinkprogress.org/alabama-pipeline-leak-prompts-gas-shortages-   22f90461bc1f#.tmqta2ual

[380] Gleb Paikachev. The Town that reveals how Russia spills two Deepwater Horizons of oil each year. The Guardian. 5th of August of 2016. Available at: https://www.theguardian.com/environment/2016/aug/05/the-town-that-reveals-how- russia-spills-two-deepwater-horizons-of-oil-each-year

[381] the 13th Chinese 5-year economic plan. A Xinghua infographic. Available at: http://news.xinhuanet.com/english/photo/2015-11/04/c_134783513.htm

[382] World Economic Forum, Ellen MacArthur Foundation and McKinsey & Company, The New Plastics Economy— Rethinking the future of plastics. 2016. Available at: http://www.ellenmacarthurfoundation.org/publications

[383] Moore, Charles (2002). "A comparison of neustonic plastic and zooplankton abundance in southern California's coastal waters and elsewhere in the North Pacific". Algalita Marine Research Foundation.

[384] 5 Gyres Institute. Available at: http://www.5gyres.org/

[385] Boyan Slat. How the oceans can clean themselves: Boyan Slat at TEDxDelft. 2012. Available at: https://www.youtube.com/watch?v=ROW9F-c0kIQ

[386] The Breakthrough Institute. Available at: http://www.thebreakthrough.org/

[387] The Breakthrough Institute, the Brookings Institution, Beyond Boom and Bust

[388] Green Scissors Campaign. Available at: http://greenscissors.taxpayer.net/reports/assets/documents/assets/documents/Green_S cissors_Report_2012_August.pdf

[389] Unburnable Carbon 2013. Wasted Capital and stranded assets. Grantham Research Institute. Available at: http://carbontracker.live.kiln.it/Unburnable-Carbon-2-Web- Version.pdf

[390] Nancy Pfund of DBL Investors and Ben Healey. What Would Jefferson Do? The Historical Role of Federal Subsidies in Shaping America's Energy Future. September 2011. Available at; http://www.dblinvestors.com/documents/What-Would-Jefferson- Do-Final-Version.pdf

[391] David Coady, Ian W.H. Parry, Louis Sears, Baoping Shang. How Large Are Global Energy Subsidies. May 18, 2015. International Monetary Fund. Available at: http://www.imf.org/external/pubs/ft/wp/2015/wp15105.pdf

[392] The Guardian. A. J. Cohen et al., The global burden of disease due to outdoor air pollution, Journal of Toxicology and Environmental Health, Part A, 68: 1301-1307 (2005)

[393] Environmental Law Institute. Estimating U.S. Government Subsidies to Energy Sources: 2002—2008. Available at: https://www.eli.org/sites/default/files/eli- pubs/d19_07.pdf

[394] David Coady. Ian Parry. Louis Sears. Baoping Shang. IMF study. How Large Are Global Fossil Fuel Subsidies? Elsevier. Available at: https://doi.org/10.1016/j.worlddev.2016.10.004 and also at: http://www.sciencedirect.com/science/article/pii/S0305750X16304867

[395] Klare, Michael. Blood and Oil: The Dangers and Consequences of America's Growing Petroleum Dependency. 2004. New York: Henry Holt.

[396] Roger J. Stern entitled "United States cost of military force projection in the Persian Gulf, 1976—2007" written in 2009. Available at: https://www.princeton.edu/oeme/articles/US-miiltary-cost-of-Persian-Gulf-force-projection.pdf

[397] Dev Millstein, Ryan Wiser, Mark Bolinger & Galen Barbose. The climate and air-quality benefits of wind and solar power in the United States. Nature Energy 2, Article number: 17134 (2017). Published online: 14 August 2017. doi:10.1038/nenergy.2017.134. Available at: https://www.nature.com/articles/nenergy2017134.epdf?referrer_access_token=yMt5 OVDiricynPA5mRnQgdRgN0jAjWel9jnR3ZoTv0O9NQQavv-jglBpgJVQy91sELwK-XsT3k9_0RWGaccyHOIqyTm wXxiyEkGvj8OSZXAss0nGufyednWpodl19vgoXAU IpICFJZXfhzctANM8M52YNvLXMRndlo4mogYvWrp0

SjQMS5Dbeme4rqQNuk   W7pkQFL7l9pk1VhJNet8luCoj8Wr-jHzJQ8Wc5LHD_tddlmdybsb2lnlEhIVeO8skMg04FWqhxmo 8xfbz4vbLZcG4bw_jUBru3eodJLDMpDd8GyD-LWaJeE8b9ztr9z0IgYE3-R_eaeQm1wNSQ5Bvqwg%3D%3D&tracking_referrer=www.independent.co.uk

[398] Alex Davies. Review: Tesla Model S P90DWired Magazine. Available at: https://www.wired.com/2015/08/tesla-model-s-p90d-review/

[399] DOE Sunshot program. Available at: http://energy.gov/eere/sunshot/about- sunshot-initiative

[400] Airbus electric airplane. Available at: http://www.airbusgroup.com/int/en/corporate-social-responsibility/airbus-e-fan-the- future-of-electric-aircraft.html

[401] NASA's electric airplane project. Available at: http://www.nasa.gov/image- feature/nasas-x-57-electric-research-plane

[402] Alfred W. Crosby. Children of the Sun. A History of Humanity's Unappeasable Appetite for Energy. Norton & Company. 2006. Available at: http://www.amazon.com/Children-Sun-Humanitys-Unappeasable- Appetite/dp/0393931536

[403] The Airbus E-Fan aircraft has made several key flights such as across the English Channel on propellers powered by electricity stored in a battery. Available at: http://www.airbusgroup.com/int/en/innovation-citizenship/airbus-e-fan-the-future-of- electric-aircraft.html

[404] Thomas, J., Huff, S., West, B., and Chambon, P., "Fuel Consumption Sensitivity of Conventional and Hybrid Electric Light-Duty Gasoline Vehicles to Driving Style". Available at: http://papers.sae.org/2017-01-9379/

[405] EIA energy consumption information for California. Available at: http://www.eia.gov/state/?sid=CA

[406] The Shelby Electric Company light bulb that has been working for 115 years. Available at: http://www.centennialbulb.org/

[407] Krajewski, Markus (24 September 2014). "The Great Lightbulb Conspiracy". IEEE Spectrum. Retrieved 12 October 2014.

[408] Wendy Wilson, Travis Leipzig and Bevan Griffiths-Sattenspiel. "Burning Our Rivers: The Water Footprint of Electricity". River Network in 2012. Available at: https://www.rivernetwork.org/wp-content/uploads/2015/10/BurningOurRivers_0.pdf

Index

A

accidents, energy-related, 389–390
mining accidents, 373, 377, 382, 385
nuclear accidents, 14, 18, 333, 335, 339–348, 351
wind turbine accidents, 131–133 acid rain, 205
Acwa Power International, 48
Adair, Red, 299
Adinolfi, Roberto, 30
agnotology, 222 agriculture
carbon capture, 155
carbon emissions from, 203
climate change impacts, 259 farmer displacement, 148
farmer suicide, 378–379
food for fuel, 152–160, 162, 394, 423
technology, consequences on, 26, 50 transportation and energy use, 158 vertical farms, 95
Air Fuel Synthesis, 46
air pollution. *See* emissions; pollution air sourced heat pumps, 88, 399

albedo effect, 194–195, 229–230, 251, 445
Aldy, Joseph, 422
al-Janabi, Rafid Ahmed Alwan, 302 Alliance BioEnergy Plus, 73, 158–160
Al Qaeda World Trade Center bombings, 300–301
Al-Saud, Abdulaziz, 282
American Petroleum Institute (API) Energy and Man symposium, 182– 183
ethanol efficiency report, 155 lobbying efforts, 191–192 peak oil awareness, 277–278 Anderegg, William, 240
Anders, Bill, 165–166
Anderson, Ray, 34
Annan, Kofi, 302
Anthropocene, 17, 137, 415 anthropogenic effects on environment carbon dioxide emissions, 168, 175–184, 187, 191, 202, 205–206, 234, 269–270, 273, 372, 399
climate change, 175, 178, 269–273, 434

Californium, 358–359
Callender, Guy S., 178–179
Canada
accidents in energy
    transportation, 389–390,
    404
climate actions, 268
climate impacts on birds, 136,
    140 oil exports to U.S., 294
oil reserves, 256, 296, 395, 399
    renewable energy use, 90
    wildfires in, 204, 214
capacitors and super-capacitors,
    70– 71
capitalism
antipathy toward, 27–28
benefits versus drawbacks,
    32–34 carbon, 196
in chain reaction, 321–322 in
    organisms, 203–204 social
    cost of, 35
stranded assets in, 96, 162, 254,
    266–268, 418
unburned, 41, 194, 279, 416,
    418, 426, 435, 438
carbon capture and
    sequestration (CCS), 44–47,
    112, 381
carbon cycle, 196–207, 227
carbon dating, 180–181 carbon
    dioxide (CO2)
anthropogenic, 168, 175–184,
    187, 191, 202, 205–206,
    234, 269–273, 372, 399
climate effects, 84, 112, 179–
    184, 276, 380
discovery of, 173–174
    extinction events and,
    205–206

global warming and, 193,
    235–236, 399 (*see also*
    global temperatures) heat
    absorption capacity, 174–
    179, 183
historical data on, 176, 178,
    180, 227 from human
    respiration, 227–228
    measurement of, 198, 201
modeling of, 43, 187–188, 273
nonradioactive carbon, 181
persistence of, 199–200
as pollution, 226–227, 368–369,
    399
ppm levels, 43, 46, 84, 178,
    180, 184, 200, 227, 399
solidifying, 46
See also carbon dioxide
    emissions; emissions
carbon dioxide emissions from
    agriculture, 203
business cost of, 265–266, 268
in China, 404
from coal burning, 179, 373–
    374 from corn ethanol, 155
payback on, 133–134
rates of, 196, 254
reducing, 43, 45–46, 422, 425,
    427
social cost of, 35
from volcanic activity, 206, 247
Carbon Engineering, 47
carbon flux, 197–202
Carboniferous period, 206–207,
    277
carbon intensity, 64–65, 90
carbon markets, 47, 193, 224
    European tax and trade
    system, 107 Regional

emissions)
external costs of, 276, 373–374
geologic formation of,
206–207 global reserves,
279
healthcare costs of, 371–372,
381
health impacts, 369–379,
381–386
land reclamation, 387 liquid
fuels from, 284–285 mining
accidents, 377
phase-down, 34–35, 279, 387
pollution from, 51, 276, 373,
375, 383
as spinning reserve, 121
subsidization of, 418
water consumption of, 386
Cockburn, Andrew, 357
Cold War
Middle East geopolitics and,
286 nuclear accidents
during, 342, 345–346
nuclear weapons development
and testing, 325, 327–328,
357–358
Colonial Pipeline EPA
violations, 403
Colorado River, 145–147
Columbia, Maryland,
commitment to renewable
energy, 91
Command and Control
(Schlosser), 324, 347
commercial building, green,
266 communications
innovations, 62 Competitive
Enterprise Institute, 209
complexity

of climate science, 216, 249
of energy markets, 107, 288,
413, 446
of modern world, 102
Comprehensive Nuclear
Test Ban Treaty (CTBT),
325, 364
concentrated solar power, 21,
115–116, 118, 125–126
Conference of the Parties
(COP), 191–193
COP15, 193, 217
COP21, 46, 78–79, 96, 104,
213, 216, 267
COP22, 78, 214
conflict, resource-related,
259–260, 305
Arab-Israeli wars, 288–295
Biafran War, 305
First Gulf War (Operation
Desert Storm), 295–300
Iran-Iraq War, 297
over water resources, 307–
308 Second Gulf War
(Operation Iraqi Freedom),
300–305
CONG (coal, oil, nuclear, and
gas) energy
carbon emissions from (*see*
carbon dioxide emissions)
carbon in, 203
consumer cost comparisons,
107–108 convenience of,
106, 111
cost of (*see* externalities of
CONG) costs to economy,
397–398 deathprint,
368–369

Earth's albedo and, 194–195 at hyrdroelectric dams, 148 reforestation, 193–194
Fork, David, 43
fossil carbon, 200
fossil fuels. *See* CONG (coal, oil, nuclear, and gas) energy
Fourier, Jean-Baptiste Joseph, 84, 174
fracking, 86, 294 coal consumption, effect on, 374 expense of, 395 impacts of, 107, 391, 402–403
Franklin, Rosalind, 189
Franks, Tommy, 303
Franta, Benjamin, 182, 191
Freese, Barbara, 370
Friedman, Thomas, 56, 107, 153, 243
friendly societies, 26
Frisch, Otto, 320–321
fuel efficiency, 62, 432, 434–435 fuels from atmospheric carbon, 46–47 carbon neutral, 152 (*see also* biodiesel and biofuels)
energy-efficiency standards, 34–36 feedstocks for, 153–156, 161 (*see also* cellulose) *See also* biodiesel and biofuels
Fukushima nuclear disaster, 30, 339, 348
fusion power, 365–366 *See also* thorium nuclear power

G

Gaddy, Bend, 39

Gaia hypothesis, 175
Galtieri, Leopoldo, 11, 312, 315
Gandhi, Indira, 362 gas. *See* natural gas
gasoline, consumer cost comparisons, 107–108
Gaston, Kevin, 134
Gates, Bill, 47
Gates, Robert M., 343
GDP (gross domestic product) carbon dioxide concentration and, 180 costs of climate change and, 194 energy use and, 68–69, 84, 96, 196, 413, 438 fossil fuel subsidies percentage of, 424–427
oil production and, 295 General Circulation Model of atmosphere, 249
General Electric, Ecomagination program, 87
geological cycle as carbon sink, 206 geological events, carbon dioxide release from, 178
geopolitics, 18, 282–365 Argentina and Falklands oil exploration, 308–318 chokepoints in oil distribution, 287 conflicts over oil wealth distribution, 305 conflicts over water resources, 307– 308 Germany's energy supply, 284– 285 global energy system instability, 306–307 of global food supplies, 56 Gulf Wars, 295–305

O

Tyndall, John, 84, 174

U

Unabomber, 29
"Unburnable Carbon" report,
194 unions. *See* labor
unions
United Nations
Austria climate conference,
188–189 Environmental
Program, 54–55
Inter-Governmental Panel
on Climate Change (*see*
Inter-Governmental
Panel on Climate Change
(IPCC)) Kyoto climate
conference, 86, 193
World Meteorological
Organization, 181–182, 188
United States
CIA-caused geopolitical
turmoil, 286 climate denial,
215–216
domestic oil production, 86,
291–295, 395
electricity consumption per
capita, 433, 435
energy consumption of military,
258–259
energy policy, 409–410,
451–452
energy production, 225
energy security, 302
energy subsidies, 319, 410,
417–418, 420–421, 423,
426
energy trade balance, 293
Gulf Wars, 295–305 Israel,
support of, 290–292

nuclear electricity generation,
332 nuclear weapons
capabilities, 328 oil
imports, 291–292, 294,
306 oil trade protection, 13
petrodollars, 307
plastics manufacturing, 405
Saudi Arabia, alignment with,
282–283, 286
transportation innovations, 60
Western Pakistan, support
of, 362 universal basic
income for workers, 31–32
uranium enriched, 338
fission reactions, 324, 333
nuclear reaction pathway,
321–322 Urban Heat Island
effect (UHI), 257–258
utilities, strategic changes in,
49–53

V

van Helmont, Jan Baptist, 160
vanHorn, Jodie, 89
van Otterloo, Grantham Mayo,
41 Venezuela
oil and gas resources, 294, 296,
306, 399
subsidization of CONG energy,
413 Venter, Craig, 39
Venus, atmospheric carbon
dioxide, 185
Verheggen, Bart, 240
Vernadsky, Vladimir
Ivanovitch, 178 Videlas,
Jorge, 310–311
Villach Conference, 188–189
Viola, Roberto, 312

quality of life (*see* quality of life) sustaining, 21

World Trade Center bombings, 300– 301

World War II
bombing of Middle Eastern oil installations, 283–284
German energy production and use, 284–285

X

xenon 135, 330

Y

Yablokov, Alexei, 357
Yeltsin, Boris, 342 Yom Kippur War, 288 YPF, 318
Yunus, Mohamed, 33

Z

Zateyev, Nikolai Vladimirovitch, 345
ZBB, 76
zero emissions goal, 64
Zhu, Chen, 369
Zipingpu Dam, 147